AF658642

The Author

Dr. S K Gupta, Born (1949), INAE Distinguished Visiting Professor, ICAR-NDRI, Karnal, is an Engineer with specialization in Soil and Water Resources Engineering and Civil Engineering. He obtained B. Tech (Agril. Engg.) from the Punjab Agricultural University (PAU), Ludhiana, in 1970 and Master of Agricultural Engineering from the same University in 1976. He obtained Ph.D. in Civil Engineering from the Jawahar Lal Nehru Technological University, Hyderabad, Telangana, India in 1984. Since his joining CSSRI in 1971, besides holding scientific positions, he has been Head, Division of Drainage and Water Management, Head, Indo-Dutch Network Project, Head, Division of Irrigation and Drainage Engineering and Project Coordinator (AICRP on Management of Salt Affected Soils and Use of Saline Water in Agriculture). After superannuation, Dr. Gupta worked as Emeritus Scientist (ICAR) and INAE Distinguished Professor at the same institute. Currently, he is professor of Agricultural Engineering (Guest Faculty) at Khalsa College, Amritsar. Dr. Gupta has been engaged in conducting researches in surface and subsurface drainage for land reclamation, irrigation water management, hydrology of salt affected soils, leaching and use of saline water in agriculture. He has been teaching at ICAR-NDRI, Karnal, Karnal Institute of Engineering and Technology, Karnal and Khalsa College, Amritsar. Dr. Gupta is Fellow of the National Academy of Agricultural Sciences, Indian National Academy of Engineering and Indian Society of Agricultural Engineering. He has published more than 150 research papers and 15 books. He has been nominated as chairperson of the working group on Problems of Water Logging and Soil Salinity in Haryana by Haryana Kisan Kalyan Pradhikaran, Government of Haryana. Dr. Gupta has been consultant to WAPCOS-Louis Berger, MoWR, MoRD, UP-Dutch Tube Well Project, ActionAid International, Synergics Hydro India-Oromia Water Works Enterprise, Ethiopia and others. He has been bestowed with Rafi Ahmad Kidwai Award by ICAR. Besides he is recipient of several other awards from NAAS, ISAE, IE (India), CBIP and MoWR among other organizations. He is the Chief Editor of Journal of Water Management of the Indian Society of Water Management.

Fundamentals of
RENEWABLE ENERGY AND GREEN TECHNOLOGIES

– *Author* –

S.K. Gupta, Ph. D.

Professor of Agricultural Engineering (Guest Faculty)
Khalsa College, Amritsar

INAE Distinguished Visiting Professor
ICAR-National Dairy Research Institute, Karnal

Former Head, Division of Irrigation and Drainage Engineering
Former Project Coordinator, Use of Saline Water in Agriculture
ICAR-Central Soil Salinity Research Institute, Karnal

2024

Daya Publishing House®
A Division of

Astral International Pvt. Ltd.
New Delhi – 110 002

ISBN: 978-93-5919-660-2 (HB)

Published by : **Daya Publishing House®**
A Division of
Astral International Pvt. Ltd.
– ISO 9001:2015 Certified Company –
4736/23, Ansari Road, Darya Ganj
New Delhi-110 002
Ph. 011-43549197, 23278134
e-mail: info@astralint.com
Website: www.astralint.com

Printed at : **Replika Press Pvt Ltd**

Preface

India's annual energy demand in 2050 is assessed at about 14,500 TWh. India is pursuing a well thought out policy of developing requisite infrastructure to overcome the anticipated mismatch between demand and supply. Such a policy also fits well in the national energy security scheme that envisages providing quality electricity services 24×7 to all. Renewable energy is India's top priority to reduce its dependence on imported fossil fuels. It will help to achieve its target of having more than 50 per cent of electricity generated from renewable sources by 2030, besides fulfilling its commitment to gradually reduce its carbon footprint and become carbon-neutral by 2070. This stupendous task requires huge investments and trained manpower for development, operation, management and use of renewable energy. To sensitize the new generation to the new technologies and ensure availability of trained manpower in future, academic courses at school, college and university levels have been aligned to include relevant issues of green energy. Agriculture is and will continue to be the major producer and user of biomass energy. Besides, farmers are now willingly tapping solar energy for various farm and domestic activities to reduce the cost of production and save fossil fuels. Therefore, a conscious decision has been taken to include bio and green energy at the undergraduate courses of agricultural engineering, agriculture, horticulture and forestry *etc.* ICAR's 5th Dean's Committee has thoughtfully designed such courses for the students pursuing these degree programs.

The book titled 'Fundamentals of Renewable Energy and Green Technologies' focuses on various aspects of bioenergy in solid, liquid and gaseous forms. Besides, it also dwells upon renewable technologies such as solar, hydro including Oceans, and wind. The book is primarily intended to serve as a textbook for various degree courses leading to B. Tech. in agricultural engineering, and B.Sc. (Hon.) in agriculture, horticulture, and forestry. The book has been divided into well-defined sixteen chapters namely: Classification of energy sources, Biomass and briquetting, Combustion process, Principles of pyrolysis, Gasifiers, Biogas production processes, Biogas plants, Vegetable oils, Biofuels, Solar energy measurement and applications, Solar collectors and their applications, Solar dryers, Photovoltaic cells and practical applications, Hydropower, Wind energy and Life cycle cost analysis. The purpose is to make the students aware of the principles, and practices of technologies to equip them for various roles while harnessing green energy resources. The following features add to the value of this book.

- ✰ The book is written in easy to understand english. It is highly illustrated with relevant diagrams and photographs for easy comprehension of issues being discussed
- ✰ Theoretical questions appear at the end of each chapter. Glossary of terms and objective type questions and their answers are included in one section towards the end
- ✰ Relevant solved numerical problems are included at relevant places in respective chapters
- ✰ Related major books and papers are referenced in the references. It may help the students and field practitioner to refer the relevant literature for more detailed understanding of a particular subject

I've had the good fortune to teach this subject at Khalsa College, Amritsar in online mode. The students asked many searching questions that helped me to improve over the study material. This book bears the stamp of these students as it owes its present form to rich contributions from the hundreds of students.

The book has heavily drawn from the literature brought out by many individuals and national and international organizations. I have attempted to refer these contributions in the references. But, I do not claim it to be a complete list as I might have inadvertently missed few. I place on record my appreciation for all the contributors who have painstakingly researched and developed the relevant processes, technologies and teaching material. I also take this opportunity to whole heartedly thank the Principal, Khalsa College Amritsar who has been kind enough to give me an opportunity to work as guest professor. I am personally thankful to Dr. Kanwaljit Singh Sandhu, Assistance Professor Agronomy, Er. Harwinder Singh and Er. Prashant Kumar, Assistant Professors Agricultural Engineering at Khalsa College Amritsar for providing logical support and resolving day to day issues whenever cropped up. They have also painstakingly gone through the teaching material and offered constructive suggestions. I take this opportunity to wholeheartedly thank them and all others who have contributed or supported me in one way or the other while writing this book. My family and friends need special thanks as they encouraged me throughout the period of writing the book.

This book as stated has been written primarily to serve as a textbook to familiarize students on the basic principles of bioenergy, solar energy, hydropower and wind energy and their application on farms and in day to day life. I believe that this book will prove to be a good resource material and will be a cherished possession of the faculty and students of agricultural engineering/ agricultural/horticulture/forestry and practicing agriculturists and engineers. Besides, it should be a valuable academic asset for libraries of colleges and universities worldwide.

Dr. S.K. Gupta

Contents

- Chapter 1 -

Classification of Energy Sources

Energy is quite important in our lives. Our body without energy will fail to perform basic functions like respiration, circulation of blood, or digestion of food. Plants will also fail to perform the process of converting carbon dioxide, water and minerals into food. Light from the Sun, a source of energy is an essential pre-requisite for this function called photosynthesis. At the dawn of civilization, humans began using wood, water, and fossil fuels as a means of heating and making machines to work. Today, we rely on one or the other form of energy for almost all types of activities in our daily life. Most of the things we see around us have been created with the help of energy. For example, most machines used for the production and manufacture of different items now operate on electrical energy. It has become so important that we cannot think of life without energy in general and electricity in particular. In fact, energy in our daily life provides comfort, increases our productivity and allows us to live the way we want to. Besides, economic growth and development of a community/sector is judged by the amount of energy used by the community/sector.

ENERGY AND POWER: DEFINITIONS

Energy has been defined as the capacity to do work or capability to produce an effort. It is expressed in N-m (Newton-meter) or Joules. The energy density is expressed as J/kg. Power on the other hand is defined as the rate at which work is performed, or energy is converted. Thus, power is the amount of energy divided by the time it took to use the energy. Its unit is the watt, which is one Joule per second of energy used. In other words, a body is said to have one watt of power if it can perform one joule of work in one second. Amount of energy is given as multiplication of power and the period of time over which it is used. For example, kilowatt is a unit of power and a kilowatt-hour (1 kilowatt times 1 hour) is the unit of energy. Similarly, the task of lifting a box take a certain amount of energy (a certain number of Joules), but the speed with which task is completed determines the power. Faster the task is completed (the smaller Δt), the more the power it takes (Joules/s or watts).

Forms of Energy

Energy exists in different forms, such as light, heat, sound, and motion (ICAR, 2016). Different forms of energy include chemical energy, nuclear energy, gravitational energy, radiant energy, thermal energy, electrical energy, motion energy, sound energy, light energy and elastic energy. All these forms of energy can be put into two broad categories *i.e.* potential energy and kinetic energy.

Potential energy: *It is stored energy or the energy of position.* Several forms of energy are known to have potential energy.

Chemical energy: *Chemical energy is the energy stored in the bonds of atoms and molecules. It is the energy that holds these particles together.* Biomass, petroleum, natural gas, propane and coal are examples of stored chemical energy.

Nuclear energy: *Nuclear energy is the energy stored in the nucleus of an atom - the energy that holds the nucleus together.* The nucleus of a uranium atom is an example of nuclear energy. The nuclear energy is released when two or more smaller atoms are combined into a larger one, known as fusion or when during splitting of an atom into two or more smaller ones, known as fission. Nuclear power plants split the nuclei of uranium atoms to produce electricity. Nuclear energy is a controversial energy source in the sense that it is not a renewable energy source, yet many people think it as a clean energy because it is not based on fossil fuels. Thus, nuclear power plants can reduce carbon emissions and help in battling climate change. Contrarily, others cite the risk of accidents and the issues related to storing nuclear wastes. They feel that these issues are too significant to warrant the development of this energy source.

Gravitational energy: *Gravitational energy is the energy of place or position, a kind of potential energy.* A rock on top of a hill contains potential energy because of its position. If a force pushes the rock, it rolls down the hill because of the force of gravity. The potential energy is converted into kinetic energy until it reaches the bottom of the hill and stops. Water in a reservoir behind a hydropower dam is an example of gravitational energy. But, it becomes rotational energy when the water is released to spin the turbines.

Elastic energy or stored mechanical energy: *Elastic energy is a form of potential energy that is stored in an elastic object - such as a coiled spring or a stretched elastic band.* Elastic objects store elastic energy when a force causes them to be stretched or squashed. It is also known as stored mechanical energy.

Kinetic Energy

Kinetic energy is the energy of motion, observable as the movement of an object, particle, or set of particles such as electromagnetic and radio waves, electrons, atoms, molecules, substances, and objects. A person walking, a thrown baseball, a crumb falling from a table, and a charged particle in an electric field are all examples of kinetic energy at work. Following forms of energy are examples of kinetic energy.

Radiant energy: *Radiant energy is electromagnetic energy that travels in transverse waves.* Solar energy is an example of radiant energy. It is also known as light energy or electromagnetic energy, Radiant energy includes visible light, x-rays, gamma rays and radio waves.

Thermal energy: *Thermal energy (or heat) is the internal energy in substances-the vibration and movement of atoms and molecules within the substances.* The faster they move, the more energy they possess. Geothermal energy is an example of thermal energy.

Electrical energy: *Electrical energy is the movement of electrons (the tiny particles that makeup atoms, along with protons and neutrons).* Electricity and lightning are examples of electrical energy.

Motion energy: The movement of objects or substances from one place to another is motion. *Motion energy or mechanical energy is the energy stored in objects.* For a moving object kinetic energy, KE is given as $(1/2)\ mV^2$ such that m is the mass of the object and V the velocity of the moving object. Thus, objects moving faster store more energy than slow moving objects. Wind and hydropower are examples of motion energy.

Sound energy: *Sound is the movement of energy through substances in longitudinal (compression/rarefaction) waves.* It is produced when a force makes an object or substance to vibrate. Normally, sound has much less energy than any other form of energy.

Light energy: *Light energy is a type of wave motion and is a form of electromagnetic radiation. It comprises photons that are produced when atoms of an object's heat up.* Light enables us to see, as objects are only visible when they reflect light into our eyes. It is the only form of energy that is visible to the human eye.

CLASSIFICATION OF ENERGY SOURCES

Any substance from which usable energy can be extracted is a source of energy. The energy sources can be classified in a number of ways.

Based on Availability in Nature

Primary sources: *Primary energy sources are either found or stored in nature.* Common primary energy sources are coal, oil, natural gas, and biomass (such as wood) Other primary sources include the sun, the wind, tides, nuclear energy from radioactive substances, thermal energy stored in earth's interior, and potential energy due to earth's gravity. Primary energy sources can be used directly or may have non energy uses, for example coal or natural gas can be used as a feedstock in fertilizer plants.

Secondary sources: *Secondary energy sources are derived from the transformation of primary energy sources. Primary energy sources are mostly converted in industrial utilities for example coal, oil or gas converted into steam and electricity.* So heat and electricity are termed as secondary energy sources. Similarly, petrol is derived from the treatment of crude oil and electric energy obtained from the conversion of mechanical energy (hydroelectric plants, Aeolian plants), chemical energy plants (thermoelectric), or nuclear plants.

Based on Transaction

Commercial energy: *The energy sources that are available in the market for a definite price are known as commercial sources of energy.* By far the most important forms of commercial energy are electricity, coal and refined petroleum products.

Non commercial energy: *The energy sources that are not available in the commercial market for a price are classified as non-commercial energy.* Non-commercial energy sources include fuels such as firewood, cattle dung and agricultural wastes, which are traditionally gathered, and used in rural households, but not bought at a price. These are called traditional fuels. Non-commercial energy is often ignored in energy accounting.

A simple model that describes the need for commercial and non-commercial energy resources is given by the following equation (Kumar *et al.*, 2021).

$$R = EN \tag{1.1}$$

Here R is the total yearly energy requirement for a population of N people. E is the per capita commercial energy-use averaged over one year. The unit of E is energy per unit time, *i.e.* power. The value of E is closely related to provision of food and manufactured goods. The worldwide average per capita value of E is about 2 kW. Regional average values vary widely, with North America 9 kW, Europe as a whole 4 kW, and several regions of Central Africa as small as 0.1 kW. The inclusion of non-commercial energy increases all these figures and has the major proportional benefit in countries where the value of E is small.

Based on Traditional Use

Conventional energy sources: The energy sources which have been traditionally used for many decades *e.g.*, coal, petroleum, natural gas and water power. Except water other sources are exhaustible in nature and cause pollution as they emit smoke and ash.

Non-conventional energy sources: The resources that are still in the process of development but have the potential of exploitation. These includes solar, wind, tidal, fossil fuels, nuclear, biomass, and geothermal. They are inexhaustible, pollution free, easy to maintain and less expensive due to local availability. These energy sources have become important sources only recently because of the oil crisis.

Based on Energy Storage or Cycling Time

Renewable energy (inexhaustible): *Renewable sources of energy are continuously and freely produced in the nature and are inexhaustible. Since these can be renewed over a relatively short period of time, these energy sources are termed as renewable.* These include firewood or fuel wood from forest, petro plants, plant biomass *i.e.* agricultural waste like animal dung, solar energy, wind energy, water energy in the form of hydro-electricity and tidal energy and geothermal energy *etc.* This book is mainly devoted to these resources of energy.

Non-renewable energy (exhaustible): *These are available in limited amount and develop over a long period of time. They are likely to be exhausted as a result of their unsustainable use.* These include coal, mineral, natural gas and nuclear power. Coal, petroleum and natural gas are also called fossil fuels being organic in origin.

Conventional or Non-renewable Sources of Energy

Conventional energy normally means the energy derived from sources which are finite in nature like oil, gas and coal. *Conventional energy sources are also termed as non-renewable energy sources because the supplies of these fuel sources are limited and cannot be replenished in a foreseeable period, if exhausted.* These sources are derived from the remains of dead plants and animals that got buried under the ground millions of years ago. The buried dead plants/organisms contained energy originating from photosynthesis. The buried and decomposed organic materials by the action of heat of the Earth's core and pressure from rocks and soils got converted into fossil fuels through anaerobic decomposition over the years. These sources of energy are used to run the vehicles, cooking, lighting, generate electricity, and for producing several industrial products such as plastics and paints *etc.* Currently, about 70 per cent of India's energy generation capacity is from fossil fuels. Coal accounts for 40 per cent of India's total energy consumption followed by crude oil and natural gas at 24 per cent and 6 per cent respectively.

India is largely dependent on fossil fuel imports to meet its energy demands. In 2009-10, the country imported 159.26 million tonnes of crude oil which amounted to 80 per cent of its domestic crude oil consumption and 31 per cent of the country's total imports. By 2030, India's dependence on energy imports is expected to exceed 53 per cent of the country's total energy consumption. Besides, domestic coal shortage is hindering the growth of electricity generation in India. As a consequence, India's coal imports for electricity generation increased by 18 per cent in 2010. Since fossil fuels contain high percentages of carbon, their use is directly linked to excessive greenhouse gas emissions and other environmental damages. The salient advantages and disadvantages of fossil fuel are as follows.

Advantages	Disadvantages
☆ Provide a large amount of thermal energy per unit of mass ☆ Easy to get and easy to transport ☆ Can be used to generate electrical energy and make products such as plastic, paints *etc.*	☆ Nonrenewable ☆ Burning produces smog ☆ Burning coal releases substances that can cause acid precipitation ☆ Risk of oil spills ☆ High cost

Nuclear Energy

The atoms of few radioactive elements such as radium and uranium act as sources of nuclear energy. The mass of nucleus serves as the source of nuclear energy which is released mainly in the form of heat. Nuclear energy normally categorized as clean energy is also a non-renewable source of energy because radioactive elements are non-renewable in nature (ICAR, 2016). The energy from these elements is produced either through the process of fission (means to split) or fusion (means to combine). In the process of nuclear fission, the heavy nucleus of radioactive atoms like uranium, plutonium, or thorium is bombarded with low-energy neutrons at an optimum speed. The neutron is absorbed such that the nucleus of these atoms becomes highly unstable. As a result, it splits into nuclei of smaller atoms. For example when uranium-235 atoms are bombarded with neutrons then the heavy uranium nucleus splits to produce barium-144 and krypton-89 (Figure 1.1 left). Simultaneously, few neutrons are also released. A lot of energy is also produced in this reaction because mass is converted into energy. The neutrons released during the process split other nuclei of the Uranium (235) in a similar fashion. The reaction, also known as the chain reaction, occurs at a fast rate releasing huge amount of energy. If all the neutrons produced during the fission of uranium-235 produce further fission, then so much energy will be produced that it will be uncontrollable. It will lead to an explosion, a process used in manufacturing atom bombs. To control the reaction rate and energy release, boron rods are used in the reactor to absorb some of the released neutrons. The controlled energy so produced is used to produce steam, which is further used to drive a turbine to generate electricity. The nuclear reactor is enclosed in a concrete chamber having a thick wall to absorb nuclear radiations.

In the process of fusion, two nuclei of light electrons are combined to form a heavy nucleus. The nuclei of atoms are positively charged and thus they repel each other. A lot of heat energy and high pressure are required to combine or fuse the two nuclei to form one heavy nucleus. Thus, nuclear fusion is carried out by heating lighter atoms to an extremely high temperature at high pressure.

The mass lost in this process gives tremendous amount of energy. For example, when two deuterium atoms or a deuterium and a tritium atom are fused then these nuclei combine to form helium which has a heavy nucleus (Figure 1.1 right). Besides, a neutron is emitted and a lot of energy is liberated. Nuclear reactions occurring at extremely high temperature are called thermonuclear reactions. This kind of reaction is used in producing hydrogen bombs which are capable of causing mass destruction.

Advantages	*Disadvantages*
✰ Very concentrated form of energy, tremendous energy from small amount of fuel ✰ Power plants do not produce smog ✰ Power plants do not produce carbon dioxide or sulphur dioxide	✰ Produces radioactive waste, which keeps on producing harmful radiations ✰ Radioactive elements are non-renewable ✰ High installation costs ✰ Risk of accidents in nuclear reactors

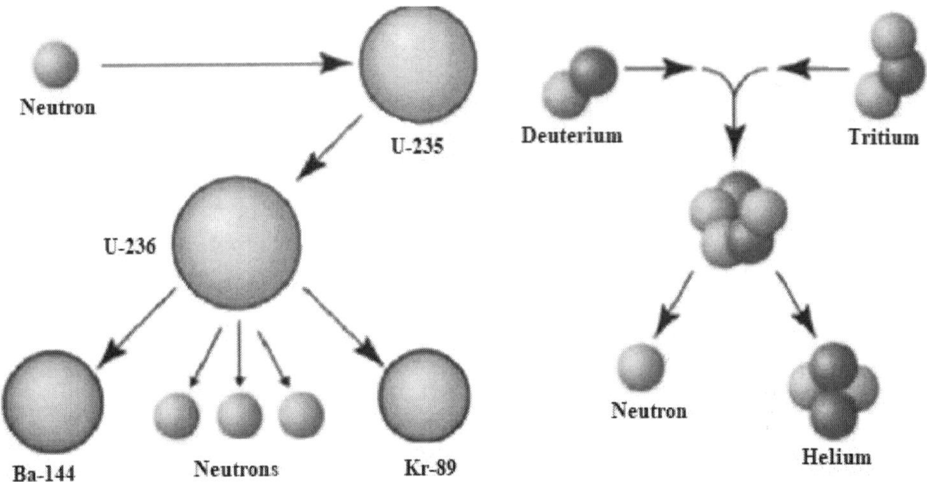

Figure 1.1. Nuclear reactions: Fission (left) and fusion (right).

Non-conventional or Renewable Sources of Energy

Non-conventional sources are also known as renewable sources of energy. Examples of non-conventional sources of energy include solar energy, wind energy, bio-energy, geothermal energy and tidal energy (Serrano, 2018). They are called renewable because it is possible to replenish these sources in finite time. These sources are not likely to be exhausted. For example, day after day, the sun shines, the wind blows, the rivers flow, and plants grow (biomass). Heat from inside the earth–geothermal energy–is continuously made by the radioactive decay of elements in the Earth's core. By 2020, electricity generation from these sources is estimated at 4444 TWh (terawatt-hour) from hydro, 2413

TWh from wind and Sun including photovoltaic, 775 TWh from other sources (IPCC, 2011, IEA, 2021). Nuclear energy, sometime included in this category because of its low carbon footprint, produces about 2658 TWh.

Sun - the Ultimate Source of Energy

The recent concerns on the decline in the world's oil supply coupled with environmental catastrophe resulting from fossil fuels forced the mankind to consider other energy options. One of the most promising options is the solar energy, which is not only renewable and sustainable but also has emissions reducing qualities (refer Chapters 10-14). Solar energy is energy derived from sun in the form of solar radiation. The output of the sun is assessed at 2.8×10^{23} KW/y out of which the energy reaching the earth is 1.5×10^{8} KW/y. Fortunately, India has high solar insolation levels with most parts of the country receiving 4-7 KWh/m^2-day and about 1500-2000 sunshine hours per-annum (5-7 hours per day for 300 days, depending upon the location). This enormous solar power can be harnessed either by direct sources (like solar cooker, solar steam systems, solar dryer, solar cells, *etc.*), or indirect sources (biomass production, wind, tidal, *etc.*). Solar power can also be harnessed to produce electricity using photovoltaic panels (Figure 1.2). National Institute of Solar Energy has assessed India's solar potential of about 748 GW assuming 3 per cent of the waste land area to be covered by Solar PV modules. States on West part of India such as Rajasthan and Gujarat have the highest potential. India has launched a solar mission in 2009 setting a target of installing 100 GW grid-connected solar power plants by the year 2022. Recently, India achieved 5th global position in solar power deployment by surpassing Italy with total installations of 30 GW by July, 2019.

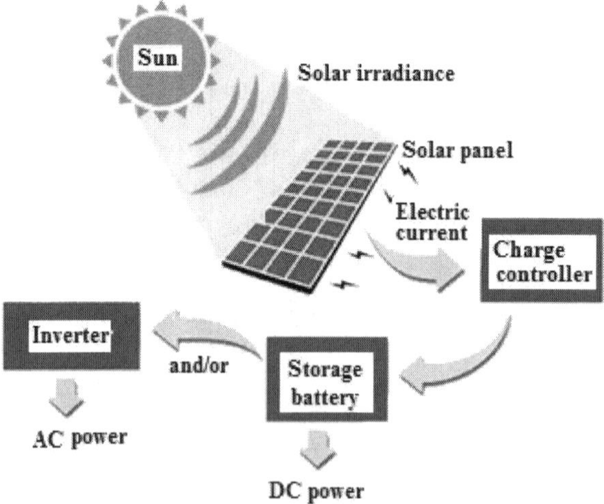

Figure 1.2. Block diagram illustrating harnessing sun energy to produce electricity.

Advantages	Disadvantages
☆ Almost limitless source of energy ☆ Does not produce air pollution	☆ Expensive to use for large-scale energy production (Now costs are reducing) ☆ Only practical in sunny areas ☆ Intermittent in nature

Wind Energy

Wind power is another clean and renewable source of energy that does not produce harmful emissions that damage the environment. Like solar power, harnessing the wind power is highly dependent on weather and location. Approximately 2 per cent of the solar energy striking the Earth's surface is converted to kinetic energy of the wind. Wind turbines are used to convert the wind's kinetic energy to electricity. Wind power exerts forces on the blades of the turbine, causing the turbine's shaft to rotate at a speed of 10 to 20 revolutions per minute (rpm). The rotor shaft is connected to a generator via a gear box that converts mechanical energy into electrical energy (Figure 1.3). High wind speeds yield more power since wind power is proportional to the cube of wind speed (Refer chapter 15). Therefore, if wind speed doubles, wind power increases eight times. The minimum wind speed required to turn the generator to produce electricity is about 12-14 km/h. Windmills work at their full capacity at wind speeds of 50-60 km/h but the turbines must be stopped, if wind speed exceeds 90 km/h as it might damage the windmill.

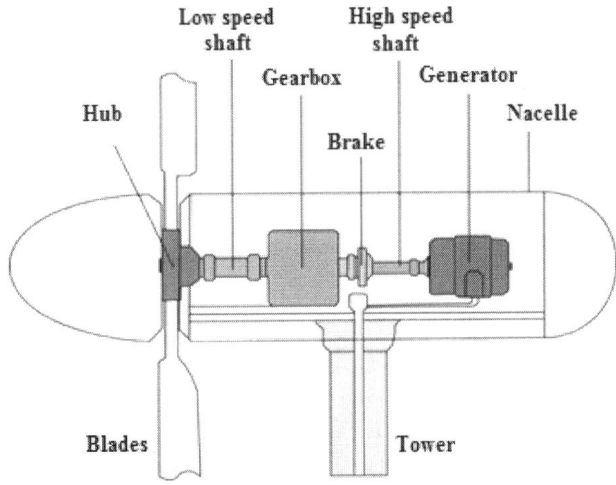

Figure 1.3. Horizontal-axis wind turbine.

The popularity of wind energy has always depended on the price of fossil fuels. For example, there were hardly any takers of wind power when oil prices were low after World War II. But the commercial use of electrical wind turbines substantially increased with dramatic increase in oil prices during 1970s. In recent years, the amount of energy produced by wind-driven turbines has increased exponentially due to significant breakthroughs in turbine technologies, making wind power economically compatible with conventional sources of energy. Today, the cost of wind-generated electricity is very close to the cost of conventional power generation utilities. India has enough opportunities to harness wind energy having around 7700 km long coastline and large exclusive economic zones. National Institute for Wind Energy observed that western states have larger potential in terms of a stable, steady and a speedy wind flow starting from Gujarat, Maharashtra, and Karnataka to Tamil Nadu and Andhra Pradesh. Advantages and disadvantages of wind power are listed as follows.

Advantages	Disadvantages
☆ Renewable source ☆ Relatively inexpensive to generate ☆ Does not pollute the environment	☆ Only practical in windy areas ☆ Produces less energy ☆ Wind mill is big, bulky and inconvenient to use as compared to other forms of energy ☆ Intermittent nature of energy produced ☆ People who live or work in close proximity have experienced symptoms that include decreased quality of life

Hydropower

Hydropower is most commercially matured technologies amongst all the renewable energy resources. The hydropower potential of India has been estimated at 84044 MW at 60 per cent load factor. The installed capacity of hydropower projects in India, at the time of its independence, was only 508 MW. It increased to about 44594 MW by the end of April 2017, making India the 7th largest producer of hydroelectric power. The most well-known hydro-power plants are dam based, where water is stored behind a dam and released in a controlled manner. Water is forced to go through a channel to *hydraulic turbines* located at lower elevations. The turbines *convert the water power* into *mechanical energy (See chapter 14). A hydroelectric generator is used to convert this mechanical energy into electricity.* This energy source is often more reliable than solar or wind power (especially if it's tidal rather than river). The power plants, depending upon the scale of operation, are grouped as large-scale > 30 MW, small-scale = 100 kW to 30 MW and micro-hydro < 100 kW. Other kinds of hydro-power plants are run-of-the-river and pumped storage schemes *(See chapter 14)*. India built dozens of dams since independence, which has played key role in increasing food production. Efforts are still required to bridge the gap between the projected potential and the installed capacity. Unevenness being witnessed in the hydropower development in major basins also needs to be addressed. Although, this activity is of non-consumptive nature, yet the quality of input water do affect the power production from these plants. Major advantages and disadvantages are listed as under.

Advantages	Disadvantages
✰ Renewable ✰ Does not cause air pollution ✰ Helps improve hygiene ✰ Generates employment opportunities	✰ Dams disrupt river's ecosystem ✰ Can be commissioned only in areas having rivers ✰ Displacement of communities from dam sites ✰ Require appropriate sites

Biomass Energy

Biomass is a kind of organic material made of plants and animals (micro-organisms). Biomass is highly diverse in nature. Biomass, on the basis of site of its origin, has been classified as follows (Figure 1.4).

- ✰ Agricultural fields and plantation biomass
- ✰ Industrial biomass
- ✰ Forest biomass
- ✰ Urban waste biomass
- ✰ Aquatic biomass

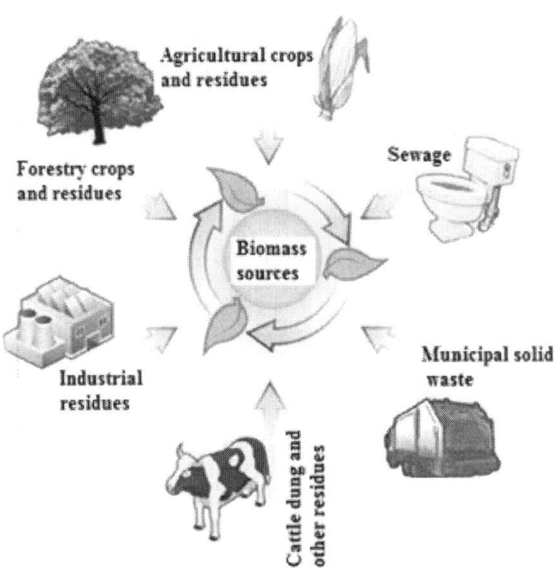

Figure 1.4. Sources of biomass.

India produces about 450-500 million tonnes of biomass per year. Although biomass can be used directly such as wood and *gobar* (dung), it can also be used indirectly. The biomass in that case is used to produce methane through anaerobic digestion and liquid and gas fuels through modern processes of pyrolysis and gasification. Methane is used for cooking, lighting and to produce steam to drive steam turbines to produce electricity. Methane and other liquid and gas fuels can be used in engines and gas turbines. India has the potential to generate around 18000 Megawatt (MW) of renewable energy using biomass and an additional 7000-8000 MW from bagasse cogeneration in sugar mills. Currently, biomass provides 32 per cent of all the primary energy used in the country having an existing capacity of over 7,000 MW. Plants like *Jatropha curcas, Neem, Mahua* and other wild plants are potential sources for biodiesel production. Goldemberg (2000) while making world energy assessment and addressing the issues of challenge of sustainability listed various kinds of biomass and processed fuels (Table 1.1).

Table 1.1. Types and examples of plant biomass

Woody Biomass	*Non-woody Biomass*	*Processed Waste*	*Processed Fuels*
☆ Trees ☆ Shrubs and scrub ☆ Bushes such as coffee and tea ☆ Sweepings from forest floor ☆ Bamboo ☆ Palms	☆ Energy crops such as sugarcane ☆ Cereal straw ☆ Cotton, cassava, tobacco stems and roots ☆ Grass ☆ Bananas, plantains and the like ☆ Soft stems such as pulses and potatoes ☆ Swamp and water plants ☆ Cereal husks and cobs ☆ Bagasse ☆ Wastes from pineapple and other fruits ☆ Nut shells, flesh and the like	☆ Plant oil cake ☆ Sawmill wastes ☆ Industrial wood bark and logging wastes ☆ Black liquor from pulp mills ☆ Municipal waste	☆ Charcoal from wood and residues ☆ Briquette and densified biomass ☆ Methanol and ethanol ☆ Plant oils from palms, rape, sunflower and the like ☆ Producer gas ☆ Biogas

Advantages and disadvantages of biomass power are listed as under.

Advantages	*Disadvantages*
☆ Renewable ☆ Waste to gold ☆ Locally available and abundant	☆ Requires large area of farmland ☆ Produces greenhouse gases, smoke and particulate matter ☆ Low energy density

Biomass conversion routes for producing energy are listed as under and will form the basis of different chapters of this book.

Physical	Densification of biomass into solid such as briquettes
Agrochemical	Fuel extraction (solid and liquid) from freshly cut plant
Thermo-chemical	Combustion Carbonization Pyrolysis Gasification Liquefaction Anaerobic digestion to methane
Biochemical	Ethanol Methanol Biodiesel Hydrogen formation cell

Geothermal Energy

The word geothermal comes from the Greek words geo (earth) and therme (heat). Thus it is the heat contained in the earth. The inside of the earth *i.e.* the core about 4000 miles deep, is very hot as much as reaching over 5,000°C (Min, 2018). It is one of the few renewable energy sources that are not directly related to the Sun as an energy source. The geothermal heat has its origin in radioactive isotopes, in the movement of the surface layers of the Earth and in the latent heat coming from inner core crystallization. Geothermal energy is contained in the rocks and fluids beneath the earth's crust and can be found as far down to the earth's hot molten rock, magma. The listed value for the outward flow of heat from the core of the earth equals 8×10^{17} Btu/yr = 8×10^{27} erg/year. It includes heat flow to the large areas covered by the earth's oceans. At places where this geothermal heat flow is usable, the conversion efficiency to electricity is not likely to exceed about 25 per cent. The early history of using geothermal energy dates back to 2000 years. Since its first commercial use in Italy in 1904, geothermal has been a consistent and expanding source of energy. According to the International Renewable Energy Agency (IRENA), geothermal energy has grown steadily from around 10 GW worldwide in 2010 to 13.3 GW in 2018. Geothermal energy relies on almost unlimited amount of heat produced under the surface of the Earth's core or heat generated in the earth itself like volcano, geysers, hot springs *etc.* Even in geothermal areas dependent on a reservoir of hot water, the volume taken out can be re-injected, making it a sustainable energy source. Many technologies have been developed to take advantage of geothermal energy of the earth.

- Hot water or steam reservoirs deep in the earth accessed by drilling
- Geothermal reservoirs located near the earth's surface
- The shallow ground near the Earth's surface that maintains a relatively constant temperature of 10-16°C

Earth energy can be tapped almost anywhere for direct-use applications or using geothermal heat pumps. In direct use applications, hot water under the earth's surface is used for heating of buildings, fish farms, spas, and greenhouses. It can also be used for cooling of buildings by pumping underground water where the temperature remains relatively stable near 16°C or less. This water is circulated into buildings, which absorbs heat resulting in cooling of the building. It can also be used to produce electricity at specialized power plants. For this purpose wells are dug few km deep into underground reservoirs to access the steam and hot water. The steam and hot water are then used to drive turbines connected to electricity generators in geothermal power plants. Although beyond the scope of this book, a dry steam power generation plant is schematically shown in Figure 1.5 for general reference of the readers. Realizing the potential of geothermal energy, Quaise energy, a US based company, is developing a rig that can dig holes up to 16 km down the earth. The company claims that it will allow tapping the inexhaustible clean energy from geothermal heat in the crust, which will be enough to meet the current and future energy demands of this planet.

Some of the attributes that make geothermal a good source of energy are listed as follows.

- It has an enormous potential – upper estimates suggest a global potential of 2 terawatts (TW)
- It is a clean energy that produces only about one-sixth of the carbon dioxide that a relatively clean natural-gas-fueled power plant produces, and very little of the nitrous oxide or sulphur-bearing gases, if any
- Geothermal energy is available 24 hours a day, 365 days a year. Geothermal power plants have average availability of 90 per cent or higher, compared to about 75 per cent for coal plants
- Geothermal being homegrown, reduces dependence on imported oil

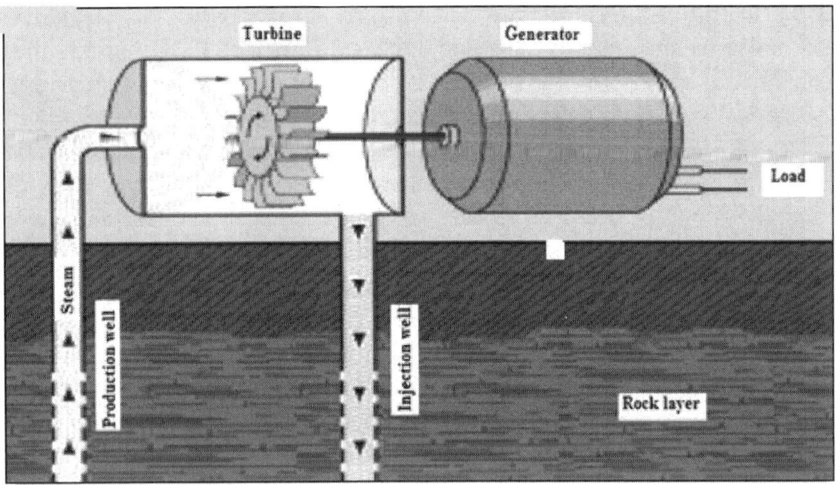

Figure 1.5. A schematic diagram of dry steam type electricity generation plant.

- ☆ Power plant require little land as these can be built partially underground
- ☆ It is very reliable for meeting the base load energy demand as compared to other renewable sources such as wind and solar. The power plants have a high capacity factor typically exceeding 95 per cent

Main limitations of geothermal energy are as follows.

- ☆ These are only practical in areas near hot spots
- ☆ Multitudes of factors affect sustainability and economics of their commercial applications
- ☆ Some geothermal plants produce some solid materials, or sludge, that may pose some problems in disposal. Some of these solids are now being extracted for sale (zinc, silica, and sulphur, for example), making this resource valuable and environmentally friendly
- ☆ Geothermal power plants have a high initial cost to build. However, the plants pay for themselves over time, because of low operating costs
- ☆ Waste water released can damage soil/aquifers and also has been the cause of mini tremors in the areas where they are operational
- ☆ Some emission of greenhouse gases during extraction of heat from ground is not ruled out
- ☆ Low concentrations of radioactive Radon may appear as a by-product of Uranium, one of the isotopes responsible for the heat accumulated below the Earth's surface

India is still at nascent stage of geothermal energy utilization. Although geothermal power potential has been estimated at 10,600 MW, no geothermal power plant has been set-up in the country for many reasons such as: high upfront cost of Rs 3000 million/MW, indicative tariff in range of Rs 10 per KWh, site specific deployment, lack of load center and power evacuation facility nearby, and high risk involved in exploration, *etc.* (Venkatachalam and Rao, 2020). Geological Survey of India (GSI) with CSIR - National Geophysical Research Institute (NGRI) carried out preliminary resource assessment for the exploration and utilization of geothermal resources during 1970-80. As per preliminary investigations undertaken by the GSI, there are around 300 geothermal hot springs in India. Most of these geothermal hot springs are in medium potential (100°C to 200°C) and low potential (<100°C) zones. The promising geothermal sites for electric power generation are Puga Valley and Chummathang in Jammu and Kashmir, Cambay in Gujarat, Tattapani in Chattisgarh, Khammam in Telangana and Ratnagiri in Maharasthra. Hot springs in the low surface temperature range from 37°C-90°C are suitable for direct heat applications. The promising geothermal sites for direct heat use applications are Rajgir in Bihar, Manikaran in Himachal Pradesh, Surajkund in Jharkhand, Tapoban in Uttarakhand and Sohana region in Haryana (Uppal, 2019). Ministry of New and Renewable Energy (MNRE) has assessed geothermal power potential of 10 GW. State-owned Oil and Natural Gas Corporation (ONGC) embarked on its maiden geothermal field development project in Ladakh in 2021. The field development of the project is proposed to be taken up in three phases. In the first phase, ONGC will drill wells and set up a pilot plant of up to 1 megawatt (MW) power capacity. In the second phase, ONGC will drill optimal number of wells and set up a higher capacity demo plant. The final phase will involve commercial development of the geothermal plant.

Ocean Tidal, Wave and Thermal Energy

Oceans occupy more than 70 per cent of earth's surface and are an inexhaustible source of renewable energy. Ocean energy is the energy harnessed from ocean tidal range (rise and fall), tidal streams, ocean waves, temperature gradients and salinity gradients.

Tides are defined as the rise and fall of sea level caused by the gravitational pull of the moon and the sun on the Earth. They are not only limited to the oceans, but can also occur in other systems whenever a gravitational field exists. Tidal power or tidal energy is a form of hydropower that converts the energy obtained from tides into useful forms of power, mainly electricity. This form of hydro energy uses twice-daily tidal currents to drive turbine generators. Although tidal flow unlike some other hydro energy sources isn't constant, it is highly predictable and can therefore compensate for the periods when the tide current is low. As per study conducted by IIT Madras, theoretical potential of tidal Energy in India is 12500 MW. Promising locations are Gulf of Khambhat and Gulf of Kutch (GJ), Sunderbans (WB), Western Ghats (MH), *etc.* A major limitation is high capital cost of tidal energy power plants due to high civil construction cost. It results in high power tariff.

Wind blowing across the surface of the ocean creates waves. These waves are also referred as energy moving across the surface of the water. *Wave energy, also known as ocean energy, is defined as energy harnessed from oceanic waves.* The technology to harness the wave energy works on the principle of rolling waves flowing through joints in a large cylindrical pipe, which pushes high pressure oil through hydraulic motors to generate electricity- which in turn is fed to an onshore grid through an underwater cable. One square kilometer of ocean interspersed with the devices would produce about 30 MW of electricity, which is enough to power 20,000 homes. Theoretical potential for wave energy in India is 41,000 MW. Promising locations are Western Coast of Maharashtra, Goa, Karnataka, Kerala, Kanyakumari, *etc.* Major bottlenecks are high upfront cost *i.e.* Rs 600 million for 1.125 MW wave energy plant at A and N islands and Rs 20000 million for 8 MW plant by Indian Navy. This result in high tariff for electricity generated through wave energy plant.

Ocean current is ocean water moving in one direction. Kinetic energy of the ocean and other tidal currents can be captured with submerged turbines that are very similar to miniature wind turbines. The movement of the marine current moves the rotor blades to generate electric power.

Ocean thermal energy (OTE) is due to the temperature differences (thermal gradients) between ocean surface water and the water at deeper depths. Ocean thermal energy conversion (OTEC) uses this ocean temperature differences from the surface to depths below 1,000 m to harness energy. A temperature difference of even 20°C can yield energy quite efficiently. OTEC technologies are based upon closed cycle and open cycle. In the closed cycle method, a working fluid such as ammonia is pumped through a heat exchanger and vaporized. This vaporized steam runs a turbine. The cold water found at lower depths of the ocean condenses the vapour back to a fluid where it returns to the heat exchanger. In the open cycle system, the warm surface water is pressurized in a vacuum chamber and converted to steam to run the turbine. The steam is then condensed using cold ocean water from lower depths.

Advantages	*Disadvantages*
☆ Running cost is low	☆ Low efficiency
☆ Continuous unlike other sources, which are intermittent in nature	☆ High initial investments

Comparison of Non-Renewable and Renewable Energy Resources

A comparative statement of non-renewable and renewable energy resources is given in the following table.

Item	*Fossil Fuels with Sequestration*	*Renewable (including Biofuels)*
Description	Fossil fuels are extracted, converted to electricity and/or hydrogen, resulting in CO_2 as by-product, which can be captured and sequestered	Renewable energy sources are converted to electricity and/or hydrogen. Biofuels may also be converted directly to mechanical power in vehicles
Sources	Non-renewable resources includes fossil fuels such as coal and petroleum	Renewable resources include sun, water, wind, geothermal such as hot springs and fumaroles and oceanic sources
Availability and depletion	200-300 years worldwide, assuming aggressive efforts to improve end-use efficiency	Indefinite. Climate change may lead to changes in output (positive or negative) in specific locations
Economic cost considerations	Combustion side is mature technology, inexpensive and reliable. Large-scale sequestration is not yet developed, its cost is unknown but likely to add significantly to life cycle cost	High cost per unit of energy, due to low capacity factors, low concentration of energy. Need for storage system due to intermittency increases cost (except for biofuels and geothermal). New technologies may reduce cost in future
Environmental cost considerations	Non-renewable energy has a comparatively higher carbon footprint and carbon emissions. Extraction of fossil fuels may negatively impact land and water resources, natural habitat, and so on	Most renewable resources have low carbon emissions and low carbon footprint. Land requirement and solidwaste issues may be there with large scale construction of infrastructure
Suitability as an energy source for transportation	Coal requires conversion to electricity or hydrogen, other fossil fuels can also be used directly as energy sources	Requires conversion to electricity or hydrogen, except for biofuels
Area requirement	Comparatively low area requirement	Requires a large land/offshore area, especially for wind farms and solar farms
Upfront cost	Non-renewable energy has a comparatively lower upfront cost. Infrastructure is available for non-renewable energy across most countries	The upfront cost of renewable energy is high. Infrastructure for harvesting renewable energy is not easily accessible in most countries

Advantages of Renewable Energy

- It is renewable, produced regularly in nature and is inexhaustible
- Available throughout the year
- The power plants based on renewable sources of energy don't have any fuel cost and therefore running cost is negligible

- Renewable sources suitable for local processing and applications. There is no need for long transmission and distribution of power
- Renewable sources have low energy density and more or less there is no pollution or ecological balance problem
- Most of the devices and plants used with the renewable energy are simple in design and construction. These can be made from local materials, local skills and by local people
- The use of renewable energy can help to save foreign exchange and generate local employment
- The rural areas and remote villages can be better served with locally available renewable sources of energy. Huge savings can be made as transporting fuels from long distances can be avoided

Disadvantages of Renewable Energy

- Intermittency and lack of dependability are the main disadvantages of renewable energy sources
- Low energy density of renewable sources requires large sized plants resulting in increased cost of delivered energy
- Low energy density also results in lower operating temperatures and hence low efficiencies
- Although renewable resources are essentially free, there is definite cost effectiveness associated with its conversion and utilization
- Much of the construction materials used for renewable energy devices are sometimes very energy intensive
- The low efficiency of these plants can result in large heat rejections resulting in thermal pollution around the plant sites
- The renewable energy plants use larger land masses

INDIA'S ENERGY SCENARIO

Energy and development have a symbiotic relationship since the energy consumption in any country is directly related to the progress of its people, its industrialization and improvement in the living standard of the people. India is now the-most populous country and is ranked third-largest energy consumer in the world in 2021 after China and the United States, using about 3.4 per cent of global energy consumption. Contrarily, world's hunger for energy combined with the decline of fossil fuels is likely to create strong vortex of energy conflict. India is aware of this impending conflict that may affect its economic growth projected at 6-7 per cent per annum. Therefore, energy has become one of the top priorities of the government making it the largest industries in India.

Energy Sources and Consumption Pattern

The energy resources of India can roughly be classified in three categories: fossil, nuclear and renewable. Coal, oil and natural gas are major fossil sources of energy. Although there is some confusion as to whether nuclear energy is non-renewable or a renewable source, yet mostly it is considered amongst the non-renewable sources. Amongst the renewable sources are hydropower, solar energy, wind energy, biofuel energy. The five sectors that consume primary energy and electricity are: transportation, industry, agriculture, residential and commercial.

India's energy consumption continues to grow exponentially as a result of population growth and modernization. Total energy consumption in 2021 was 951 Mtoe (Mtoe is *a unit of energy defined as the amount of energy released by burning one tonne of crude oil)*, which translates to around 0.7 toe per capita, which is about half the Asian average. The electricity consumption per capita reached 920 kWh in 2021, about a third of the Asian average. Coal has been the country's top energy source in 2021 with a share of 46 per cent, followed by oil (23 per cent) and traditional biomass and waste (21 per cent). In spite of the fact that India is the second largest coal producer in the world with a production of 840 million short tons on a volumetric basis, yet it experienced huge shortages being the second largest coal consumer in the world behind China. Power generation and industry are the main consumers of coal and lignite, with 76 per cent and 20 per cent, respectively. Only 3 per cent was meant for households and other services. Greater demand for power generation and industrial growth is likely to contribute to increasing coal consumption in India.

India has been the third largest consumer of petroleum products after the United States and China in 2021. To overcome the mismatch between production and demand, India imported 210 Mt of crude in 2021 representing 88 per cent of the country's oil needs (Figure 1.6). Diesel was the most-used oil product accounting for 35 per cent of petroleum products consumed. It was primarily used in commercial transportation and, to a lesser degree, in the industrial and agricultural sectors. Gasoline consumption accounted for 16 per cent of India's total oil consumption. The natural gas share of 6 per cent in the current energy mix is among the lowest in the world. Fertilizer plants were the main consumer of gas with 51 per cent. It was followed by electricity production (23 per cent) and, to a much lesser extent, transport (5 per cent) and the residential and services sector (6 per cent). Natural gas share is likely to double as gas use rises in the industrial sector and through city gas distribution for domestic sector.

According to International Energy Statistics and Estimates by US Energy Information Administration, India generated about 1628 terawatt hours (TWh) of net electricity in 2021 using various sources (Figure 1.7). Fossil fuels accounted for 76 per cent of India's electric generation. Coal had the largest share of all energy sources, at 73 per cent of total generation. Natural gas, oil, and nuclear

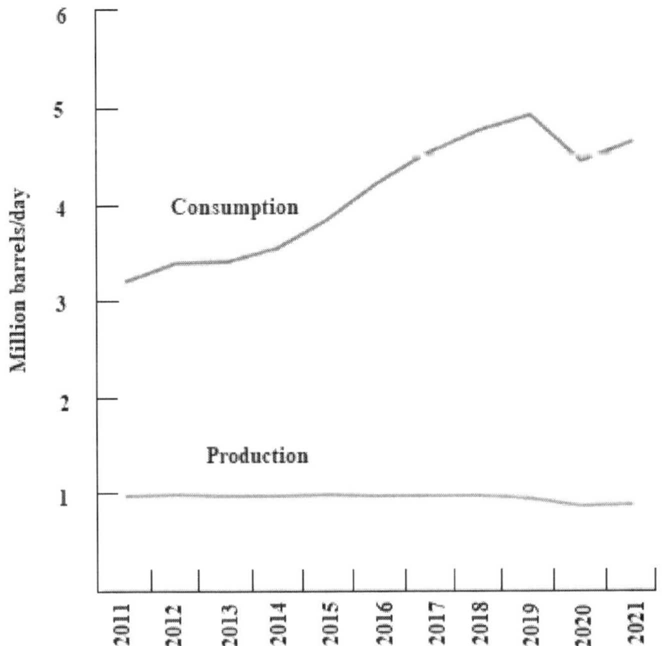

Figure 1.6. Production and consumption of petroleum products in India.

power together made up less than 7 per cent of India's power supply. India has seven nuclear power plants, with a net generation capacity of 6.8 GW, representing about 2 per cent of total utility-based generation capacity. Five nuclear reactors are under construction besides 18 additional nuclear reactors may be in place by 2025. India envisages increasing the contribution of nuclear power to overall electricity generation capacity to 9 per cent within 25 years.

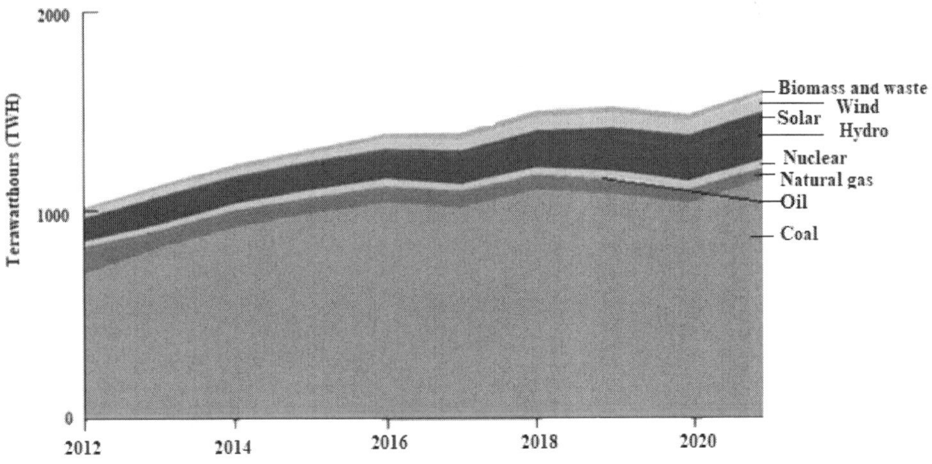

Figure 1.7. India's net electricity generation by fuel type 2012-2021.

Renewable energy made up the second largest share (21 per cent) of power generation, including hydro power (10 per cent), solar (4 per cent), wind (5 per cent), and biomass (2 per cent). The installed capacity for power generation from various sources reported in Table 1.2 shows that installed capacity is much higher than the power generated from renewable resources. At global level, renewable energy sources make up 26 per cent of the world's electricity, which according to the International Energy Agency (IEA) is expected to reach 30 per cent by 2024.

Table 1.2. Installed electrical generation capacity by source in India as on 31 January 2021

Source	*Total Capacity (MW)*	*Percentage*
Coal	200284.5	53.1
Lignite	6120.0	1.6
Hydropower (large and small)	50817.7	13.5
Solar Power	38794.1	10.3
Wind Power	38683.7	10.3
Gas	24956.5	6.6
Biomass	10314.6	2.7
Nuclear	6780.0	1.8
Diesel	509.7	0.1
Total	**338466.7**	**100**

Given India's growing energy demands and limited domestic fossil fuel reserves, the country has ambitious plans to expand its renewable energy power generation capacity. Solar power is set for explosive growth in India with addition of about 20 GW of solar power capacity by 2022. The total renewable energy capacity is currently 136 GW constituting about 36 per cent of its total capacity. The government has consciously set a target to source half of its energy through renewable sources by 2030 in accordance with India's goal to reduce carbon emissions. It means India must install 500 GW of renewable energy capacity by this time.

India continues to face several challenges to meet the country's growing energy demand. Major causes of energy crisis in India are listed as follows.

- ✰ Increasing population
- ✰ Overconsumption, wastage and theft of energy
- ✰ Aging infrastructure of power generation
- ✰ Unexplored renewable energy options such as geothermal, tidal, wave and others
- ✰ Delay in commissioning of power plants

- Poor transmission and distribution system resulting in huge losses
- Lack of skilled manpower and poor infrastructure for developing conventional and nonconventional energy sources that hampers the exploitation of full potential of national resources
- Lack of transportation infrastructure such as gas pipelines for making energy accessible
- Natural calamities resulting in disruption in power generation and distribution
- Miscellaneous factors such as tax hikes, strikes by unions, severe hot summers or cold winters causing sudden increase in demand for energy and choke supply
- Coal mining in India suffers from delays due to regulatory and environmental clearances
- Lack of awareness amongst the general masses on the importance of energy saving and climate change
- Limited investments on research and development on new potential sources of energy

The government of India has made considerable headway with energy reforms and continues to focus on greater energy security, infrastructure development, and market liberalization. But these efforts require ground support from the general public for their success. Some possible steps in this direction are:

- Move towards renewable resources making full use of government's incentives
- Use energy-efficient products
- Have lighting controls and switch-off lights as one leaves the building
- Easier grid access
- Energy simulation by big corporate to redesign the buildings and reduce running business energy costs
- Compulsory energy audit

Energy Security

India energy use has almost doubled since 2000 mainly due to rising incomes and improved standard of living making India the world's third-largest energy consumer in the world. The energy requirement will further increase several folds as India has become the world's most populous country. Moreover, coming years will see millions of Indian migrating to urban centers, households buying new appliances, air conditioning units and vehicles. Increasing pace of industrialization will put a huge burden on energy supply. These factors together will result in steep increase in per capita energy use, which is currently less than half the world average. Thus, India will continue to experience energy supply shortfall in the coming decades unless necessary steps are taken to manage supply and demand of energy.

Energy security has been defined differently by different organizations. *According to one definition energy security has been defined as the uninterrupted availability of energy at an affordable price with easy access to all as per the following equation.*

Energy security = Availability + Affordability + Accessibility + ***Acceptability***

The most comprehensive definition for energy security in India is the one provided in the report of the Expert Group of the Planning Commission on Integrated Energy Policy in India which states: *a nation is energy secure when it can supply lifeline energy to all its citizens irrespective of their ability to pay for it as well as meet their effective demand for safe and convenient energy to satisfy their various needs at competitive prices, at all times and with a prescribed confidence level considering reasonably expected shocks and disruptions.* Energy security is also categorized as short-term and long-term. *In the short-term energy security focuses on the ability of the energy system to respond promptly to sudden changes in the supply-demand balance.* For the long-term, *it should ensure energy supply in line with expanding infrastructure, provisioning of basic necessities, employment generation and manufacturing abilities and environmental needs of the nation.* This requires planning and timely investments for the future needs. Some indicators of energy security are:

- ☆ Share of fossil fuels in the primary energy supply mix
- ☆ Diversification of energy basket
- ☆ Import dependence of fossil fuels
- ☆ Diversification of import sources
- ☆ Energy intensity measured by the quantity of energy required per unit output or activity
- ☆ Average per capita commercial energy consumption
- ☆ Access to electricity
- ☆ Access to modern cooking energy fuels and technology

India, by 2025, will be the second largest consumer of global energy resources after China. By 2032, over 91 per cent of the country's energy needs will have to be imported. Rising oil demand of close to 10 per cent per year has led to sizable oil import bills. Estimates reveal that 75 per cent of total oil consumption requirement and 22 per cent of total coal consumption will be met through imports. The imports of gas and LNG (liquefied natural gas) are likely to increase in the coming years. The dependence on energy imports may further pose problems in future because of political crisis in OPEC countries, the cartelization by OPEC countries and most recent US sanctions on Iran and

Russia. Overexploitation of resources within the country is leading to loss of resources and there may be few logistical problems as monsoon affects the coal production every year. On the demand side energy demand for road transport is projected to more than double over the next two decades. Over half of the growth will be fuelled by diesel based freight transport. It is attributed to likely addition of extra 25 million trucks besides 300 million vehicles of all types by 2040. Major problems from demand point of view are over-consumption due to energy inefficiencies. It is assessed that consumption of energy in India is 3 times more for the same quantity of production than in the developed countries. The rise of the middle and high-class people has put a huge load on energy demand. Moreover, more energy is needed for cooling and warming of homes in the changing climate scenario. India's fragile energy security will therefore be under severe pressure from its rising dependence on imported oil, regulatory uncertainty, international monopolies and opaque natural gas pricing policies. Some of the strategies that can be used to overcome future challenges to our energy security are as follows.

- Notwithstanding the fact that government is giving high priority to renewable energy systems, it is lengthy long-term strategy because of its slower rate of infrastructure development. In any case, we need to prioritize the availability of fossil fuels through safe and cheap access to oil and gas either by stepping up domestic exploration or through long-term contracts. We need to build enough stockpiles of these products. India should build new refineries to maintain its ability to export refined goods to cover some of its investments on import of crude
- In order to encourage growth in the local oil and gas sector, the Hydrocarbon Exploration and Licensing Policy (HELP) should aim to minimize the government's discretion in decision-making, eliminate conflicts, reduce administrative delays, and promote the notion of revenue sharing and marketing freedom
- Diversification of energy supply sources is another alternative. Nuclear power is one of the alternatives. Since uranium reserves of India are extremely small, we need to identify partners who can help in achieving uranium needs of the country. The other alternative is blending ethanol with fossil fuels. Government in this direction is already on the right path as it has fast forwarded the deadline for mixing 20 per cent ethanol (commonly known as E20) into gasoline from 2030 to 2025
- India has some resources like cobalt, nickel, copper, and heavy rare earth metals. They are necessary building blocks for electric vehicles, solar panels, wind turbines, and batteries. We need to hasten their extraction and processing

- The *Pradhan Mantri Ujjawala Yojana* is proving to be a big milestone in household energy security. It should be supplemented with the installation of biomass pelletizing facilities and the distribution of efficient biomass *chullahs* to deliver clean fuel to rural areas. City Gas Distribution (CGD) networks for transportation and distribution of natural gas to users in the home, commercial, industrial, and transportation sectors need to be strengthened
- On the agricultural front, government subsidies and support from the National Bank for Agriculture and Rural Development's (NABARD) are needed to finance the greater use of solar irrigation pumps
- Investments on research and technology development are needed to make potential non-conventional energy sources like geothermal energy, tidal and wave energy, *etc.* economically and technologically feasible
- We have made a good headway in developing renewable energy sources. The pace of development may not only be maintained but further strengthened by providing incentives
- Energy efficient equipments both in industrial and domestic sectors should be encouraged through awareness generation. Old systems of transmission and distribution of electricity may be switched to new more efficient systems
- In order to increase the economy's competitiveness, the tax system for the import and sale of energy on a thermal value basis needs to be rationalized

India has consciously made certain policy decisions. It has already committed to provide power to every home. The provision to provide electricity on a round-the-clock basis would be an even more ambitious objective. Similarly, in line with the one hundred and thirty-three countries pledging to a net-zero carbon emissions, India has committed itself to achieve this objective by 2070. The energy transition will be long and expensive, but there is no room for compromises to be made over the matter of India's energy security. The fossil fuel-based economic system will have to be redesigned and continuous effort shall be required to increase the contribution of renewable sources in the energy basket to build a cleaner, greener and energy-secure India. The way things are happening in this sector, India is in a unique position to pioneer a new model for low-carbon, inclusive growth. If we are able to achieve this, it will show the way that robust economic expansion is fully compatible with emissions reductions and the achievement of other development goals.

RENEWABLE ENERGY SOURCES AND AGRICULTURE

Energy consumption at agricultural farms is one of the major parameters that determines the growth and progress of agriculture in any country. Farm power on Indian farms during early fifties was mainly from animate sources. Besides, non-commercial energy in the form of firewood, household dung-cakes and agro-wastes was used in various household activities mostly for preparation of food and heating. During sixties indigenous production of tractors, power tillers and stationary engines began. Besides, electricity supply to villages also improved considerably. It resulted in a significant shift in the structure of energy-use in Indian agriculture. While the contribution of human and animal power decreased, contribution of fossil fuels and electricity increased. Currently, electricity, fertilizer and diesel are the main sources of commercial energy in agriculture. The total commercial energy input in Indian agriculture has increased several folds and is likely to further increase with increasing commercialization and diversification towards high-value crops.

Energy requirements in agriculture can be divided into two groups namely direct and indirect. Direct energy is required to perform various tasks related to crop production processes such as land preparation, irrigation, intercultural operations, threshing, harvesting, and transportation of agricultural inputs and farm produce. Most farm operations are either performed using fossil fuels or electric power. Indirect energy, on the other hand, consists of the energy used in the manufacture, packing and transport of fertilizers, pesticides, seeds and farm machinery. As the name implies, indirect energy is not directly used on the farms. The indirect energy is also mainly driven by the fossil fuels and electricity. The major implications of the use of these energy sources in agriculture are:

- ✰ These energy sources are exhaustible
- ✰ These are costly and impacts the economy of agriculture
- ✰ The use of these resources contributes to greenhouse gas emissions that in turn accelerate climate change

The only realistic solution to the problems is to find alternative renewable energy sources that can replace fossil and other forms of energy. Renewable energy technologies that have potential to replace the existing sources are: solar energy, wind energy, hydropower, and biomass energy. Let us briefly look at the potential of various renewable sources in agriculture.

Solar energy is the energy derived directly from the Sun. The Sun delivers more than 10000 times the energy that humans currently use. In general, solar energy at the farms can be used in two forms. In one system, solar energy is converted into heat. Conventionally, farmers have been using the sun to dry their produce for centuries. Currently, well-designed, simple-to-use and maintain and highly efficient solar devices including solar dryers are available that can provide the energy needed at any given location and for the given time period. Modern solar crop driers are far more effective and hygienic yet very simple. Another use of solar energy is for water heating especially for livestock farming. Solar energy can also be used in greenhouse heating. So far, commercial greenhouses typically rely on the sun to supply light, but are not designed to use the sun for heating. They rely on gas or oil heaters to maintain the temperatures necessary to grow plants in the cold season. Solar greenhouses are now designed to utilize solar energy for both heating and lighting. Solar energy can easily be converted directly into electricity using photovoltaic (PV) system. The simplest and most appropriate application of this is in water pumping. PV-powered pumping systems can meet a broad range of water needs from crop irrigation to livestock watering to domestic uses. With decreasing solar modules prices as much as 70 per cent in the last 4 years, solar pumps are fast becoming a viable financial solution for irrigation. Another good use of PV is for lighting agricultural buildings and enclosures. Government of India is thinking of solarization of agriculture that might change the face of energy scene in Indian agriculture. Although less effective, but having good potential are the wind mills that can be used to pump water, grinding grains and legumes. Wind mills can be used to power both mechanical and electric pumps. Mechanical wind-powered pumps require wind speeds of 2.5 m/s minimum up to 4 m/s optimum. Leasing of lands for solar farms or wind parks is a potential area that should be exploited by the farming communities to improve their income on a regular basis.

Biomass is mostly utilized as a fuel, but now it is possible to convert it into liquid or gaseous fuels, which can also be used as feedstock for industries. Most biomass in dry state can be burned directly to produce heat, steam or electricity. On the other hand biological conversion technologies utilize natural anaerobic decay processes to produce high quality fuels from biomass. Animal waste can be easily turned into methane gas through anaerobic digestion. Anaerobic digesters are being increasingly adopted by commercial livestock farmers not only to produce energy, but also to meet new regulations for controlling animal waste. *Khadi* Village and Industries Commission (KVIC) has come out with biogas plant designs of various sizes that can be used either individually or on community scale. The gas so generated can be used for cooking, lighting or even to run generators to produce electricity. Renewable fuels such as biodiesel – a low-polluting diesel alternative fuel and ethanol – an alcohol-based fuel derived from crops can be used to run tractors or for transportation as the later alone consumes lot of fossil fuel. Ethanol can also be blended with gasoline in varying concentrations. Biomass can also be used to generate electric power by direct burning, using gasification systems, or mixing biomass with coal in coal-fired electrical generation facilities (Refer chapter 5).

In conclusion, one can say that some sort of renewable resource solar, wind, or biomass is always available at the farms. It is important to choose an appropriate source or a good combination of sources. Use of renewable energy in rural remote areas in India and in many other developing countries can help farmers to increase agricultural productivity as well as to earn more income by value addition to their products. Renewable energy and farming is a win-win combination for the farmers and the nation.

QUESTIONS (THEORY)

1. Define energy and power. Briefly discuss the various forms of energy.
2. Differentiate between conventional (non-renewable) and renewable energy sources. What are the advantages and disadvantages of conventional sources of energy?
3. Briefly discuss the various renewable energy sources.
4. List the advantages and disadvantages of renewable energy sources.
5. Briefly discuss the sources of energy and their consumption pattern in India.
6. Define energy security. Write a brief note on energy security highlighting the various strategies that can be adopted to ensure energy security.
7. Discuss the potential of renewable energy sources in Indian agriculture.

Chapter 2

Biomass and Briquetting

Renewable biomass, an energy source, is attracting the attention of the people around the globe in anticipation of fossil fuel shortages, increasing fuel prices, and global warming concerns coupled with other environmental problems. *Biomass has been defined as living or recently dead organisms and any by-products of those organisms, plants or animals. In this strict sense, biomass encompasses all living things, yet in the context of biomass energy, the term refers to those crops, residues, and other biological materials that can be used as a substitute for fossil fuels in the production of energy and other products.* Biomass feedstock include agricultural crop residues, forestry residues, grains and starch crops, dedicated energy crops, algae, wood processing residues, municipal waste, municipal solid waste (MSW) and wet waste (Figure 1.4). The term does not include coal, oil, and other fossilized remnants of the organisms. This chapter briefly introduces to the biomass and then dwells upon the densification and briquetting process to improve the characteristics of biomass fuel.

BIOMASS FEEDSTOCK

In the context of biomass energy, Biomass feedstocks comprise the following.

Grains and starch crops: Sugar cane, corn, wheat, sugar beet, industrial sweet potato, *etc.*

Agricultural residues: Corn stover, wheat straw, rice straw, orchard prunings, *etc.*

Food waste: Kitchen waste, food processing waste, *etc.*

Forestry materials: Logging residues, forest thinnings, *etc.*

Animal by-products: Cattle dung, tallow, fish oil, manure, *etc.*

Dedicated energy crops: These crops are mostly non-food crops specifically grown for biomass on marginal lands unsuitable for traditional cropping. These are either herbaceous or woody crops. Herbaceous energy crops are perennial plants/grasses that take 2 to 3 years to reach full productivity but can be harvested annually. These include switch grass, miscanthus, bamboo, sweet sorghum, tall fescue, kochia, wheatgrass, and others. The feed stocks also include short-rotation woody crops such as fast-growing hardwood trees that are harvested in about 5 to 8 years of planting. Hybrid poplar, hybrid willow, silver maple, eastern cottonwood, green ash, black walnut, sweet gum, and sycamore are some such crops/trees.

Urban and suburban wastes: Municipal solid wastes (MSW), lawn wastes, wastewater treatment sludge, urban wood wastes, disaster debris, trap grease, yellow grease, waste cooking oil, *etc.*

Algae: Algae as feedstock refers to a diverse group of highly productive organisms that include microalgae, macro algae (seaweed), and cyanobacteria (formerly called blue-green algae). The algae use sunlight and nutrients to produce biomass that contains lipids, proteins, and carbohydrates and can be converted and upgraded to a variety of biofuels and products.

Aquatic and marine wastes: Aquatic plants other than algae and waste

Any kind of biomass discussed in the preceding section can be used as a source of

- ✰ Combined heat and power (CHP) source that represent the most efficient use utilizing around 80 per cent of potential energy of biomass. The combined facilities capture the waste heat and/or steam from bio-power production and are used in nearby buildings to provide heat or to chillers for cooling
- ✰ Biofuels for various sectors including transportation sector
- ✰ Industrial products such as bio-based foams, plastics, fertilizers, lubricants, and industrial chemicals

Availability of Biomass

The total availability of biomass in India is assessed at 750 Mt. The bulk of it comes from harvested crops which generate large volume of residues both on and off-farm. Ministry of New and Renewable Energy estimated that annually about 500 Million tonnes (Mt) of crop residues are generated (Table 2.1, MNRE, 2009). Uttar Pradesh with 60 Mt of crop residues tops the list, followed by Punjab (51 Mt) and Maharashtra (46 Mt). Amongst crops, cereals generate maximum residue (352 Mt), followed by fibers (66 Mt), oilseeds (29 Mt), and pulses (13 Mt). While cereal crops (rice, wheat, maize, millets) contribute 70 per cent to the crop residues, rice crop alone contributes 34 per cent. Sugarcane tops and leaves generate about 12 Mt. Assuming a useable biomass from crop residues at 230 Mt (Table 2.1, surplus + burnt), Ministry of New and Renewable Energy assessed the power potential of 28 GW from biomass and another 14 GW through bagasse based co-generation in the sugar mills. Biomass and bagasse based projects aggregating to 10170 MW have either been installed or are in the pipeline. It is assessed that potential of crop residue biomass, if fully utilized can

Table 2.1. State-wise crop residue generated, residue surplus and burned (Million tonnes)

States	*Residue Generation (MNRE, 2009)*	*Residue Surplus (MNRE, 2009)*	*Residue Burned (Pathak et al., 2010)*
Andhra Pradesh	43.89	6.96	2.73
Arunachal Pradesh	0.40	0.07	0.04
Assam	11.43	2.34	0.73
Bihar	25.29	5.08	3.19
Chhattisgarh	11.25	2.12	0.83
Goa	0.57	0.14	0.04
Gujarat	28.73	8.90	3.81
Haryana	27.83	11.22	9.08
Himachal Pradesh	2.85	1.03	0.41
Jammu and Kashmir	1.59	0.28	0.89
Jharkhand	3.61	0.89	1.10
Karnataka	33.94	8.98	5.66
Kerala	9.74	5.07	0.22
Madhya Pradesh	33.18	10.22	1.91
Maharashtra	46.45	14.67	7.42
Manipur	0.90	0.11	0.07
Meghalaya	0.51	0.09	0.05
Mizoram	0.06	0.01	0.01
Nagaland	0.49	0.09	0.08
Odisha	20.07	3.68	1.34
Punjab	50.75	24.83	19.65
Rajasthan	29.32	8.52	1.78
Sikkim	0.15	0.02	0.01
Tamil Nadu	19.93	7.05	4.08
Tripura	0.04	0.02	0.02
Uttarakhand	2.86	0.63	0.78
Uttar Pradesh	59.97	13.53	21.92
West Bengal	35.93	4.29	4.96
Total	**501.73**	**140.84**	**92.81**

take care of the energy needs of the country. Besides the crops, we have huge potential from animal husbandry sector. The cattle population of India was nearly 305 million in 2021 against world cattle population of 996 million. The total availability of wet dung is assessed at about 740 Mt per annum assuming average wet dung per animal per day to be 10 kg and a collection efficiency of 66 per cent. This would enable to produce 28860 million m^3 of biogas, besides producing enriched manure as a by-product to the tune of 265 million tonnes. It can save huge foreign exchange on importing fertilizers. On global scale, these values can be more than 3 times these values. For example in US, biomass energy provided 4835 trillion British thermal units of energy, which constituted about 5 per cent of the primary energy use in the US.

CONSTITUENTS OF BIOMASS

The constituents of biomass are associated with a variety of physical forms. The main structural cell constituents in plant-derived biomass are cellulose, hemicellulose and lignin. The ratio of these three polymers varies from one plant to another. Other major bio-organic polymer structures such as starch and chitosan are also found in the biomass.

Cellulose

Cellulose is the structural component of a green plant's cell walls. It comprises carbon, hydrogen, and oxygen in the form of starches, proteins, and sugars and is the most abundant organic material on the earth. Cellulose makes up approximately 50 per cent of dry weight of woody biomass. It is a polysaccharide in which D-glucose is linked uniformly by β-glucosidic bonds. Its molecular formula is $(C_6H_{12}O_6)_n$. The degree of polymerization, indicated by n is quite broad ranging from several thousand to several tens of thousands. It has a high molecular weight with a molecular mass typically of the order 10^6 kg/kmol. Figure 2.1 shows the molecular structure of cellulose with repeating units consisting of two glucose units called cellobiose. Cellulose is indigestible by man but can be digested in herbivores with the help of micro-organisms present in their alimentary canal. Total hydrolysis of cellulose yields D-glucose (a monosaccharide), but partial

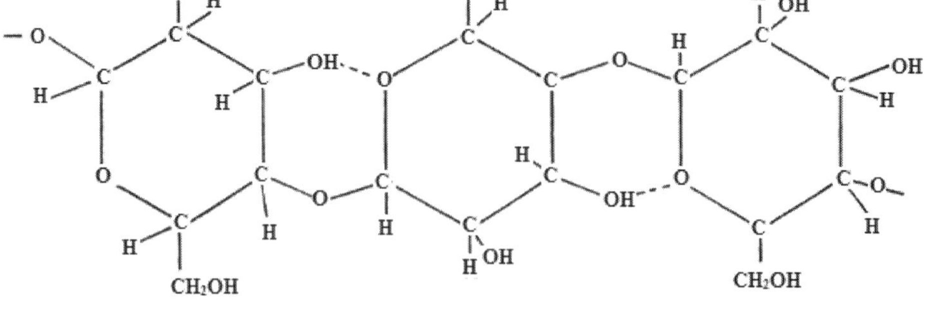

Figure 2.1. Molecular structure of cellulose.

hydrolysis yields a disaccharide (cellobiose) and polysaccharides in which n is of the order of 3 to 10. Cellulose has a crystalline structure and great resistance to acids and alkalis. Cellulose from wood is isolated during the pulping process and processed to yield ethanol, cellophane, cardboard, paper, and cellulose ethers such as acetate, rayon, and nitrates. It can also be chemically modified to yield other substances which are useful in manufacturing of plastics, rayon, photographic films *etc.*

Hemicellulose

Another component of a green plant's cell walls is hemicellulose. Unlike cellulose, which contains only one sugar (glucose), it can include a number of sugar monomers such as xylose, hexose, mannose, and galactose *etc.* It has branches with short lateral chains which consist of different types of sugars (Figure 2.2). Hemicellulose, a polysaccharide whose units are 5-carbon monosaccharides including D-xylose and Darabinose, and 6-carbon monosaccharides including D-mannose, D-galactose, and Dglucose, makes up 25 to 35 per cent of the dry weight of woody biomass. The 5-carbon monosaccharides outnumber the 6-carbon monosaccharides. The average molecular formula is $(C_5H_8O_4)_n$. They have a lower molecular weight compared to cellulose. Hemicellulose breaks down easily because the degree of polymerization n is 50 to 200, much smaller than that of cellulose. All hemicelluloses vary in amounts depending on tree species and the part of the plant. Hemicelluloses also differ in composition. Agricultural biomass such as straw and grasses are mainly composed of xylan while softwood hemicelluloses contain only glucomannan.

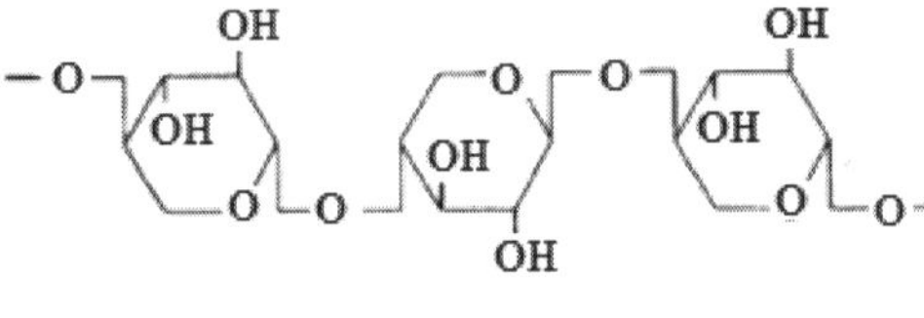

Figure 2.2. Molecular structure of a typical hemicellulose.

Lignin

Lignin is a polymer (a glue-like substance) that holds cellulose and hemicellulose together. It constitutes approximately 15 to 25 per cent of the dry weight of woody biomass and renders the biochemical conversion of woody biomass difficult. It has a large and complex molecular structure containing cross-linked polymers of phenolic monomers (Figure 2.3). Lignin has not yet been used as a raw material for industrial purposes in large quantities. It is a compound whose constituent units, phenylpropane and its derivatives, are bonded 3-dimensionally. Its complex 3-dimensional structure is difficult to decompose by micro-organisms and chemicals. In fact its function is thought to be conferring mechanical strength and protection. Cellulose, hemicellulose, and lignin are universally found in many kinds of biomass, and are the most plentiful natural carbon resources on earth.

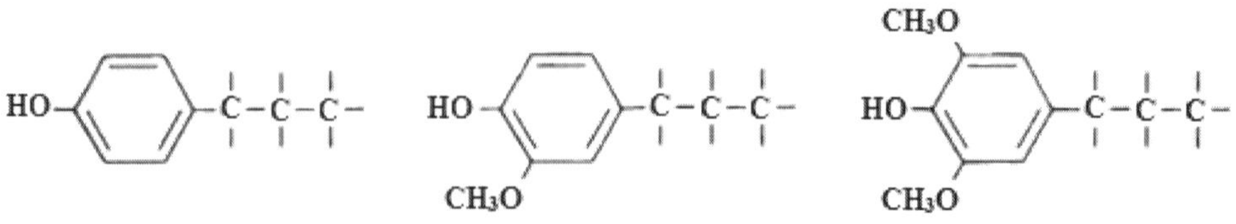

Figure 2.3. Some structural units of lignin.

Each constituent discussed in the previous sections undergo pyrolysis differently in different temperature ranges thereby producing different pyrolysis products. Differential thermogravimetry (DTG) data on some selected biomass show the following range of temperatures for the initiation of pyrolysis.

Hemicellulose: 150 to 350°C

Cellulose: 275 to 350°C

Lignin: 250 to 500 °C

While the main sources of volatiles in lingocellulosic biomass are cellulose and hemicelluloses, cellulose is the main source of condensable vapours and lignin is a major contributor to char yield.

Other Constituents

Starch

Like cellulose, starch is a polysaccharide whose constituent units are D-glucose, but they are linked by α-glycosidic bonds. Whereas cellulose is not water-soluble, part of starch is soluble in hot water (amylose, with a molecular weight of about 10,000 to 50,000, accounting for 10 to 20 per cent of starch) owing to the difference in the bond structures. A part of starch is also not soluble in water (amylopectin with a molecular weight of about 50,000 to 100,000 that accounts for 80 to 90 per cent of starch). Starch found in seeds, tubers (roots), and stems has high food value.

Proteins

These are macromolecular compounds in which amino acids are polymerized to a high degree. Properties of proteins differ depending on the kinds and ratios of constituent amino acids, and the degree of polymerization. Proteins are not a primary component of biomass, and account for a lower proportion than the previously discussed three components.

Additional Organic and Inorganic Constituents

The amounts of the other organic components vary widely depending on the specie. Some organic components of high value are glycerides (as in rapeseed oil, palm oil, and other vegetable oils) and sucrose in sugarcane and sugar beet. Other constituents are alkaloids, pigments, terpenes, and waxes. Although found in small amounts, they have very high value as pharmaceutical ingredients. Biomass besides organic macromolecular compounds also contains inorganic substances (ash) in trace amounts. The primary metal elements include Ca, K, P, Mg, Si, Al, Fe, and Na. Substances and their amounts differ according to the feedstock type.

CHARACTERISTICS OF BIOMASS AND STANDARD TESTS

Some of the characteristics and standard tests to characterize biomass are listed below.

Total Solids

The moisture content of biomass materials vary in a wide range. For example, fresh, green wood is often about half water, but many leafy crops are primarily water. A low moisture level in the fuel is usually preferable because high-moisture fuels burn less readily and provide less useful heat per unit mass. It is because much of the energy is used to heat and vaporize the water in wet fuel. Extremely dry fuel however, can also cause problems such as dust that might foul equipment or can even be an explosion hazard. Thus, moisture content of the biomass plays a large role in selecting the most suitable energy conversion process and on briquette quality. Wet conversion processes such as fermentation are often more suited to biomass with high moisture content (*e.g.* corn, sugarcane, barley straw). Dry conversion processes such as pyrolysis, gasification, and combustion are more suited to biomass with low moisture content (*e.g.* wheat straws, pine, switch grass, *etc.*). Trace moisture promotes the lignin softening and plasticizing. The effect of biomass moisture content on the suitability of briquetting using a screw extruder is illustrated in Table 2.2. It means the optimum moisture range is about 10 per cent for the two kinds of materials listed in this Table. Moisture content of the biomass sample is determined in the laboratory to work out the total solids.

Moisture content (Per cent dry basis) = 100 (Total weight of biomass – Oven dry weight)/Oven dry weight

Table 2.2. Impact of moisture content on moulding capability of two biomass materials

Moisture content (per cent)	4	6	8	10	12	14
Woody materials	No	Yes	Yes	Yes	Yes	No
Husk/stalk type materials	No	Yes	Yes	Yes	No	No

Biomass Size and Density

The size and density of the biomass particles affect the burning characteristics of the fuel as both these parameters affect the heating rate and drying during the combustion process. The improperly sized biomass will adversely impact the efficiency of the combustion process besides causing jamming or even damaging the handling equipment. Thus, fuel size dictates the type of handling equipment needed. Both bulk density and particle density of the biomass are determined in the laboratory. These values are also used to calculate the porosity and void ratio to appropriately categorize the biomass.

Ash Content

Biomass comprising grasses, bark, and field crop residues typically have high ash content (the mass fraction of incombustible inorganic or mineral material) than wood biomass. Thus, systems designed to combust wood can be overwhelmed by the volume of ash when other biofuels with high ash content are used in such systems. It will not only reduce the combustion efficiency but may also clog the ash handling mechanism. Slagging and fouling problems are encountered when ash begins to melt. It results in deposits inside the combustion equipment affecting its efficiency and life.

Flowablity

Flowability is a measure of how well a biomass flows from one point to another. This property is useful in characterizing the biomass for transportation, storage, handling and feeding.

Grindability

This property is used to classify solid biofuels in terms of their milling characteristic. *It is determined by an index expressed as the ratio of the energy consumed during the test and the amount of specified fraction of product size after milling.*

Besides these parameters, other characters of the biomass are tested/ analyzed in the laboratory. For this purpose biomass extract is prepared either in some alcohol or in pure water. Ethanol and methanol are commonly used to prepare the alcoholic extracts. The objective is to remove non-structural material from the biomass sample to prevent interferences during analysis, as well as for free sugar determination.

Structural carbohydrates: Analysis of glucose, xylose, galactose, arabinose and mannose concentrations in the biomass helps to determine cellulose and hemicelluloses concentrations in the biomass. On an average, soft and hard wood comprise about 75 per cent of cellulose and hemicelluloses.

Lignin: It is analyzed to determine the structural plant material that does not contribute to the sugar content in the sample. On an average soft and hard wood biomass comprises 25 per cent of lignin.

Acetyl content: Acetic acid concentration in the biomass determines the acid content of the feedstock. Besides acetic acid, analysis normally includes formic and levulinic acids.

Starch content: The test results represent the readily available source of sugar within the feedstock.

Ethanol content: It is a measure of total alcohol content in the fermentation broths. Analysis of fermentation broths for ethanol is mainly carried out by gas chromatography although several other equipment such as liquid chromatograph, and mass spectrometry can also be used for this purpose.

Calorific value: The heat value, or amount of heat available in a fuel (kJ/kg), is one of the most important characteristics of a fuel because it indicates the total amount of energy available in the fuel. The heat value in a given fuel type is mostly a function of the fuel's chemical composition. The sample's calorific value is determined using bomb calorimetry.

Some of the advanced instruments used to measure the characteristics of biomass are: Fourier Transform Infrared (FT-IR) Spectrometers, Fourier Transform mid-infrared (FT-MIR) Spectrometers, Fourier Transform near Infrared Spectrometer (FT-NIR), Fast Protein Liquid Chromatography (FPLC), and High Performance Liquid Chromatography (HPLC). The discussion on these instruments is beyond the scope of this book.

BIOMASS CONVERSION

Three main process technologies namely thermal, biological and extraction are used to convert biomass to energy products. These are further categorized in various sub-groups. For example, thermal conversion includes direct combustion, gasification, liquefaction, pyrolysis and partial oxidation and biological includes fermentation and anaerobic digestion (Figure 2.4). Some of these processes are discussed in the forthcoming chapters.

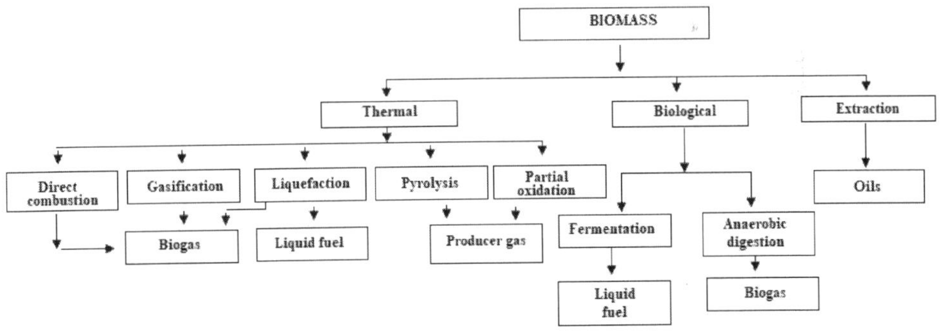

Figure 2.4. Energy conversion routes of biomass.

DENSIFICATION

Amongst all kinds of biomass, agro residues are one of most promising sources as these are locally and abundantly available as free or almost free. Although, biomass can be utilized directly as a fuel, yet its bulky nature poses some difficulties in handling. Moreover, direct burning of agro residues in domestic and industrial applications is inefficient and associated with wide scale air pollution. One of the options to achieve more efficient use of these residues is to densify them to compact pieces of definite shape and high thermal value. *The densification has been defined* as the *process of compaction of biomass into a product of higher density than the original raw material. It is also known as briquetting.* The physical properties such as moisture content, bulk density, void volume, thermal properties besides biomass composition affect the process of biomass densification.

Biomass Composition and Densification

Raw biomass has low molecular weight and macromolecular compositions. While, low molecular weight substances include organic and inorganic matter, macromolecular substances include cellulose, hemicellulose, and lignin. Cellulose may not prove to be a suitable adhesive because of the semi-crystalline structure and high hydrogen bonding of the raw materials. But heat treatment in the drying range makes the cellulose molecules quite flexible. Hemicellulose on the other hand produces adhesive products during its degradation. Thus, hemicelluloses together with cellulose present in almost all plant cell walls helps in natural bonding. Lignin content of the biomass plays the most significant role in biomass densification as it helps to build solid bridges. It begins to soften at about 180°C. It sticks to cellulose and binds the adjacent particles at about 340°C under certain pressure. The presence of lignin in plant materials allows pelletization without adding binders. It is why the briquettes made from highly lignified wood are more durable, as wood has higher lignin content than other biomass materials. Moreover, lignin upon burning yields more energy than cellulose and therefore, serves as an excellent fuel. Starch granules at high temperature and moisture content influence the binding properties of many foods and feeds. Starch gelatinization is an irreversible process and is influenced by densification process variables like heat, water, shear, and residence time. Starch not only acts as a binder but also as a lubricating agent during pelletization, helping to ease the flow of materials through the die. Proteins undergo denaturization upon heating during the densification process. It helps in the formation of new bonds and structures with other proteins, lipids, and starches improving the binding capacity. It promotes binding of straws having a certain percentage of protein content. Fat content in biomass acts as a lubricant during pelletization. It not only increases throughput but also reduces pressure requirement in the pelleting process. But too high fat content might hinder binding. Apparently, variations in feedstock composition can significantly affect densfication process as well as the quality of densified materials.

Types of Densification Processes

Densification of biomass can be performed either using the biomass directly with or without the addition of a binding material or after carbonization. Some of the processes of densification comprise the following (Vijayakumary and Desai, 2021).

Rolling/Compressing/Bailing: Rolling, compression and bailing can reduce the biomass volume to about one fifth of its loose bulk. In rolling biomass is wrapped round a rotating shaft which produces a high density roll or log. In compression, the biomass is forced through the holes in a die-plate by pressure rolls (product 10-18 cm diameter). Bailing the residues helps in easy handling of the crop residues as well as for other applications such as animal feeding, fuel and fiber for paper manufacturing. Both on-site and off-site bailers are now available. These bailers are used to prepare bales of different shapes but rectangular or rod shaped bales are more common. These processes are quite useful for agricultural residues like straw and certain type of forest biomass

Briquetting: Briquetting is a technology to reduce the volume-weight ratio of the biomass. Briquetting increases the bulk density of loose agro residues from 0.1 to 0.2 g/cm^3 to about 1.2 to 1.4 g/cm^3. In this process biomass is compacted to produce briquettes of various shapes ranging in size from 1 to 4 cm.

Pelleting: The process of pelleting involves densification of agricultural residues like straws by means of a pellet mill or pelletizer. Prior to pelleting, the straw is chopped and hammered to the required size and then steam conditioned to 60 to 82°C in a steam conditioner (Vijayakumary and Desai, 2021). Most mills have one or more steam conditioning units mounted above the main unit. The steam softens the feed and partially gelatinizes the starch to create more durable pellets. The pellet mill employs a hard steel die which is perforated with holes of 1.0 to 1.3 cm diameter. Biomass is forced through the holes of the die-plate by pressure rolls. The product size is in the range of 0.5- 1 cm.

Cutting: It is a modified form of pelleting with product size in the range of 2 -5 cm.

Extrusion: Herein biomass is forced through the holes using a screw. The resulting product is rod shaped having diameter in the range of 2-10 cm.

An alternative to densification is to convert the biomass into liquid or gaseous fuels for transportation and as feedstock for industries. This is achieved through biological conversion technologies utilizing natural anaerobic decay processes to produce high quality fuels from biomass. The processes are grouped into 3 classes:

- Thermo-chemical conversion
- Bio-chemical conversion
- Oil/biofuels extraction

These will form the subject matter of discussions in the forthcoming chapters of this book.

BIOMASS BURNING

The easiest and simplest way of utilizing the heat content of any biomass resource is to burn it as it is. Utilization of leaves, bark, wood, cakes *etc.*, for cooking food and warming the space in winter in this manner is known to human civilization for many centuries. However, it was realized that such burning of biomass has certain disadvantages.

- Most of the agro-forestry biomass contains high moisture (18-20 per cent). A large fraction of the heat generated by burning high moisture biomass is consumed in drying the biomass itself
- Most of the biomass is bulky and some of them are in powder form, generating dust and polluting the environment through production of smoke
- Difficult to handle and transport
- Storage is expensive
- Biomass has poor combustion properties. Agricultural wastes burn so rapidly that it is difficult to maintain a steady fire because of difficulties in controlling the combustion process
- Most of the biomass cannot be used efficiently in simple gasifiers and small boilers. Expensive and sophisticated furnaces are needed for this purpose
- Draft and exhaust system can get completely blocked due to ash
- Ash may insulate the heating area reducing the energy efficiency of the system
- Some of the biomass need an expensive and sophisticated equipment like fluidized bed boilers to achieve maximum efficiency

BRIQUETTING

The simplest solution to overcome all the problems discussed in the preceding section lies in densification of agricultural residues into solid fuel pellets or briquettes. This involves reducing the size by pressing the bulky mass. Unlike the loose and bulky form, combustion of briquettes is more uniform. Briquetted materials can be burned directly as fuel in somewhat similar fashion as the fuel wood and coal in domestic stoves and ovens. Some developing countries, *e.g.* India, Thailand, and few countries in Africa, have had experiences of substituting fuel briquettes for fuel wood and coal to reduce the problems of fire wood shortage and farm-waste disposal. Even development of insulation and architectural boards or wafers and cubes formation from forage materials are techniques of densification, though with a purpose different than utilizing biomass for energy production.

Briquetting of biomass is an age old technology in India, being as old as agriculture. Cow dung cakes, fuel balls made from coal-dust as well as special *chulhas* using hand compressed wood shavings are some of the prime examples of this process being used for centuries in India. Briquetting as a technology was invented/perfected in early 19th century. The Pacific Coal and Wood Co. Los Angeles, California, USA began marketing wood waste briquettes in 1923. The products obtained through briquetting could be in a solid geometrical form or in the form of hollow cylinders. It has compression ratio of approximately 7:1. The compacted fuels, known as briquettes is more or less similar to coal and has potential to replace conventional solid fuels and even diesel to meet the local needs of various sectors.

Advantages of Briquetting Biomass

- The process solves the problem of loose waste/residues from agriculture, forestry and agro-industrial processing so as to check environmental pollution
- The process increases the net calorific value per unit volume
- The fuel produced is uniform in size and quality
- No toxic gases and sulphur emission, even no odour during combustion
- Densified product with a higher bulk density (1000-1500 kg/m^3) than raw agro-wastes (50-200 kg/m^3) is easy to transport and store
- Fire risk is minimized, which is a major problem in loose storage of biomass
- The process produces high quality fuel with very low ash content (2-5 per cent) compared to 30-40 per cent in case of coal
- The briquettes are easy to burn having lower ignition temperature compared to coal
- Gases produced during burning accelerates burning efficiencies besides curtailing CO_2
- Helps to reduce deforestation by providing a substitute for fuel wood
- The process reduces biodegradation of residues
- Briquetting process consumes low energy
- Provides job opportunities
- It helps to get rid of insects that often damage the biomass and are health hazard to humans and animals

Briquetting Process

The briquetting process mainly consists of four steps (Figure 2.5).

- Collection of raw materials
- Preparation of raw materials consisting of drying, grinding and binder mixing, if required

- ☆ Densification or compaction
- ☆ Cooling, packaging, storage and marketing

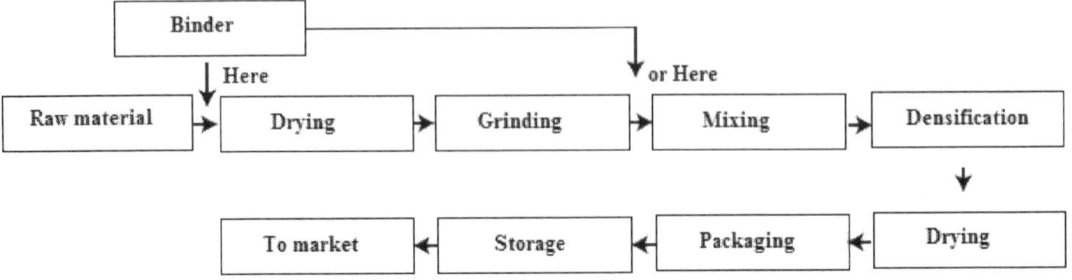

Figure 2.5. Production of non-carbonaceous briquettes: flow diagram.

Raw Materials for Briquetting

Agro residues are commonly used raw materials for briquetting in India. Almost all agro residues such as saw dust, rice husk, tapioca waste including dried tapioca sticks, groundnut shell, cotton stalks, pigeon pea stalks, soybean stalks, coir pith, mustard stalks, sugar cane bagasse, wood chips, tamarind pods, castor husk, coffee husk, coconut shell powder *etc.* can be briquetted. In general, materials that can burn but have inconvenient shape, size or form that pose problems in its direct use as a fuel can be good candidates for briquetting. Main factors that influence the selection of raw materials for briquetting are moisture content, ash content, flow characteristics, particle size and local availability. Moisture content in the range of 10-15 per cent is preferred. Biomass with high moisture content will require more energy for drying besides posing problems in grinding. Biomass feedstock having ash content up to 4 per cent is preferred for briquetting. Biomass with high ash content together with its mineral composition and operating conditions affects the slagging behaviour in the combustion units. The granular homogeneous materials which can flow easily in conveyers, bunkers and storage silos are suitable for briquetting.

Preparation of Raw Material

The preparation of raw materials includes several steps such as drying, size reduction (Figure 2.6), mixing of raw materials in correct proportions and mixing of raw materials with binder, if required. The raw materials having high moisture content than the optimum for briquetting are first dried either in open air (sun), or in solar driers, or with hot air supplied through heaters. The raw material is then reduced in size by shredding, chopping, crushing, breaking, rolling, hammering, milling, grinding, cutting *etc.* until it attains a suitably small and uniform size (1 to 10 mm). This step can be missed for materials already in the size range of 1 to 10 mm and can directly be used for briquetting. Since the size reduction process also consumes a good deal of energy, it is better to avoid excessive reduction in size. It is always preferred to use more than one raw material for briquetting mainly to have high calorific value briquettes. Mixing of these materials should be done in proper proportion in such a way that the product mix can be compacted well. Equipment used in preprocessing are drying equipment (dryer, hot air generator, fans, cyclone separator, and drying unit), cutter/clipper, shredder and grinder *etc.* Besides, material handling equipment like screw conveyors, pneumatic conveyors and holding bins may also be required.

Figure 2.6. A view of the shredding machine.

Briquette Making

Two main processes of briquetting biomass are through direct compaction and compaction after pyrolysis or carbonization. Direct compaction is further grouped under cold briquetting with or without binder and hot briquetting.

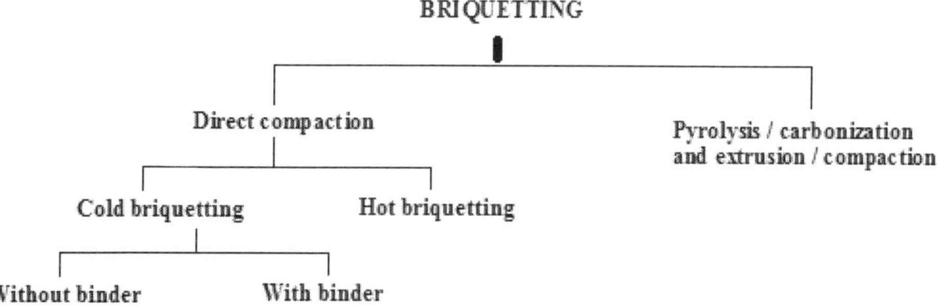

The vital requirement of briquetting of woody biomass is to destroy the elasticity of the wood, which is accomplished either by heat treatment or by applying high pressure or by a combination of the two processes. Based on the pressure used, briquetting technologies are grouped into three categories namely

- ☆ High pressure or high compaction technology (Binder less)
- ☆ Medium pressure technology (With binder)
- ☆ Low pressure technology (Pyrolysis/carbonization and extrusion/ compaction)

The densification of biomass under high pressure causes mechanical interlocking and increased adhesion/cohesion of the solid particles that helps to form intermolecular bonds in the contact area. This increase in adhesive forces at the solid-liquid interface and cohesive forces within the solid help in binding. Lignin of biomass/wood under high pressure and/or temperature helps in binding. In medium pressure technology, several additives of high viscous bonding media known as binders are used in briquetting. These binders are categorized as:

- ☆ Glucosidic (*e.g.* molasses, starch)
- ☆ Inorganic solutions (*e.g.* sodium silicate)
- ☆ Clays (*e.g.* bentonite)
- ☆ Thermoplastic (*e.g.* bitumen, pitch)
- ☆ Mortar type (*e.g.* cement, hydrated lime)
- ☆ Non-glucosidic organic solutions (*e.g.* resins)
- ☆ Fibers (*e.g.* paper fluff)

Any kind of binder must be handled properly during transport, storage and mixing. The feed must be free flowing. Last but not the least binder should be available in required quantities at reasonable cost.

BRIQUETTING MACHINES

Based on the process requirement, briquetting machines have been designed for high pressure compaction, medium pressure compaction assisted by a heating device and low pressure compaction with a binding agent.

Components of Briquetting Machines

In general, a briquetting machine consists of the following components.

Briquetting stand: It supports the drive and the bearings with the briquetting barrel assembly.

Drive mechanism: The machine is driven by variable speed drive electrical motor or some gear reduction mechanism to obtain proper speed. Power requirement ranges from 0.05 kWh/kg to 0.36 kWh/kg.

Thrust bearings: The bearings support the briquetting screw at the driven end or both the ends. It compensates the rear ward thrust resulting from the forward movement of the screw.

Feed hopper: The feed materials are supplied to the system through a feed hopper having some kind of agitator installed at the bottom to minimize bridging of the feed. Sometimes vibration of the machine alone tends to ease the flow of the feed materials. The feed hopper is often offset to the side of the barrel in the screw's rotational direction.

Barrel: Barrel, because of its huge size, is normally manufactured in number of small parts bolted together or assembled together. The construction in this manner helps to ease the repairing operation as and when required. The basic size of the barrel is given by its internal diameter. The inner surface of the barrel is hardened rendering it abrasion resistance and to bear the high shear. Normally, the pressure is in the range of 15-70 atmosphere, but under dry extrusions and plugged conditions, pressure may exceed 70 atmosphere. Heating or cooling surface is in direct proportion to the length of the barrel, being higher for bigger length of the barrel.

Die: It is used to provide shape and size to the final product.

Direct Compaction Type Machines

High pressure machines are used for residues having high lignin content. No additional binding material is required in such cases. The pressure in these machines can reach around 1200-2000 kg/cm^2 resulting in rise in temperature to about 200-250° C. This temperature is sufficient to fuse the lignin content of the residue. Fused lignin acts as a binder, provides mechanical support and repels water. The semi-fluidized biomass is densified through electrically operated briquetting machines available in the capacity range of 100-300 kg/h. The machines in this category are of two types namely piston press type (Figure 2.7) and the screw press type (Figure 2.10).

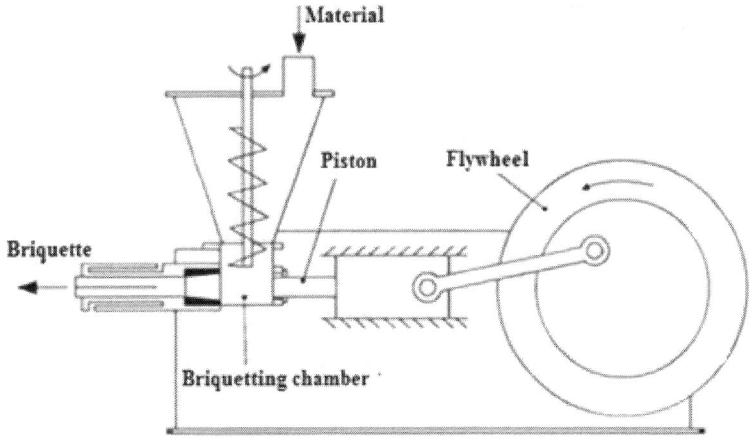

Figure 2.7. A schematic diagram of a piston press type briquette making machine.

Piston Type

The piston type machines use the rotary power of the mechanical device or the thrust of the hydraulic cylinder to reciprocate a piston (or a plunger). High pressure is generated with the movement of the flywheel (Figure 2.7). The piston plunger drives the ram that reciprocates in the forming sleeve to generate a pressing force. It punches the briquette several times in a short period raising the temperature of the raw material. As the raw material moves, it also fractionates due to generation of frictional heat. As a result of heating, lignin content of the material melts that binds the raw material with increase in its density. A simple version of piston briquetting machine is the honeycomb briquette machine. It works in a similar fashion and uses the flywheel. Briquettes in this machine are punched only once into a dye by a reciprocating ram at high pressure. Honeycomb structure of the briquettes help in better aeration during burning of the briquetts (Figure 2.8). The cost of such briquetting units depends upon their capacity and varies in the range of Rs. 3 to 20 lakhs. The main advantages of a piston press are their durability and low power consumption. The disadvantages are the need of frequent maintenance and the fact that it cannot be used to produce carbonized briquettes.

Figure 2.8. A view of honeycomb briquettes.

Pellet Press Type

These machines operate by extruding small diameter (10-30 mm) pellets through a die, which has many holes (Figure 2.9). The extruding mechanism is often 2 or 3 eccentric rollers, which move inside a large cylinder or conical die. The raw material to be palletized is dropped in the hopper. As the rollers ride over this material and rotate, they push the material through holes in the die against resistance offered by already existing pellets in the die holes. The pellets are not as hard as briquettes. Such machines were originally developed for producing animal feed and have high through puts. The power requirement varies according to the capacity of the machine. A 500 kg/h capacity machine requires about 16 kW power for its operation.

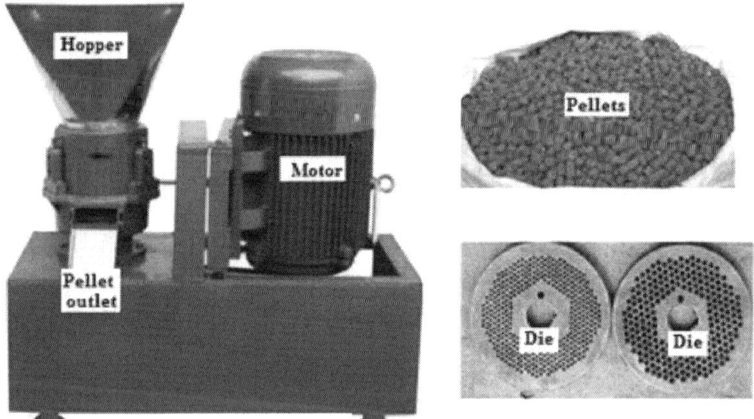

Figure 2.9. A schematic diagram of a pellet type briquette making machine.

Hydraulic Type

The principle of operation is basically the same as of the mechanical piston press. The difference is that the energy to the piston is transmitted from an electric motor via a high pressure hydraulic oil system. These machines are quite compact and light, since the forces are balanced-out in the press-cylinder and not through the frame. Almost all kinds of materials such as sawdust, peat, biomass, salt, coke, ceramic, graphite, all types of slag and dust, metals such as aluminium, brass, copper, steel, magnesium, cast iron, stainless steel, lead, sponge iron, *etc.* can be briquetted with this machine because of high pressure. Raw materials with high plasticity are also quite amenable for use with this machine. In a standard hydraulic briquetting press, the raw material less than 40 mm is fed from a hopper via an agitator and a screw conveyor. The material is transported to a pre-compacter, which pre-compresses the raw material. After that, the main piston pushes the pre-compressed material into one of the two press chambers with the die that compresses the material at pre-decided density and shape. During this process ejectors fixed parallel to the main press piston, push out the briquette through ejector holes in the front plate. At this stage, the pre-charger cylinder moves upward again. The feeding screw feeds the next charge and the process continues. Hydraulic piston press types are predominantly used to produce briquettes in India, particularly in Tamil Nadu. Cylindrical shaped briquettes with diameter in the range of 30 to 90 mm are produced. The briquettes made from these machines have high strength.

Screw Type

Screw type machine is a medium pressure machine, pressure being in the range of 50-1000 kg/cm^2. The screw type briquetting machine consists of a hopper through which material is fed continuously into a screw, which forces the material into a cylindrical die of the machine (Figure 2.10 top). This process also causes friction from the shearing of biomass. The wall friction at the barrel, internal friction in the material and high rotational speeds (600 rpm) of the screw results in an increase in temperature of the biomass. This heated biomass when forced through the extrusion die forms briquettes or pellets of the required shape. Compaction of the biomass is even greater, if the die is tapered. If the heat generated within the system is insufficient (because of the medium pressure) for the material to reach a pseudo plastic state for smooth extrusion, either band or tape heaters are used from outside the system to provide heat to the extruders (Figure 2.10 bottom). Following stages occur while processing biomass in a screw extruder.

- ✰ The biomass is partially compressed by the screw before it reaches the solids conveying zone (a zone usually formed by tapering the barrel). Thus, biomass is closely packed with increased density. Energy is dissipated to overcome particle friction
- ✰ The material is relatively softer due to high temperatures (200–250°C) as it approaches the melting and compression zone. The material loses

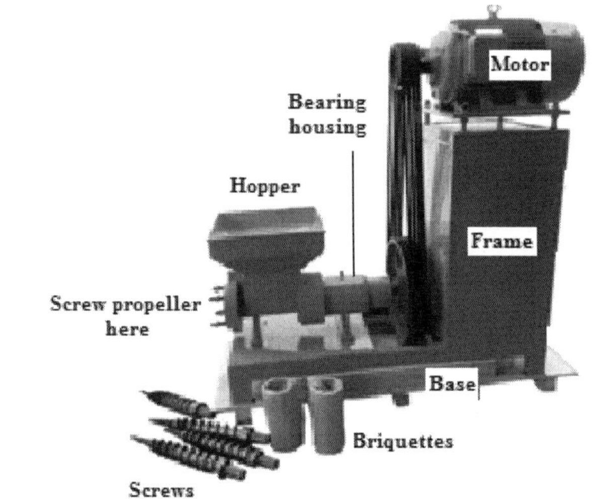

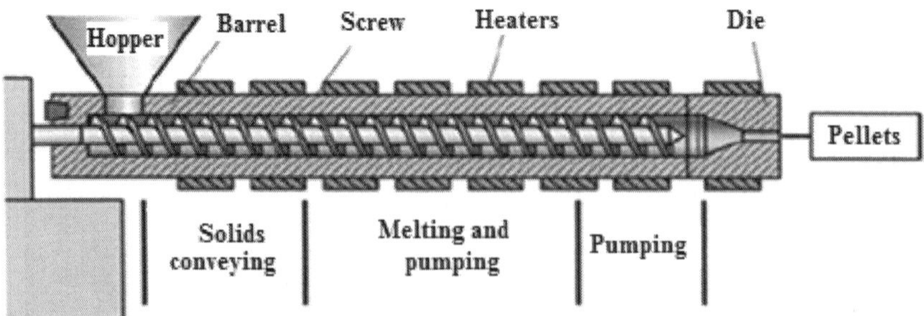

Figure 2.10. A schematic diagram of a screw type briquette making machine without heaters (top) and a machine with heaters (bottom).

its elastic nature, resulting in an increased area of inter-particle contact and local bridging. When the particles come together they form local bridges which selectively support and dissipate the applied pressure. Interlocking of particles may also occur. The moisture evaporated as steam at this stage moistens the whole biomass

- When the biomass enters the tapering die, biomass is further compressed causing temperature to rise to about 280°C. In this section, compaction and removal of steam occur simultaneously. The pressure exerted is uniformly transmitted throughout the material that forms briquettes of uniform density

The machine has certain advantages. Its operation is less noisy. It can also be used for producing carbonized briquettes. The disadvantages are the high wear and tear of the screw and relatively high power consumption. Moreover, raw material should be quite homogeneous in a particular size range. A comparative statement of a piston and a screw type machine is provided in Table 2.3.

Table 2.3. A comparison of a piston press and a screw extruder

Parameter	Piston Press	Screw Extruder
Optimum moisture content of raw material	10-15 per cent	8-9 per cent
Wear of contact parts	Low in case of ram and die	High
Output from the machine	In strokes	Continuous
Power consumption	50 kWh/tonne	60 kWh/tonne
Density of briquette	1-1.2 g/cm^3	1-1.4 g/cm^3
Maintenance	High	Low
Combustion performance of briquettes	Not so good	Very good
Carbonization to charcoal	Not possible	Makes good charcoal
Suitability in gasifiers	Not suitable	Suitable
Homogeneity of briquettes	Non-homogeneous	Homogeneous

Roller Type

Roller machine works with two closely rotating rollers having the same width and diameter. One of the rollers is driving while the other is driven. The rollers rotate at the same speed but in opposite direction (Figure 2.11). Both the rollers have ball-pockets on their surfaces. The material with or without binder is fed from the hopper at top of the rollers. When the rollers move, the two ball-pockets on different rollers coincide at the intersection of the midline and press the briquettes between them. A normal roller press uses a wedge supported roller. It is a low pressure machine having much lower output. The hydraulic pump station is used to fix the roller in advanced type machines. It provides flexible support to the roller so that it works stably even at high pressure. It also has mechanism to adjust the pressure. Forced feeding systems can be added to these machines to improve briquette quality.

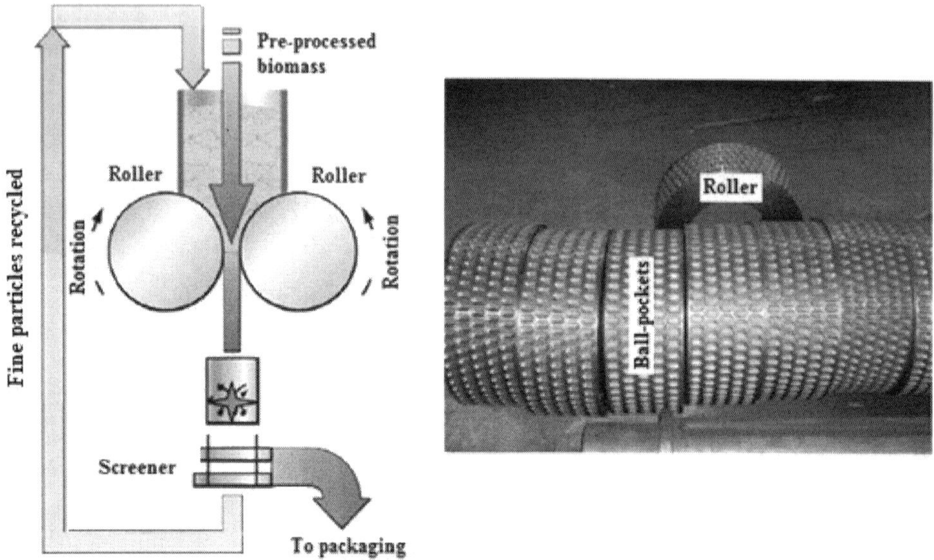

Figure 2.11. A schematic diagram of a roller type briquette making machine.

Briquetting with Binders

Most materials need some kind of binders for briquetting. In this case, some external binding material is added to assist in the formation of bonds between the biomass particles. The addition is all the more important for the carbonized materials lacking in lignin content. Briquettes can be made at low pressure. Therefore, the briquetting machines used in such cases can be operated at low pressure of about 50 kg/cm^2. Such machines are available in the capacity range of 100 to 400 kg/h. Moreover, the characteristics of the briquettes formed are better than without binders.

Screw Briquetting Machine

A screw briquetting machine works at low pressure. The process mainly comprises material extrusion by a screw extruder which also acts as a continuous feeder. As the material moves from the hopper to the die exit, its volume decreases with increase in its density. It is achieved by gradually increasing the root diameter of the threaded shaft attaining a maximum value at the die position. The screw is manufactured on a tapering shaft diameter from one side into a conical shape and then grooving the thread base on the shaft tapered surface (Figure 2.12). A heavy steel plate is welded to form the threads. The thread height remains constant but the clearance between the inner diameter of the barrel and the screw goes on decreasing. The machine is commonly used to briquette mixture of the carbonized powdered agro residues with some kind of a binder. Stick type finished briquettes in square, round, plum blossom shapes are received at the die end. The briquettes are relatively moist and need to be dried before packing for distribution. Briquettes are of better quality than piston-press unit. On the other hand, the wear and tear of screw is very high requiring frequent reconditioning.

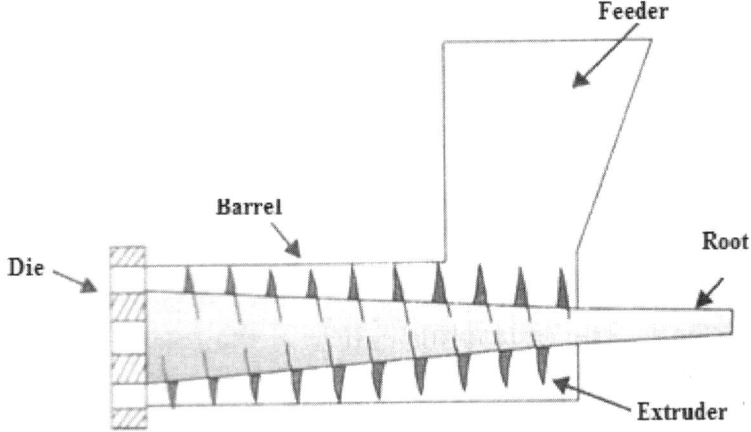

Figure 2.12. A schematic diagram of a screw type briquette making machine with binder.

Limitation of Briquetting Machines

Main limitations arise due to the following.

- Farmers and small scale manufacturers cannot afford these machines
- The high cost usually results from high power motor and engines required for their operation
- Unskilled labour in small and remote areas may find it difficult to operate these machines

QUALITY PARAMETERS

The parameters used in assessing the quality of briquettes include their calorific value, durability, strength, density, ease of ignition and smoke producing characteristics. The features may vary according to raw materials used as well as the production parameters. As an illustration, non-carbonized briquettes have the advantage of being ignited easily, but they do not last long because of their soft texture. They also produce a lot of smoke due to poor combustion characteristics compared to charcoal briquettes. The key to achieve desired quality of briquettes lies in control of the densification system variables, process variables and feedstock variables. Densification system variables included selection of manufacturing process, use of appropriate formulations, and use of additives. Process variables such as die thickness, die temperature, pressure, retention time, roller die ring gap, steam conditioning and feed rate affect the quality of briquettes in terms of their density and durability. Feedstock variables (moisture content and particle size and shape) and feedstock composition (cellulose, hemicelluloses, lignin, protein and fat) also impact the quality of the briquettes or the densified biomass.

PYROLYSIS/CARBONIZATION AND EXTRUSION

Heat treatment of the biomass can destroy the elasticity of biomass. Pyrolysis is one such process of destructive distillation of organic materials where these materials are heated at about 270°C at slow rate in the absence or minimum presence of oxygen. During pyrolysis, solid char, liquid tar, combustible gases and some organic liquids are produced. The nature and quantum of these products depend on various factors such as composition of biomass, residence time in the kiln and the temperature. Since the fiber content of biomass is broken during pyrolysis, it is easier to make briquettes from the charcoal so produced. The briquettes of charcoal are made using extrusion/compaction process. A complete flow diagram of the production of carbonaceous (Charcoal) briquettes is shown in Figure 2.13 (Asamoah *et al.*, 2016).

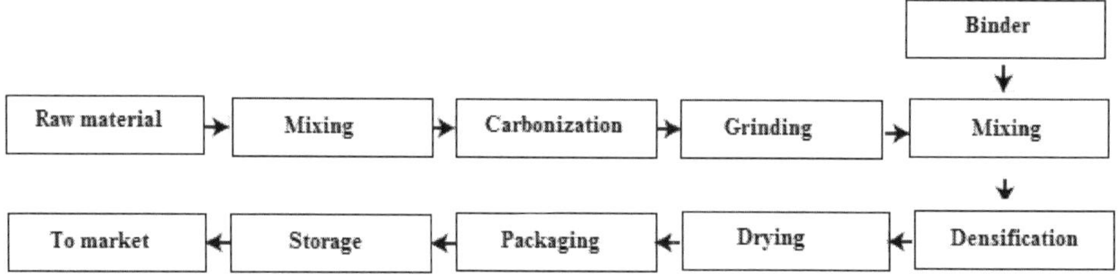

Figure 2.13. Flow diagram of the production process of carbonaceous briquettes.

Important Steps

Milling

One of essential pre-requisite for briquetting is to ground the carbonized material. This requires a hammer mill driven by either an electrical or a diesel engine. Diesel-driven mills are preferable because they can operate even in areas where electricity is yet to reach or is undependable. Contrarily, operational cost of diesel-driven mills is high. The mill should be self-ventilating without the need for any additional fan.

The Binder

The binders are materials that are added to enhance bonding and impart stable properties to the briquettes. These are added to raw materials that cannot be densified alone to form strong briquettes. The amount of binder required depends upon the binding properties of the raw material as well as that of the binding agent. One of the commonly used binders is the starch obtained from the purification of cassava roots. The binder-cost can be reduced by adopting one or a combination of the following practices.

- ☆ Rice starch may be used instead of cassava starch. It is equally suitable as a binder although slightly higher percentage may be required
- ☆ The proportion of binder used may be reduced without compromising the quality of the briquettes
- ☆ Since densifying and binding ability of the briquette machines vary widely, selection of a briquette machine with high pressure helps to reduce the amount or altogether obviate the need for the binding agent

Pressing

The presses shall ideally be of local construction requiring low investments with high productivity. The productivity of the machine should be estimated for the specific raw materials proposed to be used in the project. It is very important to set-up a chain network to achieve higher machine productivity.

Drying

The drying-cost and drying-time of briquettes will decisively impact the cost-price calculations. Therefore, final moisture content of the briquettes coming out of the machine and dried moisture content must be kept in view while selecting a machine. Need of adding a drier can also be assessed at this stage.

Cooling and Storage

Briquettes coming out of the machines are hot with temperatures exceeding 100°C. These have to be cooled and stored in a dry place.

Quality of Briquettes

Charcoal briquettes exhibit certain qualities such as low smoke emissions compared to non-carbonized briquettes. They also have low ash content, non-sparkling characteristics and long-lasting fire, and are safe against termites. In many situations, marketing of charcoal briquettes is easy as users are familiar with this type of briquettes because their colour and features resemble conventional charcoal (Figure 2.14). The disadvantage mainly arises from the fact that it is relatively difficult to start the fire than non-carbonized briquettes. The other qualities in terms of durability, strength and density vary with the raw materials used and production parameters. Charcoal briquettes are classified as durable, if the crushing strength is at least 375 kN/m^2, impact resistance index is 50 and water resistance 95 per cent.

Figure 2.14. Charcoal briquettes as they look at the time of drying.

Uses of Briquettes

The most frequent applications of briquettes are for domestic and industrial purposes, from fire places or stoves to boilers generating hot water and steam. Major sectors such as tea industries, wine distilleries, textile industries, and others are the major consumers of briquettes. Briquettes are also used in gasification process for electricity production.

REQUIREMENTS FOR BRIQUETTE PRODUCTION UNIT

Essential requirements to start a briquette production unit are as follows.

Land requirement: A minimum land area of 0.4 ha is required to start a briquette production unit. It will ensure enough space for storing the raw materials and briquettes before these are marketed.

Raw materials: First and foremost requirement to achieve high and cost-effective production is to ensure continuous availability of raw materials. It is a major determinant of a profitable briquette production unit.

Drying: Initial moisture content of raw materials is generally high. Any of the drying technologies such as solar driers/heater/hot air generator system is required to bring down the moisture content to desirable level. Provision of enough space should be made, if drying is proposed to be carried out in the sun.

Shredding machine: A shredding machine with minimum of 5 HP motor is required to powder the agro residues for briquetting.

Briquetting machine: Choose the right type of machine considering the available finances and facilities. A high pressure hydraulic piston press type briquetting machine powered by minimum of 50 HP motor may be required to produce binder less briquettes from agro residues.

QUESTIONS (THEORY)

1. What is biomass? Discuss the various kinds of biomass.
2. Discuss the various characteristics of biomass.
3. Name and describe the various components of biomass.
4. What do you understand by densification of biomass? Discuss the various densification processes.
5. List the advantages of briquetting the agro residues.
6. Discuss various steps of the general briquetting process of biomass.
7. Name various functional components of briquetting machines. Discuss any one of the machine used for briquetting without addition of binder.
8. Draw a flow diagram and discuss the various steps in carbonization of biomass and briquetting.

Chapter 3

Combustion Process

THE PROCESS

Biomass combustion is the oldest known and most widely used energy source on earth. Combustion is one of the thermal conversion processes of biomass into energy. *It is a complex thermo-chemical reaction/process consisting of several consecutive heterogeneous and homogeneous reactions*. In the process of combustion, two ingredients (biomass and oxygen) are combined in a high temperature environment to form carbon dioxide, water vapours, and heat. The basic combustion process is controlled by the fuel (the hydrocarbon) and oxidizer (air or oxygen) called the reactants. Together they undergo a chemical process releasing heat and forming the products of combustion. Mass conservation is maintained throughout the whole process. In the simplest case, all the carbon in the fuel forms carbon dioxide (CO_2) and all the hydrogen forms water (H_2O). It is known as stoichiometric combustion. The main processes of combustion under normal circumstances are drying, volatilization, gasification, char combustion, and gas-phase oxidation. The time required for each process depends on the particle size of the fuel, biomass properties, temperature, and combustion conditions. Direct conventional burning of fuel is highly inefficient process. Nonetheless, currently available advance equipment has made biomass combustion an economical, efficient, and practical energy source.

Combustion is the highest level of conversion amongst the three processes namely pyrolysis, gasification and combustion (Figure 3.1). Herein enough oxygen and heat is supplied and enough time is allowed so that the fuel is able to combust fully (https://coabnau.in/uploads/1644232046_Ag.Engg4.3Lacture-Copy(2).pdf). If the temperature of the system in the pyrolysis zone is 400°C or more, addition of oxygen will cause the biomass to leave pyrolysis and go into gasification. With the addition of more oxygen, the process will move on to combustion. The graph in Figure 3.1 also reveals that a small amount of heat is generated in pyrolysis. High heat is generated in gasification and combustion in that order. Therefore, one must go for combustion, if the major product required is heat. Notwithstanding what is stated before, it is almost impossible to separate biomass pyrolysis, gasification and combustion in their entirety. Therefore, all thermal conversions need to be optimized on the basis of the desired products. The conditions are optimized such that the primary product produced is the most in a particular thermal conversion while other products become secondary in nature.

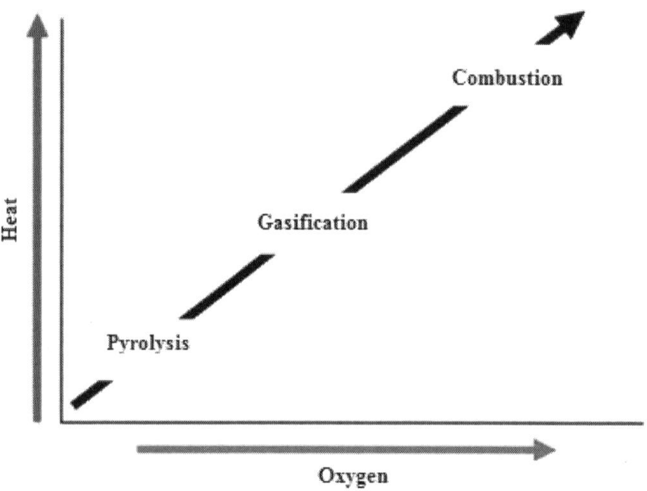

Figure 3.1. Graph illustrating oxygen and heat combinations for processes of pyrolysis, gasification and combustion.

The amount of air in excess of the stoichiometric amount is called excess air. The amount of excess air is usually expressed in terms of the stoichiometric air as per cent excess air or per cent theoretical air. Excess air ratio describes *the ratio between the locally available and the stoichiometric amount of combustion air (often denoted as i)*. For example 150 per cent excess air ratio means 1.5 × the stoichiometric air. Also, it may be referred to as 50 per cent excess air; *i.e.*, 150 per cent of stoichiometric air. For typical biomass, the combustion reaction can be described by the following equation if fuel constituents such as N, K, Cl, *etc.*, are neglected.

$$CH_{1.44}O_{0.66} + \grave{\imath}\ 1.03\ (O_2 + 3.76\ N_2) = f\ CO_2 + 0.72\ H_2O + (\grave{\imath} - 1)\ O_2 + \grave{\imath}\ 3.87 N_2 + \mathbf{f}\ C_mH_n + \text{Heat}$$

Here, $CH_{1.44}O_{0.66}$ is the approximate chemical constitution of the combustible portion of biomass *i.e.*, wood, straw, or similar material. Different types of pollutants resulting from the process can be distinguished:

- Unburnt pollutants such as CO, C_mH_n, PAH, tar, soot, unburnt carbon, H_2, HCN, NH_3, and N_2O
- Pollutants from complete combustion such as NO_x (NO and NO_2), CO_2, and H_2O
- Ash and contaminants such as ash particles (KCl, *etc.*), SO_2, HCl, the two similar classes of chlorinated aromatic chemicals namely polychlorinated dibenzo-p-dioxins and dibenzofurans (PCDD/F), Cu, Pb, Zn, Cd, *etc.*

The amount of heat produced varies depending on species, climate, and other factors, but mostly it is about 20 MJ (Mega-joules) of energy per dry kilogram of biomass.

COMPLETE VERSUS INCOMPLETE COMBUSTION

Combustion, like any other chemical reaction, can be categorized in more than one way. On the basis of rate it is categorized as rapid and slow. *Rapid combustion is a form of combustion in which large amounts of heat and light energy are released, which often results in a fire. This also happens in machinery such as internal combustion engines. Slow combustion takes place at low temperatures.* Cellular respiration is an example of slow combustion. Since combustion does not always proceed to completion with 100 per cent efficiency, it has also been categorized as complete and incomplete combustion.

Complete Combustion

Combustion is said to be complete when the reaction produces a minimal number of products. It is also called clean combustion, i.e. oxidation of a hydrocarbon that produces only carbon dioxide and water. An example of clean combustion is burning of candle wax, where the heat from the wick vaporizes wax (a hydrocarbon), which reacts with oxygen in air to release carbon dioxide and water. Ideally, all the wax burns so nothing remains once the candle is consumed. The water vapour and carbon dioxide dissipate into the air.

Incomplete Combustion

Incomplete combustion occurs when there is insufficient oxygen for the fuel to completely convert to carbon dioxide and water. Incomplete combustion may yield several additional products, including char, carbon monoxide, and acetaldehyde. It is why; it is also called dirty combustion. Oxidation of hydrocarbon producing carbon monoxide and/or carbon (soot) in addition to carbon dioxide is an example of incomplete combustion. Another example is burning of coal, where lot of soot and carbon monoxide is released.

BIOMASS TYPES

It is possible to utilize major biomass feedstock such as wood, agricultural waste, municipal solid and residential waste to produce heat, steam or electricity through direct combustion. Although, it is possible to burn any biomass but in practice combustion is feasible only for biomass with moisture content less than 50 per cent. For biomass having high moisture content, it needs to be dried before it is used for combustion. Ideally, biomass should have no more than 20 per cent moisture. Biomass having high moisture content is better suited to biological conversion processes.

Characteristics of biomass from combustion point of view have been classified as macroscopic or microscopic. The macroscopic properties are its heating value, moisture content, particle size, bulk density, and ash fusion temperature. Microscopic properties include thermal, chemical kinetic and mineral data. Some of the values vary greatly with physical properties such as density, porosity, and internal surface area. Thus, biomass can often benefit from a certain amount of pre-processing that makes it quite suitable as a combustion fuel. This includes sizing, drying, and/or densification. In general, combustion of biomass produces hot gases at temperatures around 800-1000°C. Carbon based biomass will always produce CO_2 and H_2O. Non-carbon based biomass can produce a variety of other oxidation products like sulphur oxides, nitrogen oxides, and chlorides. The energy flow in a fire-wood system is depicted in Figure 3.2.

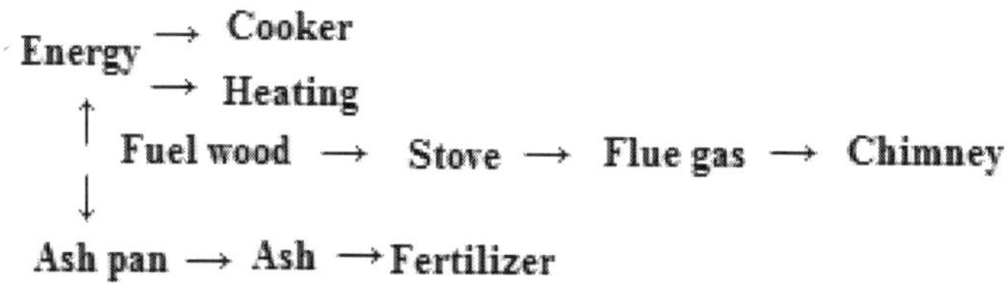

Figure 3.2. Material and energy flow in a stove fire wood system.

Utilization of Fire-wood

Fire-wood can be used for combustion with simple equipment wherein the combustion proceeds in stages as per the following sequence:

Drying → dry-distillation → flame combustion → char combustion

Initially some heat is consumed in drying the biomass. If supply of air is less, then some hazardous tar is produced at the dry-distillation stage. At flame combustion stage, CO and soot (carbon particles) are produced, and part of tar is converted into carcinogenic PAH by pyrolysis. In order to prevent these pollutants from getting in to the flue gas, high temperature and oxygen content has to be maintained by using a little bit excess air compared to stoichiometry air requirement. Air ratio of 1.25-1.40 is usually employed for safe combustion of firewood. Too high air ratio is also undesirable as it will dilute the flame and the flame temperature will be relatively low. Thus, in many cases excess air is supplied as secondary air (Figure 3.3). Besides, devices to place a combustion catalyst over the flame are now commercially available, which help to achieve complete combustion even at air ratio close to 1.0. However, it is necessary to replace the catalyst on yearly basis. Besides, there are possibilities of flow distortion. Char combustion normally comprises 10-20 per cent of the whole combustion.

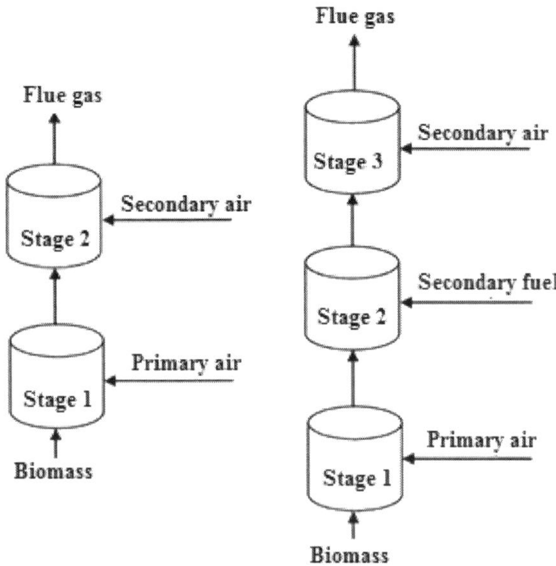

Figure 3.3. Staged combustion with air staging (left) and fuel staging (right).

IMPROVED DEVICES FOR COMBUSTION

Staged Combustion

Staged combustion method is devised to reduce emissions of nitrogen oxides (NO_x). Air staged supply and fuel staged supply are commonly used techniques in these equipment (Figure 3.3). Both air and fuel staging is also employed. Functionally, both air staging and fuel staging help to reduce NO_x in biomass combustion.

Air Staging

Combustion air is divided into two streams. The first stream called the primary air stream is mixed with fuel in a ratio that produces a reducing flame. Primary air is normally kept below stoichiometric level (ì < 1). Combustion operation at low excess air is possible due to good mixing, thus enabling high efficiency on one hand and high temperature with complete burnout on the other. The second stream is injected in the combustion chamber downstream of the flame (Figure 3.3 left). Secondary air enables a good mix of combustion air with the combustible gases formed during devolatilization and gasification in the fuel bed. The sum of the air in two stages or the net ratio is slightly in excess compared to the stoichiometric ratio. The concentrations of unburnt pollutants in air staging can be reduced close to zero (*e.g.*, CO < 50 mg/m^3 and C_xH_y < 5 mg/m^3). Enlarged furnace volume is necessary for the successful air staging operation since appropriate residence time is needed between the fuel bed and the secondary air inlet.

Fuel Staging

This is staging of combustion using fuel instead of the air. Fuel is divided into two streams. The first stream feeds primary combustion that operates in a reducing fuel to air ratio. The second stream is injected downstream of primary combustion, causing the net fuel to air ratio to be only slightly oxidizing. Excess fuel in primary combustion dilutes heat to reduce temperature. The second stream oxidizes the fuel while reducing the NO_x to N_2.

Air-fuel Staging

Both air and fuel are divided into two streams and fed into the furnace at two different levels. The primary fuel is combusted with excess air (i > 1). A consecutive reduction zone is achieved by feeding secondary fuel. Final combustion air for the secondary fuel is supplied through an inlet after the fuel staging (Figure 3.3 right).

EMISSION OF POLLUTANTS

Unburnt Pollutants

Incomplete combustion happens when the supply of air or oxygen is poor. Incomplete combustion of fuels releases carbon monoxide, carbon, excessive water vapours, aldehydes, and soot. Amongst them, carbon monoxide is the most potentially dangerous product. Less energy is released than during complete combustion. Complete burning of biomass depends upon the temperature, time, and turbulence (mixing). While the temperature (around 850°C) and residence time (around 0.5 s) can easily be controlled, the mixing between combustible gases and air is the main limiting factor that determines the burnout quality. To achieve sufficient mixing in fixed bed combustion, previously described two-stage combustion can be adopted. In fluidized bed, good mixing is achieved in the bed and the freeboard.

NO_x Emissions

In the combustion processes, NO and NO_2, normally summarized as NO_x, are formed during the following 3 reactions.

- Thermal NO_x is formed from nitrogen in the air at high temperatures
- Prompt NO_x is formed in the presence of hydrocarbons
- Fuel NO_x is formed from nitrogen-containing fuels

Fuel bound nitrogen is the main source of NO_x emissions in biomass combustion. Fuel nitrogen is converted to intermediate components such as HCN and NH_i with i varying from 0, 1, 2, 3. These are further oxidized to NO_x in the presence of oxygen, which is normally the case in conventional combustion. In the absence of oxygen, intermediates can interact in the reduction zone and form N_2.

$$NO + NH_2 \rightarrow N_2 + H_2O$$

Two kinds of approaches namely primary and secondary measures have been devised to reduce environmentally undesirable NO_x formation. Staged combustion technologies such as air staging and fuel staging comprise primary measures for NO_x reduction. Both the techniques help to reduce NO_x up to 50 per cent for wood with low and up to 80 per cent for biofuels with high nitrogen content. However, some specific conditions have to be ensured to achieve this reduction potential. As an example, in air staging, a primary air excess around 0.7, a temperature in the reduction zone of 1150°C and a residence time of 0.5 s are needed. Practically, relatively high temperature can limit the application due to undesirable ash softening and deposit formation. For fuel staging, similar results are achieved at lower temperature as low as 850°C. The furnace design and operation in this case is quite complex due to the need of two independent fuel feeding systems.

Considering the limitations posed by primary measures, several secondary measures have been devised for NO_x abatement. The most relevant techniques are selective non-catalytic reduction (SNCR) and selective catalytic reduction (SCR) using the same reaction as mentioned for staged combustion, *i.e.*, $NO + NH_2 \rightarrow N_2 + H_2O$. In SNCR, urea or ammonia is injected as reducing agent and as source of NH_2. It has to be applied in a narrow temperature window of around 820°C to 940°C. It can reduce NO_x up to 90 per cent. SCR is typically applied in the flue gas in a temperature range of around 250° to 450°C. It enables NO_x reduction of more than 95 per cent. High concentrations of undesired side products such as HNCO, N_2O, NH_3, HCN, and others can be formed in both types of secondary measures under unfavorable conditions. It is why; primary measures are preferred over secondary measures provided it is possible to achieve desired emission reduction.

Particulate Emissions

Combustion of biomass fuels in small scale heating appliances results in considerable amounts of particle emissions. Mass concentrations of particles in the flue gas may be in the range of 7-67 mg/Nm^3 as reported in several studies in the literature. The particle numbers may be in the range of 0.8×10^7 - 5.5×10^9 particles/cm^3. The dominating chemical compositions of the fine particle are potassium fraction followed by sulphur, chlorine and sodium. Coarse particles are of Ca, Si, K, S, Na, Al, P, and Fe. The main fuel constituents for aerosol formation are typically K, Cl, S, Ca, Na, Si, P, Fe, and Al. Majority of the particles emitted directly to the ambient air from the combustion devices are less than 1 μm (micrometer) or particulate matter 1 (PM 1). Primary measures used to control NO_x emissions can be used to reduce particulate emissions by at least a factor of 10.

COMBUSTION TECHNOLOGIES

Biomass combustion is mainly used for heat production in small and medium scale units such as wood stoves, log wood boilers, pellet burners, automatic wood chip furnaces, and straw-fired furnaces. Heating systems are often in the size range from 0.5 MW_{th} (Megawatt thermal) to 5 MW_{th} with some systems going up to 50 MW_{th}. The biomass combustion systems can be categorized by the flow conditions in the furnace. The 3 types of systems are: fixed bed combustion, fluidized bed, and entrained flow or dust combustion. For small scale combustion, downdraft boilers with inverse flow that use two-stage combustion principle described previously have been introduced. Use of log wood furnaces at very low loads should be avoided as it will result in high emissions of unburnt pollutants. In such cases, log wood boilers should be coupled to heat storage tanks to achieve high efficiencies. Wood pellet furnaces can be used for small heat outputs usually required for heating buildings. They can be used both as stoves and as boilers. Stoker (A mechanical device to feed the biomass fuel in furnace) furnaces are mostly used with wood chips and similar fuel having relatively low ash contents (Figure 3.4). Grate furnaces are used for biomass with high ash and water content.

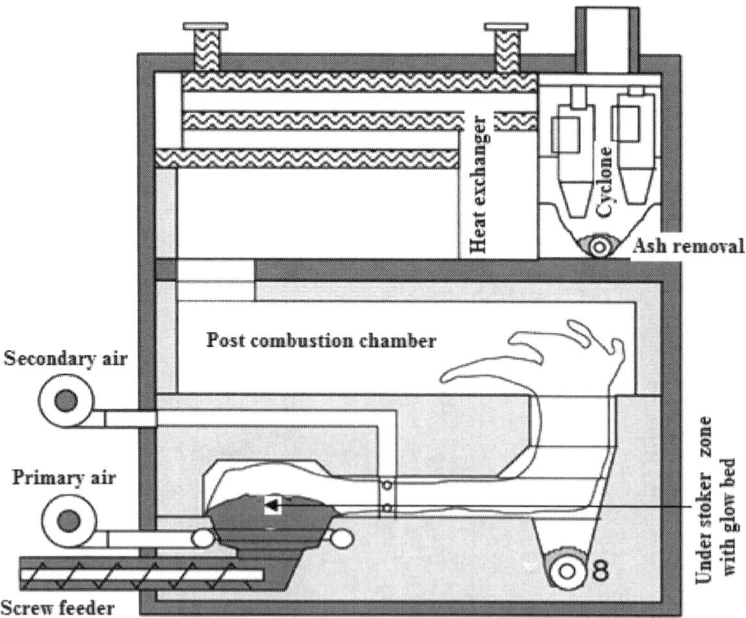

Figure 3.4. Stoker furnace with primary and secondary air, mixing zone, and post combustion chamber.

Special types of furnaces have been developed for straw stored in the form of bales having very low density. For this purpose, conventional grate furnaces operated with whole bales, cigar burners and other specific furnaces are used. Stationary or bubbling fluidized bed (SFB) and circulating fluidized bed (CFB) boilers are used for large-scale applications with waste wood or mixtures of wood and industrial wastes from the pulp and paper industry. Nearly homogeneous conditions of temperature and emission concentrations can be achieved in CFB boilers enabling high burnout quality even at low excess air. Additional opportunities of catalytic effects can also be achieved because a wide choice of bed materials can be used in CFB. The option of heat removal from the bed in CFB boilers is quite useful as it helps to control the combustion temperature enabling the operation at low excess air without excessive ash sintering. NO_x emissions are relatively low because conditions similar to air and fuel staging are attained for nitrogen conversion.

Combined heat and power production (CHP) from a biomass plant essentially consists of an electrical generator combined with equipment to recover and use the heat produced by the generator. The generator may be a prime mover such as a gas turbine or a reciprocating engine. Alternatively, it may consist of a steam turbine generating power from high-pressure steam produced in a boiler. In some cases, a CHP scheme may be a combination of prime mover(s), boiler(s) and steam turbine(s). Purely biomass based plants are in the range of 0.5 MW to 10 MW. However, co-firing of biomass is possible in fossil-fired large power stations (>100 MW). Such big size plants are difficult to operate with dedicated biomass combustion system because of limitations of biomass availability.

CO-COMBUSTION

Co-combustion also known as co-firing or co-utilization is the burning of more than one fuel to produce power. Normally, it refers to the blending of biomass with coal in the furnace of a conventional coal-fired plant. Currently it is one of the simplest ways of using biomass to replace fossil fuels with no new investment or specialized technology. This section is devoted to biomass co-combustion where up to 20 per cent of total fuel weight or energy of biomass is used in a coal or gas fired power station. Opportunity to replace up to 20 per cent of the coal fuel with biomass makes co-firing an attractive proposition as it can substantially reduce CO_2 emissions. In addition, SO_2 emissions will also be much less in co-firing as biomass contains much less sulphur or metal compounds than coal. The reduced SO_2 emissions may also result from special interactions between biomass and coal during combustion.

Advantages of Biomass Co-combustion

A wide range of usable biomass fuels are available for co-firing. Co-firing rate can be adjusted considering the quantity of biomass available. The rate to a great extent can also partially compensate the variations in the quality of biomass materials. Use of co-combustion is especially useful when a coal or fossil-fuel plant is located near an agro-industrial or forestry product processing plant where large quantities of low-cost biomass residues are available. The advantages and limitations of plants using co-combustion can be grouped under technical, economical and environmental heads.

Technical

- ☆ Co-utilization of biomass with conventional fuels can reduce cost, improve efficiency, and reduce emissions
- ☆ An agro-industrial or forestry processing plant can make quite efficient use of its residues in generating electricity. But in most cases availability of agro-residues is seasonal in nature. Co-firing with a fossil fuel can generate electricity all the year round. For example, a sugar industry can produce large amounts of electricity during the harvesting and processing season. During the off-season the plant has to remain idle. This has two drawbacks. First, it is an inefficient use of equipment having limited lifetime. Second, electrical distribution utilities will not pay the full premium for electricity supplies as they have to source the electricity from other sources in the off season. Thus, an agro-processor can get both the advantages of making efficient use of its equipment and getting premium payments for the electricity by the distribution facility
- ☆ A major advantages of co-combustion also emerge from the fact that pre and post treatment facilities of the existing plants can be utilized
- ☆ Overall high efficiencies are obtained with power generation from biomass in plants having co-combustion facilities compared to the use of biomass in small stand alone plants
- ☆ In some cases, higher sulphur to chlorine ratio resulting from co-combustion may reduce the risk of corrosion

Economical

- ☆ The additional investments for the co-combustion equipment are significantly lower
- ☆ The reduced CO_2 emissions can be used to avail CO_2 credit certificates for economic gains

Environmental

- ☆ It is now widely accepted that fossil-fuel power plants are highly polluting in terms of sulphur, CO_2 and other GHGs. Using the existing equipment, perhaps with some modifications, and co-firing with biomass may represent a cost-effective means for meeting stringent emissions targets. Low sulphur (relative to coal) content of biomass and nearly zero net CO_2 emissions allow biomass to offset the higher sulphur and carbon contents of the fossil fuel. Low SO_x emissions occur because of low S content of the biomass

- ☆ High volatile matter content of the biomass results in improved combustion, better burn out and low unburned carbon in the ash

Limitations of Biomass Co-combustion

Biomass co-combustion has following technological, economic and environmental problems.

Technical

- ☆ A combustion system is always designed for a certain fuel. The burners are designed for certain volatile components and ash. The fuel handling system is designed for particular moisture content, size distribution, dust *etc.* In co-firing, it will be necessary either to adapt the existing or build a new combustion system for the fuels. The same is valid for storage and conveying. Cost of operation may increases due to boiler retrofitting, higher operation cost due to increased fouling and corrosion, and a possible decrease in efficiency
- ☆ Co-firing in pulverized coal-fired boiler requires milling. To use biomass in such systems, biomass has to be dried and sometimes pelletized, or milled to sufficiently fine size. Additional cost for fuel preparation may offset some of the economical benefits. This issue does not apply to fluidized bed boilers
- ☆ Some biomass fuels (sewage sludge, some cereals) high in chlorine can prove to be corrosive due to increased HCl formation. Moreover, corrosion rates of some components may also increase due to relatively higher temperature in co-firing in comparison to stand-alone facilities
- ☆ Many biomass fuels containing large amounts of alkalines especially potassium may aggravate the fouling problems
- ☆ Additional slagging problems may be encountered as a result of accelerated reducing atmosphere
- ☆ Catalyst used in Selective Catalytic Reduction (SCR) of NO_x can be blocked by ash particles or deactivated by potassium, chlorine, and may also be poisoned by some heavy metals (As, Zn) especially in case of sewage sludge
- ☆ Efficiencies of electrostatic precipitators may be reduced in biomass co-firing systems

Economical

- ☆ Operating costs may be typically higher for biomass than for coal because of higher fuel cost. Even if the fuel is nominally free at the point of its generation (as many residues are), its transportation, preparation and on-site handling typically increase its effective cost per unit energy

- Competition with standalone plants may be difficult because some incentives are especially meant only for standalone bio-energy

Environmental

- Possibility of increased deposit formation in the boiler cannot be discounted
- Critical parameters for the utilization of ash in the cement and concrete industry are the concentrations of alkali metals, P_2O_5, SO_3, Cl and unburned carbon. The ash quality can be negatively influenced mainly by alkali metals, chlorine contained in the biomass, and the unburnt carbon content. It may limit ash utilization in building materials. Thus, the share of biomass has to be limited to approximately 10 per cent of the fuel input to avoid undesirable changes in ash composition

METHODS OF CO-COMBUSTION

Following two options are applied for co-combustion of biomass and coal.

Direct co-firing: The biomass is directly fed to the boiler furnace (fluidized bed, grate, or pulverized combustion). Although it can be done directly yet in most cases It is done after physical preprocessing of the biomass such as drying, grinding, or metal removal

Indirect co-firing: The biomass is first gasified and the product gas is fed to the boiler furnace (a combination of gasification and combustion)

Co-firing helps to achieve the following objectives.

- A small amount of biofuel or waste is fired together with coal to get rid of waste
- Biomass is used to replace some amount of coal to reduce carbon emissions
- A small amount of biomass fuel with a high heating value is fired together with a fuel having a low heating value (such as sludge) that needs thermal support to attain the desired combustion temperature
- Spontaneous use of biomass with fossil fuels in different ratios depending on price, availability and local supply conditions

Some of the methods of co-firing are described in the following section.

Pre-mixing

Biofuel can be fed with coal and burned together in the burners when the proportion of biofuel to coal is low. It is the simplest option and involves minimal investments. The biomass in this case is mixed with the coal in the fuel yard and is directly fed to the boiler furnace after being passed through the same mills *i.e.* crushers, bunkers and pulverizers as the coal. Efficiencies of multi fuel fluidized bed boilers can exceed 90 per cent, besides lowering of flue gas emissions compared to coal alone.

Joint Direct Injection

This option involves separate handling, metering and comminuting of the biofuel. It is injected into the pulverized fuel upstream of the burners or directly in the burner (fluidized bed, grate, or pulverized combustion) after physical pre-processing of the biomass. This option needs a number of biofuel transport pipes across the boiler front (Figure 3.5A), which might already be congested. Moreover, it may also be difficult to control and to maintain the burner operating characteristics over the normal boiler load curve.

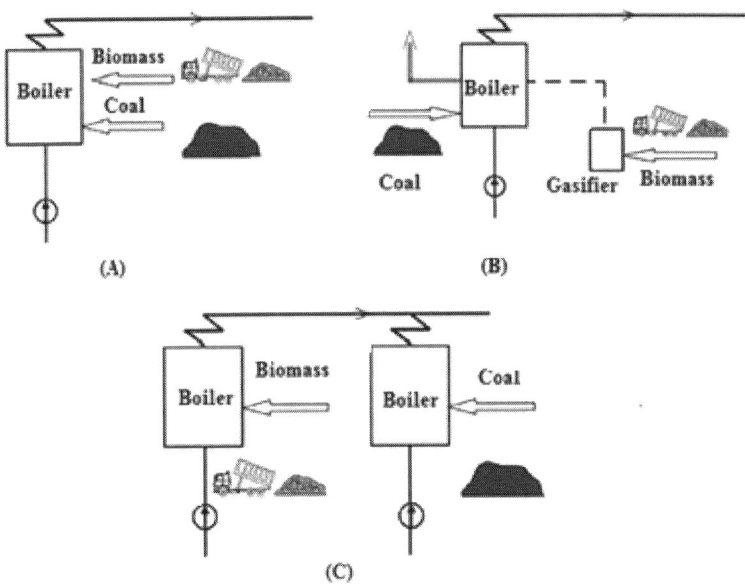

Figure 3.5. Various methods of co-combustion.

Separate Burning of Biomass and Coal

Direct co-firing of biomass results in several problems because of its high alkaline and chlorine contents. Corrosion, slagging, fouling in the boiler, heat exchanger and piping, poisoning of catalysts, and relatively low performance of electrostatic precipitators are some such problems. To overcome these problems, indirect and parallel co-firing has been introduced. In this case biomass is gasified and the product gas is fed to the boiler furnace (Figure 3.5B). The capital cost of such a system is more resulting in higher production cost than direct co-firing. Moreover, cooling and cleaning of gas may be required, which in itself is quite challenging and implies higher operational costs. Nonetheless, the approach has certain advantages making it a useful proposition.

- It presents minimum risk to boiler operation and offers a high degree of fuel flexibility
- Since the gasification takes place separately, the ash from coal and biomass can be separately stored
- A wide variety of biomass fuels can be used as the potential problems from various kinds of biomass of differing chemical composition and physical properties are dealt with before the gas fuel enters the main combustion chamber
- Only one flue gas cleaning system is required because flue gases are not separated. Thus, investment cost can be reduced in comparison to parallel combustion

Parallel Co-firing

Parallel co-firing is most popular in the pulp and paper industries as dedicated biomass boilers are used for bark and waste wood, bio-residues and by-products resulting during production of paper. In this system, biomass is burnt in a separate boiler for steam generation (Figure 3.5C). The steam is used in a power plant together with the main fuel. Parallel combustion enables complete separation of the ashes and flue gases from different fuels. Hence, disadvantages or limitations resulting from undesired alkali metals or contaminants in the ash are avoided. Moreover, flue gas cleaning equipment can be optimized for individual fuels although it may increase investments.

Re-burning of Biomass in Upper Furnace

A new option that is still in the development stage is re-burning of biomass. This method can be associated both with direct or indirect co-firing. It involves the use of biofuel as a re-burn fuel mainly to control NO_x emissions. The combustion of biofuel is carried out in a specially-designed re-burn system located in the upper furnace or alternatively on a separate grate at the bottom of the pulverized coal boiler. The main advantage of this option results from the fact that biomass having high water content and large in size can be burnt. Thus, it is possible to minimize the costly and energy-consuming fuel pre-treatment.

Applications of co-combustion of biomass with coal have been studied in greater details. Bubbling fluidized beds (BFB), circulating fluidized bed (CFB), cyclone, and stoker boilers, have been designed and developed. These can accept a much wider range of fuel sizes, composition, and moisture content than burners in pulverized coal boilers. But many possible effects of the interaction between fuels called synergy effects have not been sufficiently treated in the literature. This is an area of research that needs attention. It will make the selection of fuel combinations safer and allow better utilization of the positive effects that may arise in co-combustion by combining certain fuels.

TECHNOLOGIES FOR CO-COMBUSTION

Grate Combustor

It is the simplest and oldest design for combusting solid fuels. In this case, fuel is directly combusted over a grate with no further processes or circulation of air. It is the least efficient method having high flue gas emissions. On the contrary, combustion temperatures in the chamber in the range of 800 to 1400°C are higher than other technologies. Wood can be easily used in the grate combustor without causing any problems. For other biomass types, grate combustor is generally sensitive to changes in fuel quality and moisture content. Studies have shown that in small plants recycled fuels are relatively safe in terms of emission levels but in other cases there could be ignition and poor combustion problems.

Atmospheric Fluidized Bed Combustors (AFBC)

It is a well established commercial technology based around a fluidized movement of fuel and air for coal co-combustion plants. The combustion chamber has sand bed that maintains high combustion temperatures despite possible impurities, low energy content or high moisture content of the injected fuel. Therefore, this technology is suitable for a variety of fuel qualities and moisture levels making it fit for wood and other biomass types. While 90 per cent of the bed within the chamber is sand and ash, the rest 10 per cent is fuel. SO_2 and NO_x emissions are considerably less than grate combustors. Only little modifications may be required in this system to operate it for biomass combustion alone. Some variations of this technology are circulated, bubbling and stationary fluidized bed boilers.

Pressurized Fluidized Bed Combustor (PFBC)

The principles are exactly the same as the AFBC except that the combustion takes place at higher than atmospheric pressure. Therefore, some problems of biomass fuel crossing across the pressure boundary into the reactor may be encountered.

Pulverized Combustor

The coal and the biomass fuels to be injected into the burners of such combustors need preprocessing. They need to be reduced into fine material (generally < 1 mm). This requirement of preprocessing limits the choice of fuels. In spite of this limitation, it is the most widely used technology in the world for power generation. The low energy density of the fuel means a higher volume flow to the boiler and use of high volume of locally available biomass. This option is a preferred technology because of low emissions. Biomass in pulverized combustors can be used with coal in any one of following three ways.

- Mixing of prepared biomass fuel upstream with coal before firing
- Pre-mixing biomass in the coal preparation plant
- Dedicated separate coal and biomass burners

Examples of Co-combustion

The main application of direct co-firing is in coal-fired power stations. The typical size range of these plants is in the range of 50 MW to 700 MW with a few units in the range of 5 to 50 MW. The majority of the plants are equipped with pulverized coal boilers in which co-combustion is applied in different ways. An example is 635 MW power plant in the Netherlands that burns local urban waste wood and demolition wood. The waste wood replaces 3-4 per cent of the coal in this plant. Thanks to the scale and the flue gas cleaning processes, the urban waste wood is used with high efficiency having little adverse environmental impacts. With an overall electric efficiency of the plant at 43 per cent, the net efficiency for wood is estimated at 36-38 per cent. In comparison, a modern 25 MW wood fired plant may reach an efficiency of only about 30 per cent.

Co-firing of biomass has also been applied in cement kilns especially for the disposal of contaminated waste wood. Additionally, biomass fuels are co-fired in municipal solid waste (MSW) incinerators. From logistics and efficiency point of view, it is quite advantageous as biomass can be stored while municipal waste needs to be burnt immediately. The combination of biomass with oil or natural gas also offers specific advantages. Topping with natural gas enables a significant increase of the plant efficiency, because steam from a biomass boiler can be superheated to higher temperatures with natural gas.

CO-GASIFICATION

Co-gasification is defined as the gasification of a mixture of waste/biomass and coal. The combination protects the environment by reducing GHG emissions from process equipment. In recent years, co-gasification of biomass and coal has been broadly investigated by researchers because it creates opportunities in industries such as forestry, agriculture, and food processing to manage large quantities of combustible agricultural and wood wastes. In addition, the cost of adapting an existing coal power plant to co-fire biomass is significantly lower than the cost of building new systems dedicated only to biomass power. The biomass in the range of 3 -5 per cent on energy basis can be directly co-fired. However, this rate may be much higher at 20 per cent with cyclone boilers. The syngas produced from co-gasification is hydrogen-rich and contains CH_4, which can be effectively used in power plants. The volatiles readily decompose in the co-gasification process and form free radicals which react with the organic matters of the coal. In this way, while conversion rate increases, CO_2, SO_2 and

NO_x emissions are reduced. Presence of wastes in the feedstock on the contrary may result in higher concentrations of hydrocarbons and tar in the product gas. Co-gasification opportunities are increasing over time as many industries are adopting co-gasification of biomass and plastic wastes, petroleum coke, and tires.

QUESTIONS (THEORY)

1. What do you understand by combustion? Give the general equation of the combustion process. How it differs from pyrolysis and gasification in terms of oxygen requirement and heat production?
2. Name and describe 3 major pollutants resulting from combustion of biomass.
3. What is co-combustion? List its advantages and limitations.
4. What is direct co-firing? How it is useful. Discuss the various direct methods of co-firing.
5. Discuss the various methods of indirect co-firing.
6. Differentiate between: complete and incomplete combustion, air staging and fuel staging in staged combustion, and atmospheric fluidized bed combustors (AFBC) and pressurized fluidized bed combustor (PFBC).

Chapter 4

Principles of Pyrolysis

Thermo-chemical conversion of biomass can be achieved through combustion, pyrolysis, gasification, and hydrothermal liquefaction (Table 4.1). A look at this table reveals that the temperature range and products formed during these processes are quite different. While combustion has been discussed

Table 4.1. Operating conditions and product range generated during various thermal processes

Thermal Process	*Temperature (°C)*	*Atmosphere*	*Products*	*Mean Overall Yield (Per cent)*
Combustion	> 900	O_2 (air)	CO_2 +H_2O +N_2 + ashes to be treated	≈ 65
Pyrolysis	< 500	Inert gas or low pressure	Char + Tars + Gas, pyrolysis parameters control the relative proportions	≈ 45
Gasification by fast pyrolysis	> 700	Inert gas or low pressure	Mainly gas (CO_2 +H_2 +C_2H_4 + CO +….) with low quantity of char	≈ 75
Gasification	> 800	Air or H_2O vapours	Gas (CO_2 +H_2 +CH_4+ N_2 + CO) + ashes to be treated	50-60
Liquefaction by fast pyrolysis	< 550	Low pressure	High viscosity liquid (phenols)	≈ 75
Direct liquefaction	300-350 slurry in water	CO, high pressure	High viscosity liquid (phenols) non-soluble in water	≈ 80

in Chapter 3, pyrolysis process is covered in this chapter, and gasification process is covered in Chapter 5. Direct liquefaction process is not included being beyond the scope of this book.

PYROLYSIS

Pyrolysis is thermal decomposition of biomass occurring in the absence of oxygen producing a range of useful products. It involves breaking down of large complex hydrocarbons into relatively simpler molecules of gas, liquid and char. Biomass in the pyrolysis process is heated to moderate temperatures of 400-600°C as against gasification where biomass is heated to high temperatures >700°C. Pyrolysis is the fundamental chemical reaction, a precursor to the combustion and gasification processes. It occurs naturally in the first two seconds. The products of biomass pyrolysis include biochar, bio-oil and gases including methane, hydrogen, carbon monoxide, and carbon dioxide. Depending on the thermal environment and the final temperature, pyrolysis yields mainly biochar at low temperatures less than 450°C at slow heating rates, and mainly gases at high temperatures, greater than 800 °C, with rapid heating rates. The main product is bio-oil at intermediate temperatures and under relatively high heating rates (Figure 4.1). Although, bio-oil can be used as a feed stock in existing petroleum refineries, oxygen-containing compounds (aldehydes, ketones, phenolics and organic acids) make the oil too unstable and acidic for introduction into existing pipelines, tankers and refineries.

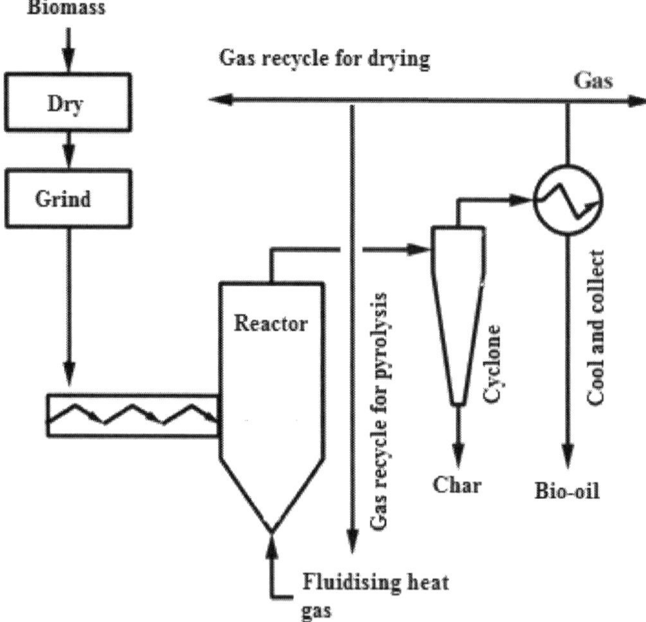

Figure 4.1. Biomass pyrolysis cycle.

Pyrolysis Techniques

Pyrolysis process, depending on the operating conditions, can be divided into three sub-classes namely conventional slow pyrolysis, fast pyrolysis and flash pyrolysis. The range of important operating parameters for these processes is reported in Table 4.2. Slow pyrolysis takes several hours to complete and gives biochar as the main product. On the other hand, fast pyrolysis yields 60 per cent bio-oil and takes seconds to complete the process. In addition, it gives 20 per cent biochar and 20 per cent syngas. Therefore, it is preferred technology for biomass pyrolysis.

Table 4.2. Operating parameters of various pyrolysis types

Operating Parameters	Slow Pyrolysis	Fast Pyrolysis	Flash Pyrolysis
Pyrolysis temperature (°C)	300-700	600-1000	800-1000
Heating rate (°C/s)	0.1-1	10-200	> 1000
Particle size (mm)	5-50	< 1	< 0.2
Solid residence time (s)	300-550	0.5-10	< 0.5
Major products	Gases Char Bio-oil (tar)	Bio-oil (thinner) Gases Char	Gases Bio-oil

Slow Pyrolysis

This process has been used for the past thousands of years primarily for charcoal production. It is why; it is also known as conventional pyrolysis. In slow wood pyrolysis, biomass is heated to ~500°C. The heating rate is slow approximately 0.1-1°C/s. At this heating rate, char yield is higher than the liquid and gaseous products. The char yield is about 50 per cent. The higher vapour residence time in the reactor allows the gas-phase products to continue to react with other products to form char.

Fast Pyrolysis

Fast pyrolysis is a rapid thermal decomposition of carbonaceous materials in the absence of oxygen in moderate to high heating rates. Fast pyrolysis uses much faster heating rates of about 10-200°C/s and produces relatively higher quantities of liquid and gases. High liquid yield is attributed to fast heating rates that allow the conversion of thermally unstable biomass compounds to a liquid product before they form undesired coke. Typically, fast pyrolysis produces 60-75 per cent of liquid bio-oil, 15-20 per cent of solid char and 10-20 per cent of non-condensable gases by weight depending on the feed stock used. Fast pyrolysis occurs on the time scale of a few seconds or less. The feedstock in this case must be dry, have smaller particles (< 3 mm), and typically done at atmospheric pressure with rapid quenching of the products. Fluidized bed reactors are best suited for this process as they offer high heating rates, rapid devolatilization and are easy to operate. Other reactors such as entrained flow reactors, circulating fluidized bed reactors, rotating cone reactors *etc.* can also be used for this purpose.

Flash Pyrolysis

An improved version of fast pyrolysis is flash pyrolysis also known as ultra-fast pyrolysis. It is an extremely rapid thermal decomposition process having a high heating rate exceeding 1000°C/s and reaction times of few to several seconds. Present reactors for flash pyrolysis include fluidized bed reactor, vacuum pyrolysis reactor, rotating cone reactor, entrained flow reactor, ablative, vortex or blade, and twin screw reactors. Entrained flow or fluidized bed reactors are considered to be the best reactors for this purpose. Compared to the others, this process requires smaller particle size for higher yields due to rapid heating rates and short reaction times. Four essential features of this process are:

- ✰ Very high heating and heat transfer rates requiring finely ground feed
- ✰ Carefully controlled reaction temperature of around 500°C in the vapour phase
- ✰ Residence time of pyrolysis vapours in the reactor less than 1 s
- ✰ Quenching (rapid cooling) of the pyrolysis vapours to give the bio-oil

Typically, flash pyrolysis produces >75 per cent of liquid bio-oil, 15 per cent of solid char and 10 per cent of non-condensable gases. A comparative statement of the three processes is given in Figure 4.2.

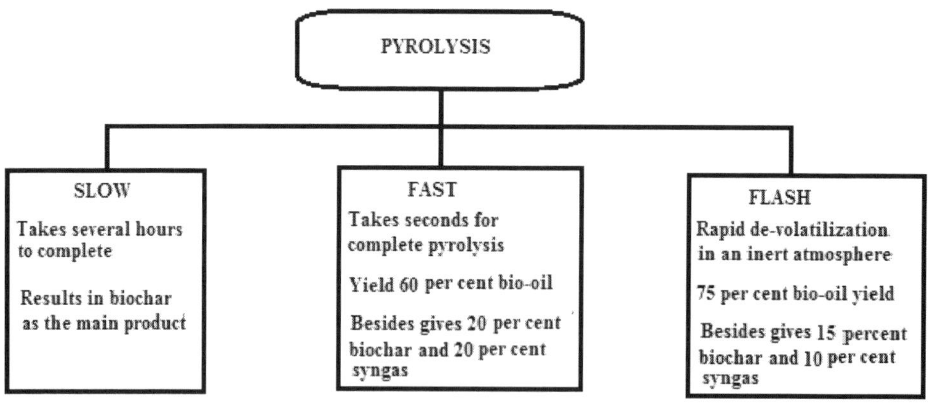

Figure 4.2. Comparative statement of 3 pyrolysis processes.

Economical and environmental advantages and disadvantages of pyrolysis are as follows.

Advantages

- ☆ Pyrolysis offers a flexible and attractive way of converting solid biomass into an easily stored and transportable liquid, which can be successfully used for the production of heat, power and other chemicals
- ☆ Helps to utilize waste renewable resources such as lumber processing waste (barks, sawdust, forest thinning, *etc.*), and agricultural residues (straws, manure, *etc.*) through a carbon neutral route
- ☆ Self-sustaining energy
- ☆ Pyrolysis can be performed at relatively small scale and at remote locations resulting in reduced transport and handling costs as well as enhanced energy utilization of the biomass resources
- ☆ Potential to produce chemicals from bio-based resources
- ☆ The need of external fuel is minimized by using the produced gases as fuel
- ☆ Toxic components and pathogens are degraded at high temperatures

Disadvantages

- ☆ The pyrolysis process is complex and requires high operational and investment costs
- ☆ An air purification system is necessary to further treat emanating flue gases during pyrolysis
- ☆ Ash produced may have high heavy metal content depending on the concentrations of various metals in the biomass. The ash is regarded as harmful/dangerous waste and must be carefully handled and disposed

PROCESS CONSIDERATIONS

Pyrolysis comprises of words Pyro = heat and Lysis = break down. When biomass is heated, products in the form of liquid gases and solids are produced as per the following reactions.

$$\text{Biomass} + \text{Heat} \rightarrow \text{Liquid} + \text{Syngas} + \text{Solid}$$

The products liquid, syngas, and solid are actually all valuable fuels and are defined as biooil (or bio-oil), bio-syngas (or bio-syngas), and biochar (or bio-char), respectively. Chemically, the process can be described by the relation:

$$CH_xO_y \rightarrow n_1C + n_2H_2 + n_3\,H_2O + n_4\,CO + n_5CO_2 + n_6\,CH_4 + \text{Char} + \text{Tar}$$

Molecular formula CH_xO_y in the above equation describes the nature of biomass, and n_1 to n_6 are constants of reaction determined by state of gasification. The main product of the pyrolysis process is carbon (C), hydrogen (H_2), carbon monoxide (CO), carbon dioxide (CO_2) and methane (CH_4), tar and other hydrocarbons. As an example, pyrolitic reaction for cellulose can be written as:.

$$3\,C_6H_{10}O_5 \rightarrow C_6H_8O + H_2O + 2CO + 2\,CO_2 + CH_4 + 7C$$

Herein $C_6H_8O + H_2O$ represent liquid bio-oil, $2CO + 2\,CO_2 + CH_4$ represent combustible gases and C is biochar.

The pyrolysis-gasification of biomass occurs in four stages; drying process (50-150°C), pyrolysis (above 160° C), the oxidation reduction (1000-1200°C) and reduction reaction (600-900°C). The reaction of pyrolysis occurs only at temperatures above 160°C. Within the biomass components, the pyrolysis of hemicellulose happens between 210 and 310°C. The pyrolysis of cellulose starts from 300–315°C and ends at 360–380°C, with a peak at 342–354°C. Lignin starts to decompose at about 200°C and continues until 1000°C. Volatile matters are separated from dry biomass during the pyrolysis process. Several units of pre-processing and post processing are necessary additions to any pyrolysis unit (Figure 4.3).

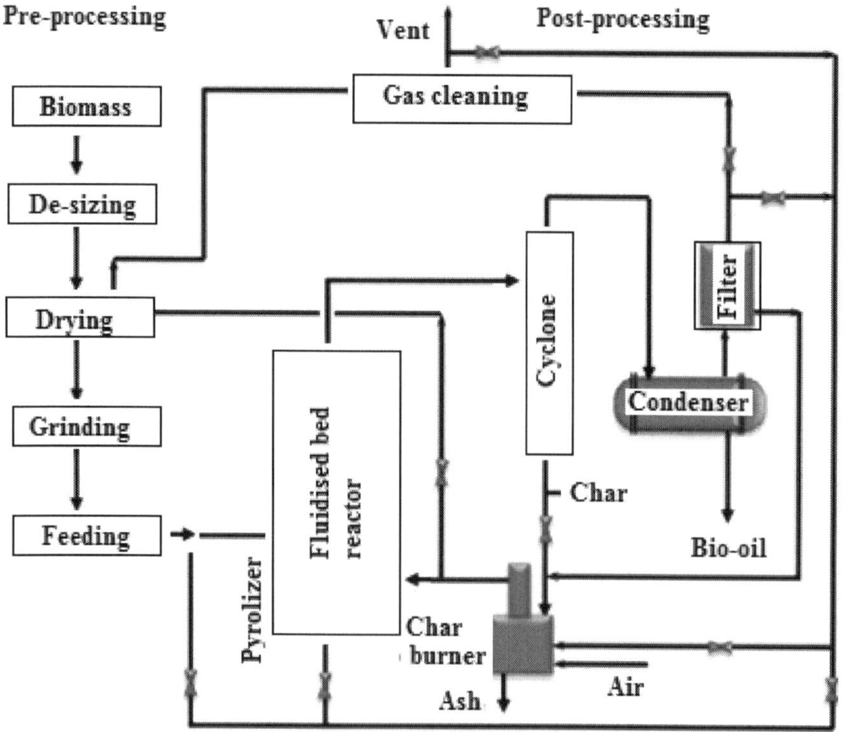

Figure 4.3. General components of pre and post processing in pyrolysis.

Raw Materials

A wide range of biomass feedstock can be used in pyrolysis. The broad spectrum of organic waste materials includes vegetable wastes, wooden wastes, agricultural residues, sludge, waste and polluted oil and synthetic wastes. Two major characteristics of the feedstock that affect the product outputs and their quality are the moisture content and the particle size. The moisture content of the feedstock should be around 10 per cent. The products produced from biomass having high moisture content will have high moisture levels. On the other hand, the process may only produce dust instead of oil at too low moisture contents. Therefore, high-moisture waste streams such as sludge and meat processing wastes require drying before using them for pyrolysis. The efficiency and nature of the pyrolysis process is also dependent on the particle size of feedstock. Most pyrolysis technologies can only process small particles to a maximum of 2 mm keeping in view the need for rapid heat transfer through the particle. It means the feedstock has to be size-reduced before it can be used for pyrolysis.

Pyrolysis Products

Three major products of pyrolysis of biomass are in the form of liquid, solid and gases.

Liquid: The liquid bio-oil is derived from the gases which have condensed in the form of liquid. It can be used as low grade diesel oil. It may also have tars, water and heavy hydrocarbons.

Solid: The solid material produced during pyrolysis is char. It consists of carbon and has applications in improving soil fertility.

Gases: These are non-condensable gases referred as syngas or biogas. The biogas is a mixture of gases that include CO_2, H_2O vapours, CO, C_2H_2, CH_4, C_2H_6, C_6H_6, *etc.*

The heating rate and residence time of a pyrolyzer are adjusted to maximize the production of desired product. The design norms for maximizing each of the pyrolysis products are given below.

1. A low heating rate (< 0.01 to 2.0°C/s), a low final temperature and long gas residence time maximizes char production.
2. A high heating rate, a moderate final temperature (450 to 600 °C) and a short gas residence time maximizes liquid production.
3. A slow heating rate, a high final temperature (700 to 900 °C), and a long gas residence time maximize gas production.

Besides several other physical, chemical and operating parameters control the whole pyrolysis process. Some of them are listed below.

Physical factors: Particle size, density

Chemical factors: Biomass composition (*e.g.*, the biomass resource)

Operating parameters: Heating rate (length of heating and intensity), residence time, pyrolysis temperature and pressure, the kind of ambient atmosphere and the type of catalysts used, if any

These parameters can be regulated by selection among different reactor types and heat transfer modes (Figure 4.4), such as gas–solid convective heat transfer and solid–solid conductive heat transfer.

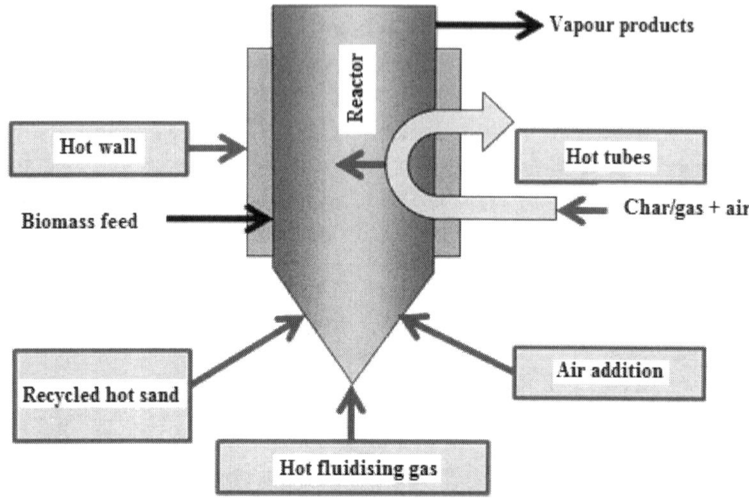

Figure 4.4. Methods of heat transfer to a pyrolysis reactor.

PYROLYSIS REACTORS

Laboratory Scale

Varieties of reactors have been designed and developed for continuous feed fast pyrolysis. They vary primarily by solids handling, mixing, and heat transfer mechanisms (Figure 4.4). Some reactor types have been successfully tested in the laboratory scale to realize fast pyrolysis conditions. These include Resistively heated micro-furnace or tube pyrolyzer, Resistively heated element pyrolyzer, Curie-point filament pyrolyzer, Laser pyrolyzer, radiative pyrolyzer, molten salt pyrolyzer, plasma pyrolysis reactor and vacuum pyrolyzer/reactor. These technologies still await successful demonstration on pilot scale. Therefore, these are not being covered in detail except the micro wave and plasma pyrolysis reactors where some headway has already been made.

Micro-furnace or Tube Pyrolyzer

Micro-furnace provides a constantly heated isothermal pyrolysis zone into which samples are introduced through a liquid syringe, solid plunger syringe, or in a little cup. Currently, lack of control on temperature/time characteristics of the sample has made the continuous-mode pyrolyzer less attractive to perform precise analytical works. These control problems can be overcome with the combination of a pulsed molecular beam source and the continuous mode pyrolysis oven. A comparative statement of microwave assisted and conventional electrical pyrolysis is given in Table 4.3.

Table 4.3. Comparison between microwave-assisted and conventional electrical pyrolysis

Microwave-Assisted Pyrolysis	*Conventional Electrical Pyrolysis*
Conversion of energy	Transfer of energy
Noncontact heating	Contact heating
Hot spot	No hot spot
Selective	Non-selective
Lower thermal inertia and faster response	Higher thermal inertia and slower response
Lower energy consumption	Higher energy consumption
Rapid heating	Slow heating
Shorter reaction times	Longer reaction times
Volumetric heating	Superficial heating
Higher level of control	Lower level of control
Improved product yields	Lower product yields

Plasma Pyrolysis Reactor

Gas molecules in plasma pyrolyzer are ionized by electric discharges (Figure 4.5). It is highly reactive due to the electrons, ions and the high energy density in the gas. While cold plasmas are generated at vacuum pressure, thermal plasmas are generated at atmospheric pressure. The temperature of thermal plasmas can

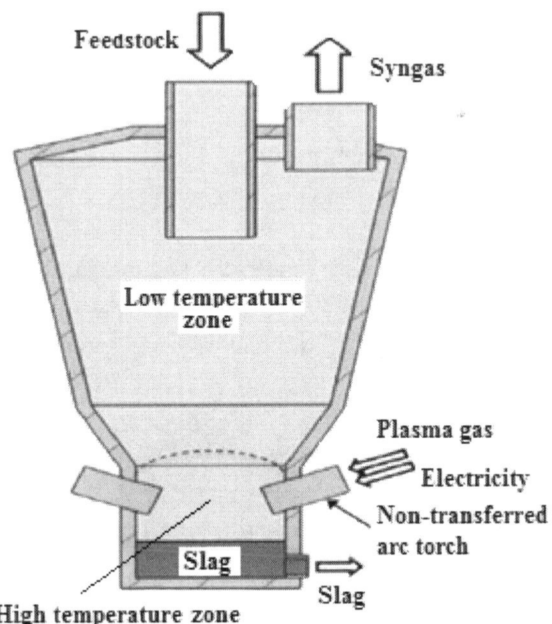

Figure 4.5. Schematic view of a plasma pyrolytic reactor.

be as high as 5000 K or even higher. Plasma pyrolysis reactor offers some unique advantages in comparison to conventional pyrolysis at low temperatures and slow heating rates. The high energy density and temperature associated with thermal plasmas and the corresponding fast reaction times provide a potential solution for the problems that occur in conventional pyrolysis processes, such as low gas productivity and the generation of heavy tarry compounds. In spite of this, plasma pyrolysis of biomass for energy and chemical production are seldom used because electric power consumption is quite high. In fact, the temperatures in thermal plasma (usually 2500-9500°C) are too high for biomass pyrolysis. This method is more practical and of interest for the utilization of biomass material for the purpose of syngas and char production.

Pilot Scale

Pilot scale reactors are classified either on the basis of biomass movement or on the basis of heat supplied. On the basis of biomass movement (gas-solid contacting mode) pyrolyzers are classified into fixed bed, moving (fluidized) bed and entrained bed. Moving bed reactors have further been sub-classified into sub-groups and sub-sub-groups (Figure 4.6).

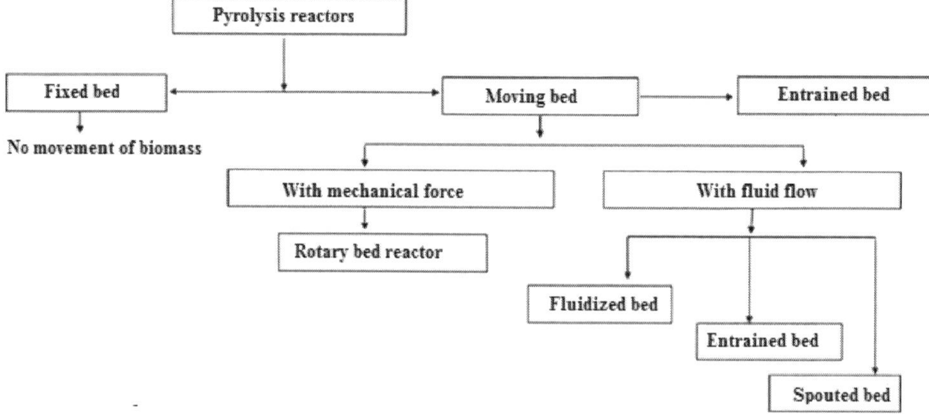

Figure 4.6. Classification of pyrolytic reactors on the basis of biomass movement.

On the basis of heat supplied, the reactors have been categorized in four groups.

- ✰ Part of the material burned inside the reactor to provide the heat to carbonize the remaining material
- ✰ Direct heat transfer from the gases produced by combusution of the pyrolysis product or any other fuel outside the reactor
- ✰ Direct heat transfer from inert hot material (hot gases or sand introduced into the reactor)

☆ Indirect heat transfer through the reactor walls *i.e.* external heat source due to combustion of one or more pyrolysis productes or any other fuel

KINDS OF REACTORS

Pyrolysis reactors have been categorized as:

Slow pyrolysis: Fixed bed pyrolyzers, heated kiln reactors, fluidized bed reactors, vacuum pyrolysis reactors, and screw feed/auger reactors.

Fast pyrolysis: Fluidized bed reactors (bubbling and circulating), rotating cone reactors, ablative pyrolysis reactors, twin screw reactors and ultra-rapid pyrolyzers.

Fluidized bed and circulating bed reactors are the most favorable configurations because of their easy operation and good quality products.

Distinction between Fixed and Fluidized Bed Reactors

The bed section of a pyrolyzer consists of a sand mass where drying, pyrolysis and gasification take place. In this section, organic particles of the waste are converted into gas under low oxygen conditions (Basu, 2010). A number of flow regimes can be established depending upon the velocity of the fluidizing gas (Figure 4.7). As the flow of the fluidizing gas passing through the bed of particles increases, particles vibrate but there is no change in the height of the bed. It is called fixed bed reactor (Figure 4.7A). With further increase in the gas velocity, a point arrives when the drag force by the upward flowing gas equals the weight of the particles. The bed voids also increase slightly. It is the onset of fluidization and it is called the minimum fluidization (Figure4.7B). The velocity at this point is called the minimum fluidizing velocity. Increasing the velocity of the gas beyond the minimum fluidizing velocity leads to the formation of bubbles (Figure 4.7C). With further increase in velocity, the bubbles coalesce and grow in size as they rise. If the ratio of the bed height to the bed diameter is sufficient, the diameter of the bubbles may be the same as the diameter of

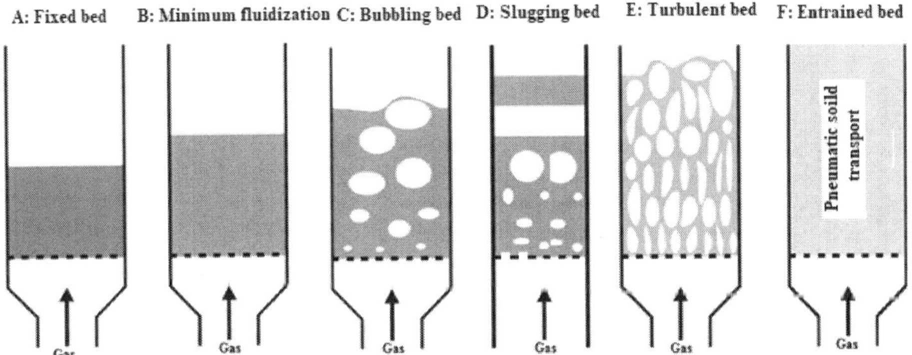

Figure 4.7. Schematic representation of the different regimes of fluidized bed.

the bed. This is known as slugging bed (Figure 4.7D). With further increase in gas flow rate, the velocity may exceed the terminal velocity (the velocity required to transport particles out of the bed). In this case, upper surface of the bed disappears and a turbulent motion (Figure 4.7E) of solid clusters and voids of gas of various shapes and sizes form in the bed. Further increase in the gas velocity creates an entrained bed with disperse, lean or dilute phase fluidized bed which results in pneumatic solid transport (Figure 4.7F).

FIXED BED REACTORS

Beehive oven, a kind of fixed bed reactor, is traditionally used for charcoal production. It typically works in batch mode. The heat is supplied by an external source or through limited combustion in a beehive oven (Figure 4.8). The most critical factor in the efficient conversion of wood to charcoal is the careful operation of the kiln. Wood must be dried and carefully stacked to allow an even flow of air through the kiln and sufficient time for reactions to take place. Improved beehive ovens are made of brick including the oven's base, floor and its sides. Sometimes, outer surface is sealed with a layer of mud or lime. The oven is heated by fire inside for a period long enough to heat all the bricks. Although wood is the most common fuel source many other sources have been tried including agricultural waste such as millet stems and corn cobs as well as coconut shell. The process is divided into the following stages.

Combustion: Oxygen supply is high and temperature rises from ambient to over 500°C. After the fire is established, the oxygen supply is reduced, and the firing point is closed so that temperature drops to about 120°C.

Dehydration: Free water is driven out at a reduced temperature of about 100°C and the kiln gives out thick, white and moist steam.

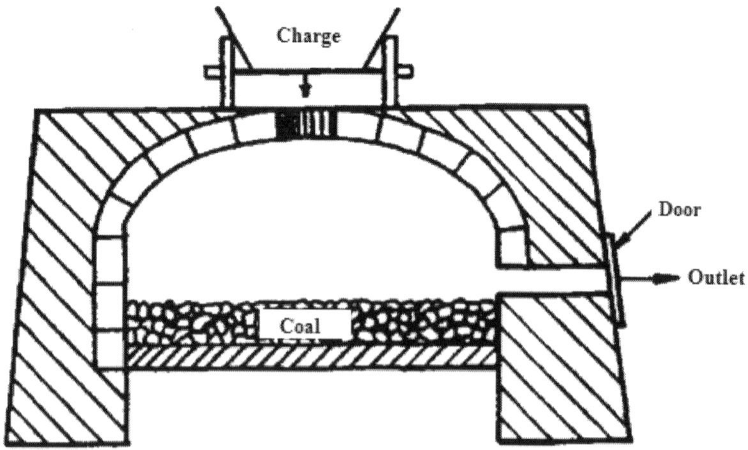

Figure 4.8. Beehive oven, a fixed bed pyrolyzer, for charcoal production.

Exothermic reaction: Once the wood has dried, temperature rises to about 280°C and the wood begins to break down into charcoal, water vapours and other chemicals. The temperature is maintained by controlling the air flow through holes and vents that helps more wood to burn.

Cooling: When carbonization is complete, the kiln cools to below 100°C. At this time, charcoal is removed for further cooling.

Heated Kiln Reactor

Heated kiln reactor is the most commonly used reactor for slow pyrolysis. The residence time in this reactor is much longer compared to vacuum pyrolysis and fluidized bed reactors. The reactor can be heated externally by a heating jacket surrounding the kiln or internally by the combustion of product gases. It can be operated in batch mode that does not require agitation. It can also be operated in continuous mode where some kind of mechanical agitation is required. Several methods such as conveyors or sweepers are employed for mixing and transport in the heated kilns. Same objective can also be achieved with rotating or inclined kilns. All types offer simple cost-effective designs. The product output may differ depending on factors like the residence time and temperature.

Other Reactors

Other kinds of fixed bed reactors are grouped into downdraft and updraft fixed bed reactors like the gasifiers discussed in the next chapter. In a down draft fixed bed reactor, the solid and product gas move downward in a co-current mode. Solids move slowly down a vertical shaft and reacts with introduced air at the throat that supports the gasifying biomass. With high carbon conversion, a relatively clean gas is produced with low tar. In contrast, solids move down a vertical shaft and contact the air in a counter-current mode in an updraft fixed bed reactor. The product gas is very dirty with high levels of tars. Tar crackers are used to alleviate this problem. The technology is simple, reliable and proven for fuels of relatively uniform size. A sweep gas can be used for effective removal of the product gas from the pyrolysis reactor.

FLUIDIZED BED REACTORS

Bubbling Fluidized Bed Reactor

Bubbling fluidized bed reactor comprises a plenum chamber, bubbling zone and disengagement zone (Figure 4.9). Plenum zone with distributors results in uniform distribution of the gas. Rate of fluidized gas passing through the reactor is adjusted so that the solid fluidization is in the bubbling regime. It means the bed has fully expanded and is bubbling aggressively but without reaching the turbulent flow regime. The pyrolysis vapours coming out are led to post processing units. The fluidized beds offer some advantages in terms of uniform temperature, good heat transfer, high reaction rates with bio-oil yield going up to 75 per cent, a medium level of complexity in construction and operation, and ease of scaling up. They have greater tolerance to particle size range but perform well when particle size is small. The major disadvantages are: the requirement of large quantity of inert gases, and high operating costs. Chances of product dilution from fluidizing gas and trapping of some sand by the char is also not ruled out. They also have higher particulates and moderate tar levels in product. Gas tar cracking catalysts can be added to the bed to alleviate this problem.

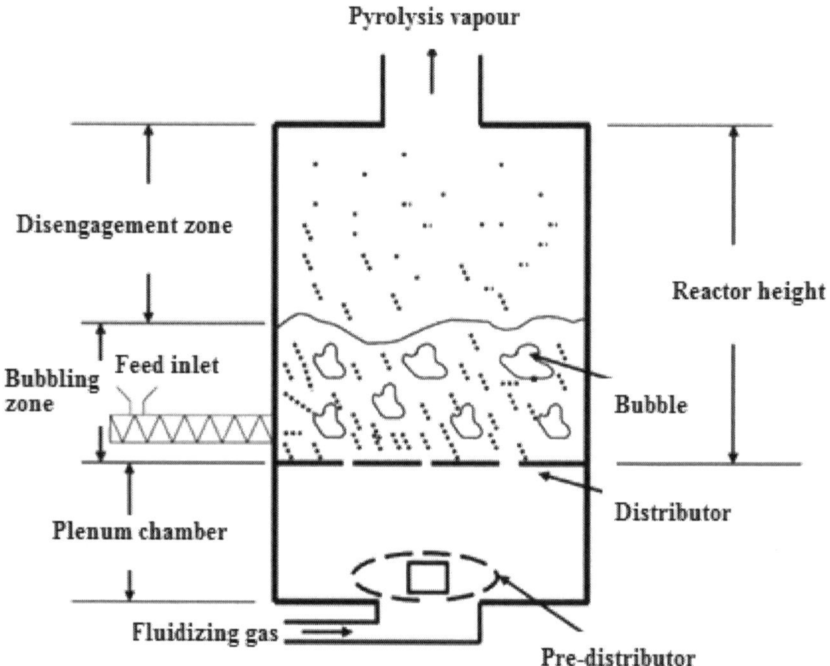

Figure 4.9. Bubbling fluidized bed reactor.

Circulating Fluidized Bed (CFB) Reactor

The circulating fluidized bed reactor works on the same fluidization principle as the fluidized bed reactor except that the bed is highly expanded and the bed particles circulate in a constant loop comprising a cyclone (Figure 4.10) and loop seal. Although several configurations of circulating fluidized bed exist, yet all of them generally include a reactor and a cyclone (Figure 4.10). Gas and solids move up the reactor with some degree of internal refluxing. The average residence time of solid particles is longer than the gas residence time though the difference is not as high as in a bubbling fluidized bed. High heat transfer rates from sand ensure rapid heating of biomass particles. A special characteristic of this pyrolyzer is uniform heating of the bed particles around the entire height of the unit. A major advantage of CFB is that the entrained char from the reactor can be easily separated and burnt to resupply heat back to the reactor. Ablation is more prevalent than with regular fluidized beds. Disadvantages include need of large quantities of heat carriers (sand); more complex operation, and high operating costs. Moreover, erosion problems may occur inside the reactor vessel due to circulation of hot solids.

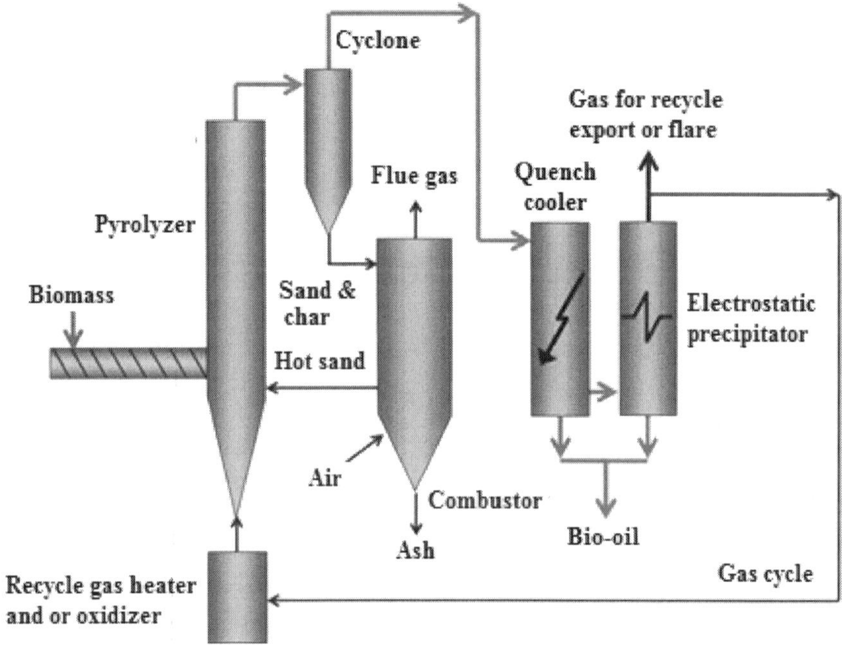

Figure 4.10. Circulating fluidized bed reactor.

Entrained Flow Reactor

Entrained flow fast pyrolysis is a simple technology and is favoured option because of its similarities to coal gasification system. Entrained flow pyrolysis utilizes some type of mixture or slurry as the feed having small particle sizes depending on the operating conditions (Figure 4.11). Very high heating rates of the particles, residence time varying from milliseconds to a few seconds, and the pressure as high as 5 atmospheres can be achieved in this reactor. A major issue with entrained flow pyrolysis is the gas pressure effects on the heat transfer. Since the potential increase in pressure affects the temperature profile within the reactor, it may not always suit the operating conditions. This issue can be taken care of by using much shorter residence time in the pyrolysis reactor. Major limitations arise from low heat transfer rates, particle size limit < 2 mm, and limited gas/solid mixing. Besides, feed preparation for woody biomass may be quite costly. The experience with these reactors has not been satisfying as anticipated mainly due to poor heat transfer between the hot gases and soil particles.

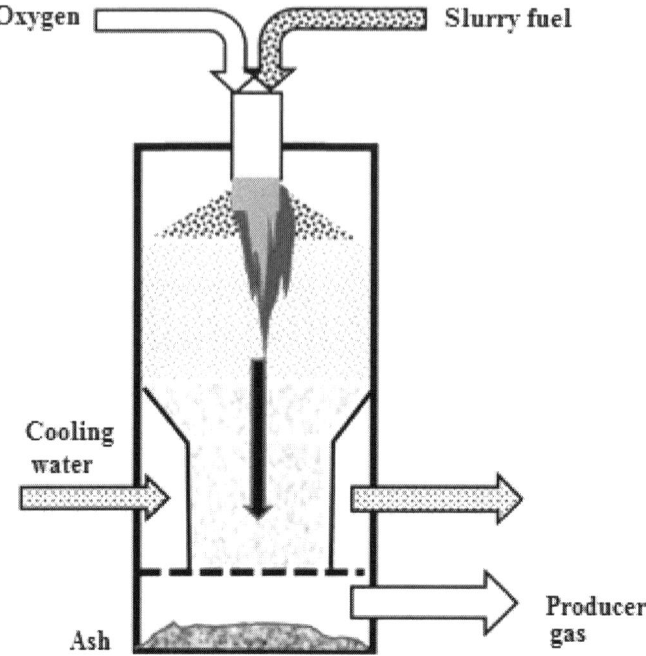

Figure 4.11. Entrained flow reactor.

Vacuum Furnace Reactor

Biomass is thermally decomposed under low absolute pressure inside the vacuum furnace reactor. The biomass fed into the reactor is mechanically stirred by a conveyor during the pyrolysis process. The pyrolysis in this case is slow with a low heating rate and residence time of approximately 40 s. The vapours produced are quickly removed from the vacuum and condensed to recover bio-oil. Two important features of this reactor are: i) it is capable of handling larger particles than most fast pyrolysis reactors and ii) It produces much less char in the liquid product than other reactors due to lower gas velocities. No carrier gas is needed in this pyrolyzer. Mechanically, the process is complex as the product (charcoal) is expelled from the reactor through a pressure seal. The typical liquid yields for dry biomass feed vary from 35 to 50 per cent. A multiple hearth vacuum pyrolysis set-up is shown in Figure 4.12. High capital and maintenance cost and high sealing/gasket requirement often limit the use of these pyrolizers.

Ablative Reactor

The biomass in an ablative pyrolysis reactor is pressed against a hot moving surface either by centrifugal or mechanical forces (push/press). The high pressure between the biomass particle and the hot reactor wall so created ensures rapid heat transfer to the biomass particles resulting in liquid to melt

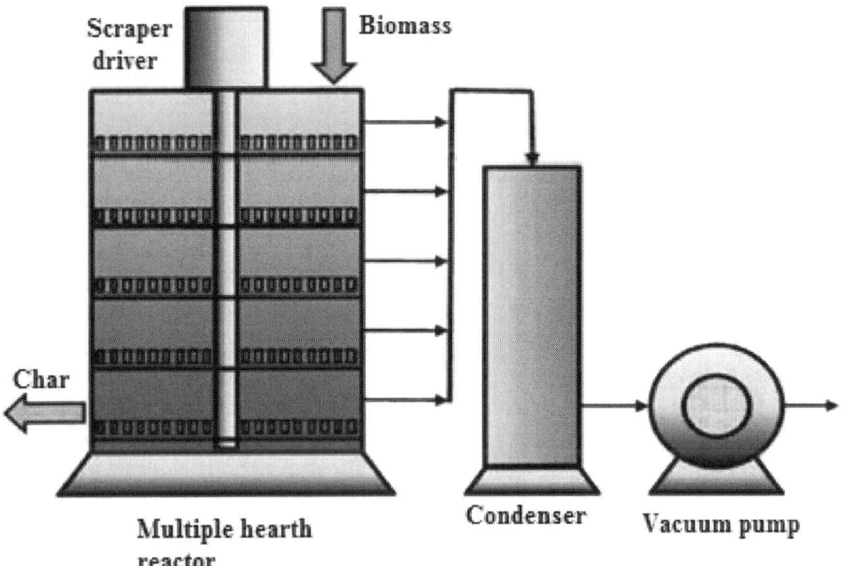

Figure 4.12. A multiple hearth vacuum pyrolyzer.

out of the biomass. The final liquid yield may be as high as 80 per cent. High pressures are normally required in these reactors as the reaction rate is found to increase with increasing pressure. Temperatures are usually under 600°C but very high heat is needed at the reactor walls (Figure 4.13). The ablative pyrolysis reactors have several other advantages over conventional fluidized bed reactors.

- ✰ No milling of the biomass is required, because the heat transfer is through the direct contact with the hot surface
- ✰ Because of the compact design, these reactors have good heat transfer with high heating rates and relatively small contact surface

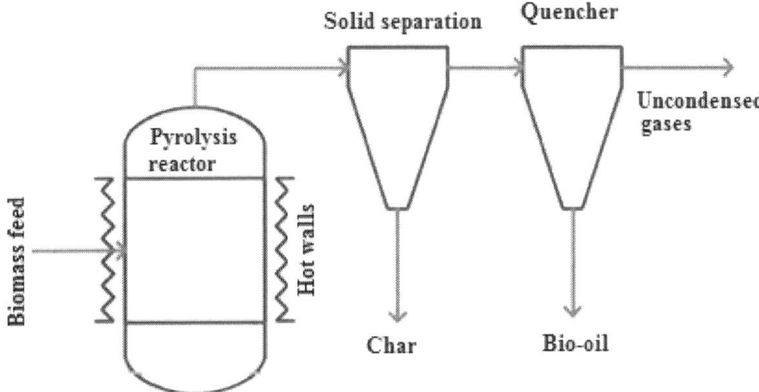

Figure 4.13. An ablative reactor for pyrolysis.

- They have high energy and cost efficiency as no heating and cooling of fluidizing gases is involved
- Condensation units with small volume requiring less space at lower costs can be installed

Note that scale up of this process has not yet taken past the laboratory phase. Yet the process has shown the potential of adoption on commercial scale.

Cyclone Reactor

The cyclone reactor is another version of ablative pyrolysis reactor. Here, carrier gas and particles enter the reactor tangentially and centrifugal forces press the particles against the hot reactor wall for pyrolysis. The pressure against the hot wall is comparatively lower than an ablative reactor. Therefore, particle sizes have to be much smaller. There are no mechanically moving parts and no heat carrying medium (such as sand). The cyclone reactor also automatically separates the products, as solid residues leave the reactor at the bottom while gases and vapours leave at the top. In principle, the process can be designed in a similar manner as an ablative reactor shown in Figure 4.13 except that there is no need for solids separation as it takes place inside the reactor itself.

Rotating Cone Reactor

The rotating cone reactor is used in flash pyrolysis of biomass with rapid heating of about 5000 K/s, a short residence time of the solids and negligible char formation. The distinguishing features of this pyrolyzer are very short solids and gas-phase residence times of about 0.5 s and 0.3 s respectively. The rotating cone pyrolyzer reactor differs from the fluidized bed reactor in the sense that a centrifugal force is used to transport the biomass particles and products in a spiral through the reactor either in upward or downward direction. The biomass is fed at the bottom of the rotating cone rotating at about 360 to 960 rev/min (Figure 4.14). Relatively fine particles are required to obtain good liquid yield. The outside walls of the cone are heated. A motor spins the cone. The biomass is pyrolyzed within a small annular volume. The gas/condensable comes out of the top. The sand and the char/ash spill over the upper rim of the cone into the fluidized bed surrounding it. The condensable biogas leaves through a tube. The char burns in the fluidized bed and the heat so produced is used to heat the cone and the solids. The solids are recycled to supply the heat for pyrolysis. Typical liquid yield is 60 to 70 per cent. Although, carrier gas required is much less than for a fluid bed and transported bed systems, additional gas is needed for char burn off and for sand transport. This type of pyrolyzer is compact, has a relatively simple operation, and has a low heat carrier/sand requirement. However, it has limited capacity and may be difficult to scale up. Integrated operation of three sub-systems namely rotating cone pyrolyzer, riser for sand recycling, and bubbling char combustor may pose some design complexity.

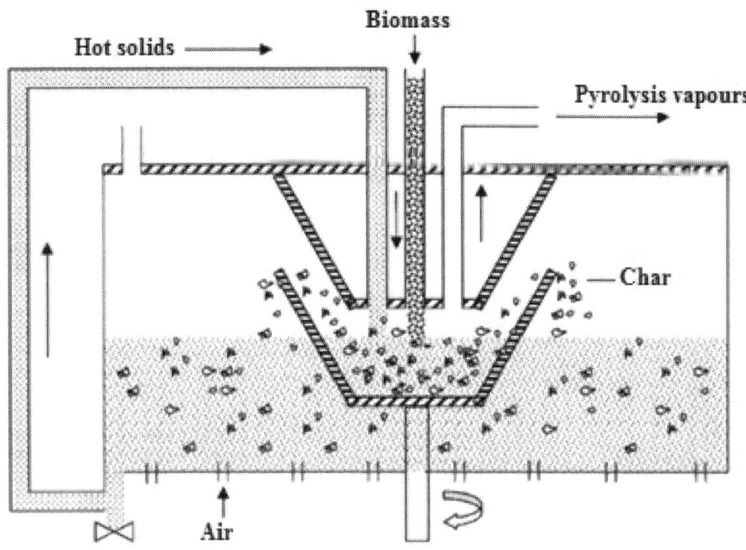

Figure 4.14. Rotating cone pyrolyzer reactor.

Auger Reactor

An auger pyrolyzer looks like a cylinder with corkscrews inside depending upon whether it is slow (single screw, Figure 4.15 left) or fast pyrolysis (double screw, Figure 4.15 right). A motor is attached to the rotating corkscrew. Hot sand and biomass particles are fed at one end of the screw. The screw mixes the sand and biomass and conveys them along. It provides a good control of the biomass residence time. The condensable comes out of the top and char/ ash from the bottom.

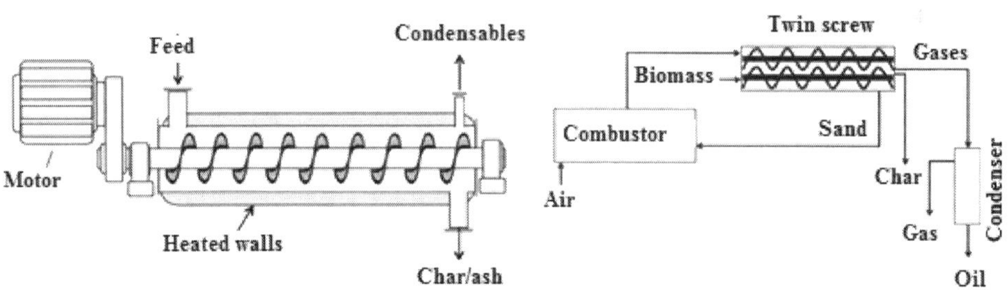

Figure 4.15 Auger pyrolizers: single screw for slow (left) and twin screw for fast (right) pyrolysis.

The advantages of this reactor are:

- ✰ Compact size
- ✰ Efficient mixing and transport of the biomass particles through the reactor

- No carrier gas required and hence no fear of dilution of final products
- Lower processing temperatures (400°C)

The disadvantages include long residence times, lower bio-oil yields, high char yield, and limits in scaling up due to heat transfer limits. Besides, sand must be reheated in a separate vessel. Its mechanical reliability may be a concern.

The twin screw pyrolysis reactor uses two offset screws encased in the reactor (Figure 4.15 right). They rotate in the same direction and have intertwining flights to dislodge particles caught in the thread. The products obtained are 53 to 78 per cent liquid, 8 to 20 per cent gas and 12 to 34 per cent char.

Ultra-rapid Pyrolizers

Ultra-rapid pyrolyzer is characterized by high heating rate and short residence time resulting in high liquid yields. Inert gas (such as nitrogen) is heated to 100°C above the temperature of the reactor and sent into the reactor at a very high velocity. The nitrogen bombards a stream of biomass injected into the reactor. A heat carrier (such as sand) can also be heated externally and used to bombard the stream of biomass through multiple jets. This creates an exceptionally high heating rate which heats the biomass to the pyrolysis temperature in a few milliseconds. The products obtained from pyrolysis leaves the reactor from the bottom and are immediately cooled off to maximize liquid yield and to prevent secondary reactions.

BIO-OIL

Bio-oil is a dark brown liquid. It has quite similar composition as the biomass, but much higher density than woody materials resulting in reduced storage and transport costs. Bio-oil is not suitable for direct use in standard internal combustion engines. But it can be upgraded to either a special engine fuel or through gasification processes to syngas and then biodiesel. Bio-oil is particularly attractive for co-firing because it can be more readily handled and burned than solid fuel. Bio-oil offers major advantages over solid biomass and gasification due to the ease of handling, storage and combustion in an existing power station as no special start-up procedures are required. In addition, bio-oil is also a vital source for a wide range of organic compounds and specialty chemicals. The low quality of oil produced through pyrolysis can be improved by physical and chemical processes, which might drive up production costs, but as circumstances change it may make some economic sense.

QUESTIONS (THEORY)

1. What is pyrolysis? Show the biomass pyrolysis cycle through a neat diagram.
2. Name and describe the various pyrolysis techniques.

3. List the advantages and disadvantages of pyrolysis.
4. Name and discuss the various kinds of products produced during pyrolysis. List the factors that affect the product mix of pyrolysis.
5. Write short notes on pyrolysis reactors: Plasma pyrolysis reactor, fixed bed reactor, bubbling fluidized bed reactor, entrained flow reactor, ablative reactor, cyclone reactor, and bio-oil.
6. Distinguish between fixed and fluidized bed reactors using neat line diagrams.

Chapter 5

Gasifiers

Increasing use of energy has been one of the major contributing factors for high productivities in agriculture. So far, increase in energy use has mainly come from fossil fuels. Increasing scarcity of fossil fuels, rising costs and import bills, governments around the globe are pumping huge investments and efforts to popularize use of alternative energy sources like solar, wind, and other non-conventional sources. One amongst these sources is biomass energy. Since solid biomass as a fuel is inconvenient to use, is suited to certain limited applications, and has low utilization efficiency, it needs to be converted to liquid and/or gaseous fuels. Conversion to a combustible gas mixture called producer gas can overcome most of the problems associated with the use of solid biomass fuels. The gasification technology, based on incomplete combustion of biomass, was extensively used for transportation and on-farm systems during World War II. Gasification has proven to be quite reliable and beneficial with many attractive features.

- Ability to process a wide range of feed stocks including coal, heavy oils, petroleum coke, heavy refinery residuals, refinery wastes, hydrocarbon contaminated soils, biomass, and agricultural wastes. Therefore, gasification process is much more attractive than ethanol production or biogas where only selected biomass materials can be used
- Although conversion to gas usually results in loss of energy up to 25 per cent, yet the overall efficiency is very high because use of gas is highly efficient
- It can be employed at any scale
- It is ideally suited to decentralized applications

- It produces a consistent product that can be used for the generation of electricity or as primary building blocks for manufacturers of chemicals and transportation fuels
- The ability to remove contaminants in the feedstock and to produce a clean syngas product
- The ability to minimize the amount of solid waste requiring landfill disposal. Solid by-products are nonhazardous and have a market value as these can be used as fuel or construction material
- Economic factors are quite favourable as biomass feedstock is easily available in agricultural systems

GASIFICATION PROCESS

The thermo-chemical decomposition of hydrocarbons of the biomass in a reducing (oxygen deficient) atmosphere into a combustible gas mixture normally called producer gas is called gasification. Other names of the producer gas are syngas or synthesis gas, generator gas, wood gas, coal gas, biogas or others. Syngas is strictly defined having only CO, H_2, CO_2 and H_2O as its constituents. Whereas complete combustion of biomass produces only carbon dioxide and water vapour, partial combustion produces carbon monoxide and hydrogen with traces of methane all being combustible gases. It also contains non-useful products like tar and dust (Figure 5.1). Gasification is carried out in a reducing environment at relatively high temperature in the presence of an oxidizing agent also called a gasifying agent, a substance that promotes the chemical reaction. Reducing atmosphere in the gasification means that only 20 per cent to 40 per cent of stoichiometric amount of oxygen (O_2) for complete combustion is allowed during the reaction.

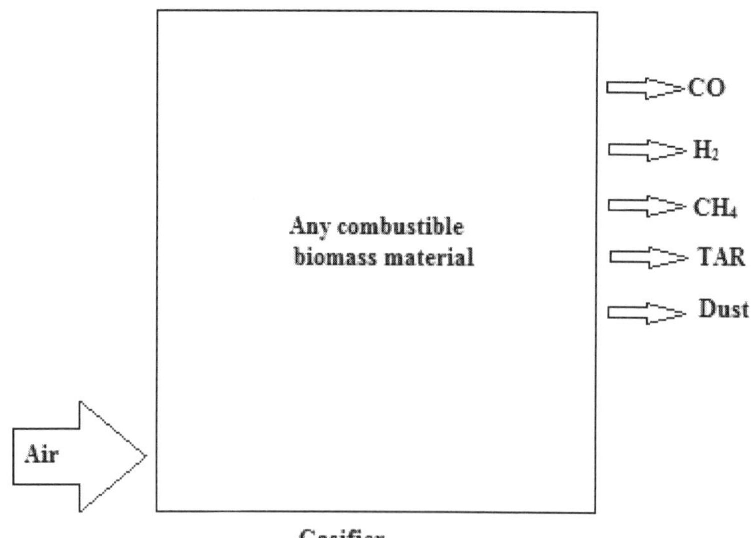

Figure 5.1. Gasification products in deficient oxygen environment.

Gasifying agents are mainly oxygen (O_2), steam, CO_2 or an appropriate mixture of air (N_2, O_2). Heat is supplied either directly or indirectly to raise the temperature of the gasifier to 600-1000°C. The combustible gases in the gasifiers are produced as a result of reaction of water vapours and carbon dioxide passing through a glowing layer of charcoal. Clearly, the key to design an efficient gasifier is to create conditions such that a) biomass is reduced to charcoal and, b) charcoal is converted at suitable temperature to produce CO and H_2.

A comparative statement of gasification and combustion processes is given in Table 5.1, which brings out the differences in the gasification and combustion process discussed in Chapter 3.

Table 5.1. Comparison of gasification and combustion technologies

Features	Combustion	Gasification
Purpose	Generation of heat or destruction of waste	Creation of valuable, usable products from waste or lower value material
Process type	Complete combustion using excess oxygen (air)	Thermal and chemical conversion using no or limited oxygen
Raw gas composition (before gas cleanup)	CO_2, H_2O, SO_2, NO_x, and particulates	H_2, CO, H_2S, NH_3, and particulates
Gas clean-up	Flue gas clean-up at atmospheric pressure Treated flue gas is discharged to atmosphere Any sulphur in the fuel is converted to SO_2 that must be removed using flue gas clean-up systems, generating waste that must be land-filled. Clean flue gas primarily consists of CO_2 and H_2O	Syngas cleanup at atmospheric to high pressures depending on the gasifier design Treated syngas used for chemical, fuels, or power generation Recovers sulphur species in the fuel as sulphur or sulphuric acid Clean syngas primarily consists of H_2 and CO.
Solid by-products/ products	Bottom ash	Char or slag
Ash/char or slag handling	Bottom ash and fly ash are collected, treated, and disposed in most cases as hazardous waste	Low temperature processes produce a char that can be sold as fuel High temperature processes produce a slag, a non-leachable, non-hazardous material suitable for use as construction material Fine particulates are recycled to gasifier. In some cases fine particulates may be processed to recover valuable metals

Features	Combustion	Gasification
Temperature	800°C to 1000°C	700°C – 1500°C
Pressure	Atmospheric	Atmospheric to high

Wide range of gasifiers is available for the wide range of biomass including forestry products such as wood, agricultural cellulosic residues, aquatic biomass and municipal solid wastes. The key parameters affecting the gasification are: equivalence ratio (ER) the actual air (supplied in the gasification) to the stoichiometric air required for combustion, biomass characteristics, moisture content, superficial velocity, operating temperature, gasifying agent, residence time, pressure, use of catalysts, and biomass/steam ratio. Fundamental gasification processes in the presence of an oxidizing agent at high temperature are as follows:

Evaporation of surface moisture: Surface moisture evaporates from the raw material at the boiling point of water (depends on pressure). Inner moisture may still remain if the raw material is huge.

Evaporation of inherent moisture: Inherent moisture normally evaporates at 110-120°C.

Volatilization: Thermal decomposition is a heat generating reaction which is a characteristic phenomenon of biomass commonly designated as $C_mH_nO_p$. Since cellulose is the main component of biomass (typically 50 per cent of the weight), biomass is usually represented as $C_mH_nO_p$ having similarity with the chemical formula of cellulose $C_6H_{10}O_6$. Thermal decomposition of biomass begins at 200-300°C.

Volatilization and gasification reaction: The temperature is further raised during volatilization, and the volatile matter of the lightweight hydrocarbons (C_mH_n, where m and n are integers of at least 1; a low value of m indicates lightness and a high value of m indicates heaviness) is transformed into heavy C_mH_n with a high boiling point. Subsequently, the C_mH_n reacts with the gasifying agent and are converted to lightweight molecule clean gas. Tar and soot can also form in case diffusion of the gasifying agent is slow resulting in condensation of C_mH_n.

Char gasification: Following volatilization of the volatile content in the raw biomass, the fixed carbon and ash become char, which gets heated to the surrounding temperature. The subsequent reaction with the gasifying agent transforms the carbon into CO and CO_2. A wet gas reaction occurs ($C+H_2O \rightarrow CO+H_2$), when the gasifying agent contains excess steam and the surrounding temperature is over 750°C. The gas produced in such an environment is mainly composed of CO, CO_2 and H_2.

Char residue: The reaction rate of the wet gas being slow, it results in the formation of char residue. The formation of tar, soot and char on the one hand reduce efficiency while on the other pose some troubles to the equipment.

Although there is a considerable overlap amongst these processes, yet each can be assumed to occupy a separate zone where fundamentally different chemical and thermal reactions take place.

Pyrolysis Zone

Wood pyrolysis discussed in chapter 4 is an intricate process that is still not completely understood. The products produced during pyrolysis mainly depend upon temperature, pressure, residence time and heat losses. Following general remarks can be made about the process. Only water is driven off up to a temperature of 200°C.

$$\text{Moist fuel} \rightarrow \text{Fuel} + H_2O$$

Carbon dioxide, acetic acid and water are given off from 200 to 280°C. The real pyrolysis, takes place from 280 to 500°C. While the active pyrolysis reflects the loss of the hemicellulose, cellulose and part of lignin, passive pyrolysis reflects the slow and continuous loss of residual lignin. The temperature ranges of these stages and the kinetic parameters of the degradations depend primarily on the rate of heat transfer, the composition of the biomass, and the degree of the oxidizing environment. Large quantities of tar and gases containing carbon dioxide are produced during the process. Besides light tars, some methyl alcohol is also formed. The gas production at 500 to 700°C is small and contains hydrogen. Addition of oxygen at such a stage causes gasification. The pyrolysis products react with oxygen as an oxidant to form CO, CO_2 and H_2. These products also react with water as an oxidant to form CO, CO_2, and H_2. The process is called steam reforming. The pyrolysis products also react with CO_2 as an oxidant to form CO, CO_2 and H_2 in a process known as dry reforming. The breakdown reaction of the biomass is written as:

$$C_mH_nO_p + O_2 \text{ (21 per cent of air)} + H_2O \text{ (steam)} \rightarrow aCO_2 + bH_2O \text{ (unreacted steam)} + cCH_4 + dCO + eH_2 + fC_2 + C \text{ (char)} + \text{tar}$$

Combustion Zone

The combustible portion of a solid fuel is usually composed of elements carbon, hydrogen and oxygen. During complete combustion, carbon dioxide is obtained from carbon in the fuel and water is obtained from the hydrogen, usually as steam. The combustion reaction is exothermic and yields a theoretical oxidation temperature of 1450°C. Oxidation of part of the biomass is necessary to generate thermal energy required for the endothermic reaction and to maintain the operative temperature. Reaction temperatures are around 600–900°C for partial oxidation and around 800–1400°C for complete oxidation. Partial oxidation releases 111 kJ/mol of heat while complete oxidation releases 394 kJ/mol.

Complete oxidation reaction: $C + O_2 \rightarrow CO_2$

Partial oxidation reaction: $2C + O_2 \rightarrow 2CO$

$$2H_2 + O_2 \rightarrow 2H_2O$$

Reaction Zone

The products of partial combustion (water, carbon dioxide and non-combusted partially cracked pyrolysis products) now pass through a red-hot charcoal bed. The char and tar produced during oxidation release gases at around 600–950°C via several reactions during the reduction process. The typical reactions include the heterogeneous water-gas shift (HWGS) and water-gas shift reactions via the endothermic HWGS reaction. The char and tar react with water vapours (derived from biomass and air) to produce CO and H_2 at temperatures exceeding 750°C.

Boudouard equilibrium: $C + CO_2 \leftrightarrow 2CO$

Heterogeneous water-gas shift: $C + H_2O \rightarrow CO + H_2$

Water-gas shift reaction: $CO + H_2O \rightarrow CO_2 + H_2$

Steam reforming reaction: $CH_4 + H_2O \rightarrow CO + 3H_2$

Hydro-gasification reaction: $C + 2H_2 \leftrightarrow CH_4$

$$CO_2 + H_2 = CO + H_2O$$

The reduction reactions reduce the gas temperature being endothermic in nature. Consequently the temperatures in the reduction zone are normally 800-1000°C. Lower the reduction zone temperature (~ 700-800°C), lower is the calorific value of the gas.

Tar decomposition: $C_nH_m \rightarrow C_{n-x}H_{m-y} + H_2 + CH_4 + C$

Steam-tar reforming: $C_nH_m + 2nH_2O \rightarrow (2n + m/2)\ H_2 + nCO_2$

Two important terms in gasification are equivalence ratio, ϕ defined earlier and the efficiency of the gasifier. Equivalence ratio, ϕ is expressed as:

ϕ = Actual air supplied to the system for combustion/Theoretical or stoichiometric requirement of air for combustion

Typical values of ER range from 0.2 to 0.4. Total energy in the gaseous phase increases with the increase in equivalence ratio, ϕ.

The efficiency, h of a gasifier is defined as the chemical energy output in the dry producer gas at15°C to energy input from the biomass.

h = Energy output in the dry producer gas/Energy input from the biomass

The efficiency is usually around 75 per cent.

Example 5.1

1 kg of biomass produces 2.5 m^3 of gas with an average calorific value of 5.5 MJ/m^3 at 15°C. Calculate the gasification efficiency, if the average calorific value of biomass (dry) is 18.8 MJ/kg.

$\eta_{Gas} = 2.5 \times 5.5 \times 1 \times 100/18.8 = 73$ per cent

RAW MATERIALS

The biomass including agricultural residues and other common solid wastes can be used for gasification. Solid biomass may include wood (construction wastes and timber off-cuts and unutilized woods such as forest thinning, remaining timbers, and damaged trees), paper, peat, lignite, and coal including coke derived from coal. Biomass used for gasification is broadly grouped as dry, wet, and others (Table 5.2). For example, dry biomass is classified as woody or herbaceous and wet biomass as sludge/excreta, common kitchen food waste and other. Each group is further categorized as waste, unutilized, and produced (Table 5.2). For example unutilized category comprises unutilized herbaceous biomass, crop residues such as rice-wheat straw and rice husks, and produced such as short-rotation woody crops (*Eucalyptus*, willow, *etc.* and herbaceous biomass, grasses such as Napier grass, sorghum, and *Miscanthus*). Solid fuels are composed primarily of carbon with varying amounts of hydrogen, oxygen, and impurities, such as sulphur, ash, and moisture. The aim of gasification is the almost complete transformation of these constituents into gaseous form so that only ashes and inert materials remain.

Table 5.2 Classification of biomass feedstock

Moisture Level	*Classification*	*Waste*	*Unutilized*	*Produced*
Dry	Woody	Construction wastes, timber off-cuts	Forest thinning, damaged trees, remains of forest timbers	Short duration woody crops
	Herbaceous		Crop residues such as rice-wheat straw and rice husk, sugarcane wastes, aquatic biomass such as algae, water plant	Grasses and crops
Wet	Sludge/ excreta	Sewage sludge/ livestock excreta		
	Common food	Food processing waste, kitchen waste		
Others	Other	Molasses, waste food oil	Land-fill gas	Cultivated maize, sugarcane

Ultimate and proximate analysis are used to fully characterize the biomass. Ultimate analysis is used to quantify chemical constituents of biomass namely carbon, hydrogen, and oxygen and reported using the $C_mH_nO_p$ formula. Here m, n, and p represent the elemental fractions of C, H, and O, respectively. Proximate analysis gives the composition of the biomass in terms of gross components such as moisture (M), volatile matter (VM), ash (ASH), and fixed carbon (FC). The quality of biomass as a fuel in the gasifier can be classified as good or bad as per the following parameters:

- Energy content
- Bulk density
- Moisture content
- Dust content
- Tar content
- Ash and slagging characteristic

Keeping the above characteristics of fuel in mind, only two fuels namely charcoal and wood have been thoroughly tested and proven to be reliable. Some of the characteristics of other biomass types are listed in Table 5.3. Note that millet and rice husks have about 20 per cent and wheat straw around 10 per cent ash content. The ash content in sewage sludge may be as high as 42 per cent and municipal solid waste as 12 per cent (Bhavanam and Sastry, 2011).

Table 5.3. Typical components of different types of biomass

Plant	*Lignin (wt. per cent)*	*Cellulose (wt. per cent)*	*Hemicellulose (wt. per cent)*	*Ash (wt. per cent)*
Subabul wood	24.7	39.8	24.0	0.9
Wheat straw	16.4	30.5	28.9	11.2
Bagasse	18.3	41.3	22.6	2.9
Corn cobs	16.6	40.3	28.7	2.8
Groundnut shell	30.2	35.7	18.7	5.9
Coconut shell	28.7	36.3	25.1	0.7
Millet husk	14.0	33.3	26.9	18.1
Rice husk	14.3	31.3	24.3	23.5

Preprocessing of Raw Materials

Preprocessing of some agricultural and forestry products and other solid wastes is essential before these are used for gasification. Some of the common pre-processing operations are listed as follows.

Biomass	Pre-processing Processes
Agricultural products and residues	Chipping, drying and densification
Forestry products	Hogging, chipping, drying and densification
Municipal solid wastes	Separation of ferromagnetic and non ferrous metals through magnetic separation and/or by eddy current or any other equivalent devices. Removal of glass pieces by air classification, trammeling and screening. Size reduction of the MSW through shredding.

The advantages of pre-processing are:

- It is easy to store and handle the feedstock
- Ensures uniform distribution of the pre-processed feedstock in the gasifier preventing channeling action in the flow paths
- The top sealing device of the gasifier can be operated without any accident
- Produced gas is superior and is of uniform quality
- Steady rate of ash or char production

The disadvantages include additional capital investment for the pre-processing equipment and maintenance costs resulting in escalated cost of the fuel.

Gasifying Agents

Gasification can be accomplished under different atmospheric conditions such as air, steam, steam-oxygen, and carbon dioxide. The gasifier atmosphere governs the calorific value of the syngas produced. Syngas produced with air as the gasifying agent has low heating value. It is mainly due to dilution of the syngas by the nitrogen contained in air. Syngas of medium calorific value is produced with steam or a combination of steam and oxygen. Besides, combination of steam and air gives much higher H_2 yields than with air alone. This combination also helps to reduce the energy required in the process, which is normally provided by combusting a fraction of the biomass.

TYPES OF GASIFIERS

Gasification as already stated is a process of conversion of the fuel or organic wastes/matter into a gas involving a sequence of chemical and thermal reactions such as oxidation and reduction. *The reactor in which this sequence of processes takes place at high temperatures in presence of limited oxygen is called gasifier.* Gasifiers, on the basis of the physical conditions of the feed stock in the reactors, are grouped into two kinds.

- Fixed bed gasifiers
- Fluidized bed gasifiers

FIXED BED GASIFIER

A fixed bed gasifier is generally a vertical reactor (furnace). The biomass is fed either from the top or from the side at a certain height. Feedstock in the gasifier is supported either on a fixed grate or on a sand bottom. The fixed bed gasifiers are further categorized as updraft, downdraft and cross draft (cross flow) gasifiers according to the way air or oxygen is introduced in it.

Updraft Gasifier

It is the oldest and simplest type of gasifier. It is a cylindrical vessel in which height to diameter ratio is usually kept at 3:1. The feed is introduced from the top while air is introduced from the bottom through a grate. Thus, feed and air move counter currently in the gasifier (Figure 5.2 left). While the drying zone is at the top, the combustion or the oxidation zone lies at the bottom (Figure 5.2 right). Gasification occurs through zones of decreasing temperatures as the gas rises through the reactor-fuel bed. Hot gases pyrolyze the biomass above the combustion and reduction zones. Pyrolysis of biomass releases volatiles and a sizeable amount of tar is formed. Some of the tar may leave with the outgoing gases. Since the reaction gases flow counter to the path of the incoming cool feedstock, gases exit at a relatively low temperature. The major advantages of this type of gasifier are its simplicity, high charcoal burn-out and internal heat exchange leading to low gas exit temperatures, small pressure drop, good thermal efficiency and little tendency towards slag formation. The gas flowing through the biomass also gets filtered, and a part of the particulate matter entrained in it is captured by the material. This helps in lowering the particulate content of the outgoing gases. This kind of gasifier can be operated with several types of feedstock (sawdust, cereal hulls, *etc.*). Possibility of channeling in the equipment is its major drawback. It can lead to oxygen break-through and

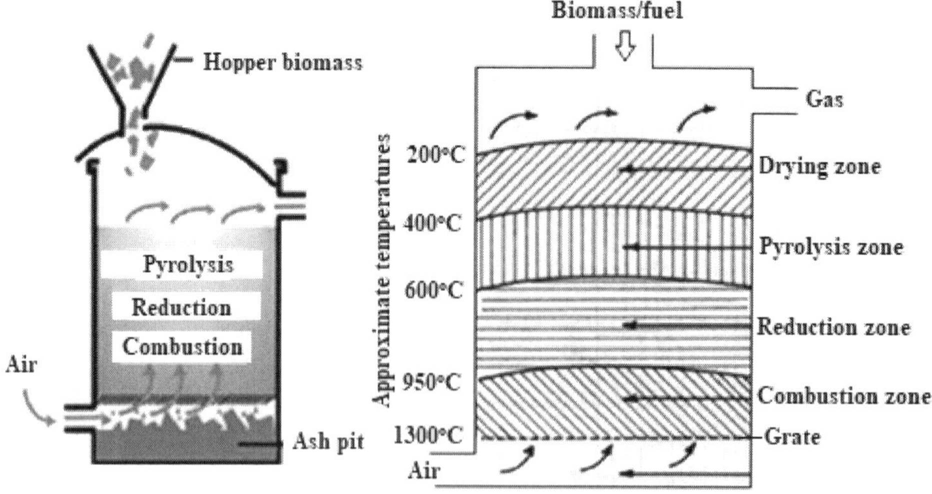

Figure 5.2. Updraft gasifier: schematic (left) and process zones (right).

dangerous explosive situations. Therefore, it is necessary to install automatic moving grates. Besides, the fuel gas containing high tar content may pose problems in its disposal or in gas cleaning operations. Moreover, its direct use in internal combustion (IC) engine may pose problems because of high tar and moisture content. One of the common problems is that the engine takes relatively long time to start-up.

Downdraft Gasifier

Downdraft gasifiers are available mainly in two designs (Reed and Das, 1988). One, the Imbert gasifier (Throated or closed top gasifier) and other, the stratified gasifier (Throat less or open core gasifier). These gasifiers have been used to gasify bark, wood blocks, chips and pellets, straw, maize cobs, refuse derived fuel, and waste pellets with various gasifying media like air, oxygen, and steam. Interest among stakeholders and researchers worldwide has increased in favour of downdraft gasifiers mainly due to its potential to produce mechanical and electrical power from biomass on small- scale at affordable prices. In general, a downdraft gasifier has a drying zone, where moisture content in the upper layers of the biomass is removed by evaporation. The temperature in this zone is about 120°C acquired by heat transfer from the lower zones which are at much higher temperatures. The dried biomass moves down to the second zone called the pyrolysis zone. The temperature here ranges from 200 to 600°C from top to bottom. Throughout this zone, the biomass loses its volatiles. In the lower part of this zone where temperature is around 400°C, an exothermic reaction takes place in which structure of biomass breaks down. As a result, water vapours, methanol, acetic acid and significant amounts of hydrocarbon tar are evolved. The remaining solid is char (carbon). The third zone is called the oxidation (or combustion) zone with temperatures ranging from 900 to 1200°C. A pre-determined quantity of air is drawn into this zone through nozzles. A portion of char and pyrolyzed gases coming from the second zone are burnt in this zone. The principal reactions are exothermic and oxidizing in nature, and the resultant products are carbon dioxide and water vapours. These products along with un-burnt pyrolysis gases and char pass on to the fourth and last zone called the reduction zone. This zone is at temperatures ranging from 900 to 600°C, the highest temperature being near the oxidation zone. These reactions are endothermic and consequently the temperature in the zone progressively decreases. At the end, the char is fully consumed and the final products are producer gas and ash. The volumetric composition of biomass based producer gas is: CO 20-22 per cent, H_2 15-18 per cent, CH_4 2-4 per cent, CO_2 9-11 per cent, and N_2 50-54 per cent. The gas also contains measurable amounts of particulate material and tar. The heating value of the producer gas ranges from 4000 to 5000 kJ/m^3.

Imbert Gasifier

The constricted hearth downdraft gasifier is called the Imbert gasifier after its inventor, Jacques Imbert. It is being commercially manufactured under various names. Recent modifications include single throat and double throat. Single throat gasifiers are mainly used for stationary applications whereas double throat are used for varying loads and automotive purposes. The upper cylindrical portion of the gasifier unit is simply a storage bin or hopper for wood chips or other biomass fuel. The spring-loaded, airtight cover is opened to refill the fuel hopper every few hours as needed (Figure 5.3 left). It remains closed and sealed during gasifying operation. The spring also permits the cover to function as a safety valve because it can pop open whenever excessive gas pressure builds-up. In a gas generator for vehicle use, the down stroke of the engine's pistons creates the suction force which moves the air into and through the gasifier unit. A blower is also used to create the proper airflow. About one-third of the way up from the bottom of the gasifier unit, a set of radially directed air nozzles are used to inject air into the wood as it moves downward.

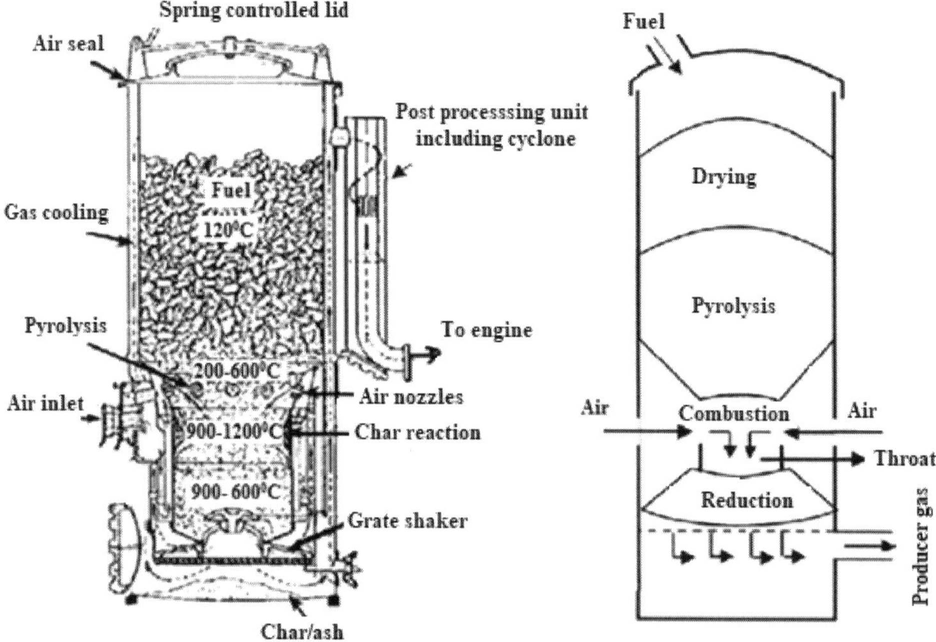

Figure 5.3. Imbert gasifier: line diagram (left) and process zones (right).

During normal operation, the incoming air burns and pyrolyzes some of the wood, most of the tars and oils, and some of the charcoal that fills the gasifier below the nozzles. Most of the biomass is converted to gas within the combustion zone (Figure 5.3 right). Below the combustion zone lies the gas-reduction zone usually consisting of a Imbert hearth. Now V-hearth or a flat plate hearth constriction has been introduced that allow formation of high quality insulation resulting in low tar production and higher efficiencies. The resulting hot combustion gases - carbon dioxide (CO_2) and water vapour (H_2O) - pass through the hot char where they are chemically reduced to combustible fuel gases: carbon monoxide (CO) and hydrogen (H_2). It results in cooling of the gas that helps to remove most of the charcoal and improve the quality of the gas.

Usually, wood contains less than 1 per cent ash (by weight). But as the charcoal is consumed, it eventually collapses to form a powdery charcoal/ash mixture which may represent 2 to 10 per cent (by weight) of the total fuel mass. This fine char and ash dust can eventually clog the charcoal bed and reduce the gas flow unless this dust is removed. Since the charcoal is supported by a movable grate, it can be shaken at intervals to remove the ash. Ash build-up below the grate can be removed during cleaning operations.

Imbert gasifier requires wood with low moisture content (less than 20 per cent by weight) or a uniform, blocky fuel to allow easy gravity feed through the constricted hearth. It is not possible to use twigs, sticks, and bark shreds. The constriction at the hearth and the protruding air nozzles may obstruct the passage of the fuel and create ridging and channeling. It may result in poor quality gas output as non-pyrolyzed fuel falls into the reaction zone. The capacity of the Imbert downdraft gasifiers is limited to 500 kW because of physical limitations of biomass particle size.

Since, the gas from the lmbert gasifier can be directly used in the engines a post processing unit consisting of a cooling unit, water filled precipitating tank and an automotive radiator type gas cooler is provided (Figure 5.3 left). The precipitating tank removes all unacceptable tars and most of the fine ash from the gas, while the radiator further cools the gas. A second filter unit containing a fine mesh filtration material is used to remove the traces of any ash or dust that may have survived while passing through the cooling unit. Once out of the filter unit, the wood gas is mixed with air in the vehicle's carburetor and is consumed within few seconds after it is produced.

Stratified Downdraft Gasifier

The stratified downdraft gasifier has been developed to overcome the problem of bridging and channeling in throated (Imbert) downdraft gasifier. It employs a balanced, negative-pressure concept in which the old type of sealed fuel hopper is no longer necessary. A closure is only used to preserve the fuel as and when the engine is stopped. This new version has several popular names such as stratified downdraft gasifier, open top gasifier and throat less downdraft gasifier. The gasifier consists of a cylindrical vessel. During operation both the air and biomass move downwards. The name stratified has come from the fact that air during operation passes uniformly downward through four zones. The uppermost zone contains non-reacted biomass through which air and oxygen enter. This region serves the same function as the fuel hopper in the older World War II designs. In the pyrolysis zone, the wood fuel reacts with oxygen (Figure 5.4). Most of the volatile components of the fuel are burned in this zone and

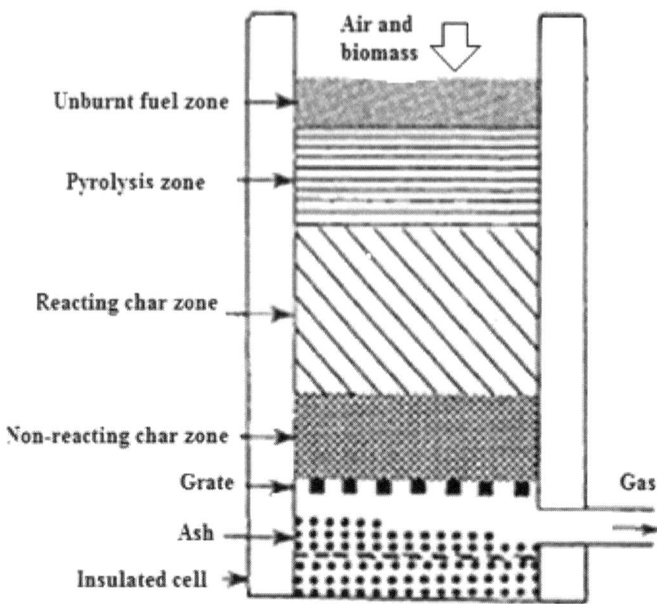

Figure 5.4. Line diagram of stratified downdraft gasifier.

provide heat for continued pyrolysis reactions. All the available oxygen in the air is completely consumed in this zone. The third zone is made up of charcoal from the second zone. Hot combustion gases from the pyrolysis region react with the charcoal to convert carbon dioxide and water vapour into carbon monoxide and hydrogen. The inert char and ash, which constitute the fourth zone, are normally too cool to cause further reactions. This zone also serves as a buffer and as a charcoal storage region. Below this zone is the grate. The presence of char and ash over the grate protect it from excessive temperatures. The foremost question about the operation of the stratified downdraft gasifier concerns char and ash removal. As the charcoal reacts with the combustion gases, it eventually attains a very low density and breaks up into dust containing all of the ash as well as some percentage of the original carbon. This dust may be partially carried away by the gas and might eventually plug the gasifier. Hence, it must be removed by shaking or agitation. The gasifier is automatically shaken by the vehicle's motion when the gasifier unit is used to power vehicles. An important issue in the design of the stratified, downdraft gasifier is the prevention of fuel bridging and channeling. High-grade biomass fuels such as wood blocks or chips will flow down through the gasifier because of gravity and downdraft air flow. However, other fuels (such as shredded chips, sawdust, and bark) can form a bridge, which may obstruct continuous flow and cause very high temperatures. Bridging can be prevented by stirring, shaking, or agitating the grate with some means or by the vehicle's movement. Some changes have been made in the design of this gasifier that allow varying air distribution in the gasifier, improve insulation of gasifier and re-circulation of the gas within the gasifier (Bhavanam and Sastry, 2011). These changes have improved the performance of throat less gasifier for its use in power and heat applications.

The stratified downdraft design has a number of advantages as listed below.

- The cylindrical shape is easy to fabricate and permits uniform flow of fuel
- The open top ensures uniform access of air and permits fuel to be fed easily and uniformly, which keeps the local temperatures in control
- No special fuel shape or pre-treatment is necessary as any blocky fuel can be used
- The open top throat less gasifier can be used for biomass having high ash content up to 20 per cent
- The gasifier has good scale up properties

Cross-draft Gasifier

Cross-draft gasifier is mainly an adaptation for the use of charcoal as a fuel although it can also be used for other biomass types. Charcoal gasification results in very high temperatures (1500°C and higher) in the oxidation zone, which may pose material selection problems. Insulation against these high temperatures in this gasifier is ensured by the fuel (charcoal) itself. Biomass in a cross-draft gasifier is introduced from the top and the air from the side of the gasifier through a horizontal nozzle (Figure 5.5). The air enters at a very high velocity inducing substantial circulation and flows across the bed of fuel and char. The biomass moves down as it gets dried, devolatilized, pyrolyzed, and finally gasified. The resulting fuel gas is discharged through a vertical grate on the opposite side of the air injection site, more-or-less at the same level as the air entrance. Thus, the producer gas leaves the gasifier at almost the same temperature as gasification (~ 800-900°C). The combustion and gasification zone is located near the entrance of the air while the devolatilization and pyrolysis zones are at a higher level than the entrance and exit. The thermal efficiency of the gasifier is usually low because of high heat losses. Secondly, the overall residence time of the producer gas in the high temperature zone is small (as the gas enters and exits from opposite ends), resulting in limited tar cracking. This leaves a significant amount of tar in the outgoing gas. The reactions in the cross draft gasifier are similar to the downdraft gasifier. Major advantages are its short height, very fast response time to attached loads and flexible gas production. It can even be operated at very small scales. Major disadvantages are: very high sensitivity to slag formation and high pressure drop. It has minimal tar-converting capabilities and the consequently need high quality (low volatile content) charcoal/fuel.

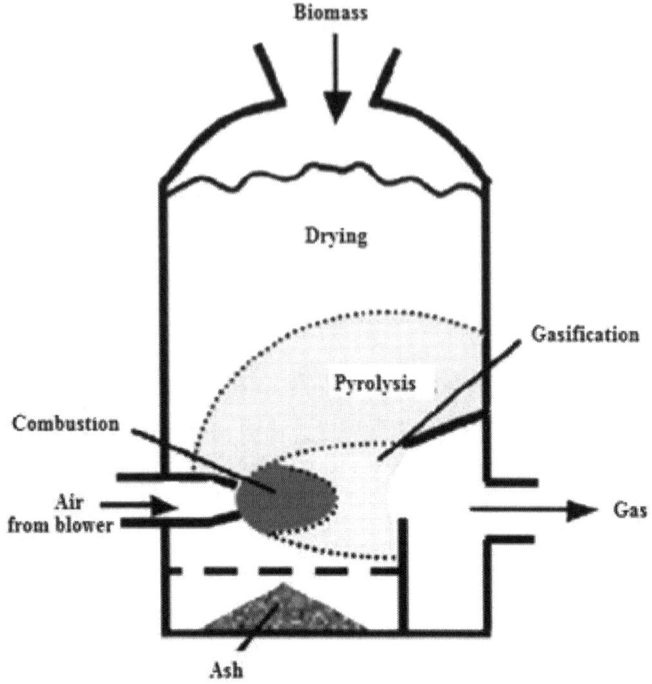

Figure 5.5. Schematic diagram of cross draft gasifier.

FLUIDIZED BED GASIFIERS

Fluidized bed (FB) gasification has been used extensively in coal gasification. A fluidized bed gasifier is a tall refractory lined unit. The height to diameter ratio of a fluidized bed reactor is normally kept at 10:1. Fluidizing medium comprises inert materials like sand typically silica sand, or ash or catalytic material such as CaO or olivine. The material is supported on a perforated plate at the bottom of the reactor, which ensures uniform distribution of the fluidizing medium. Fluidization of the medium is achieved by applying controlled flow of air or oxygen at a fluidization velocity through the perforated supporting plate. As a result the whole bed including sand is kept in suspension and separated. The temperature in a fluidized bed reactor is quite uniform throughout the bed. The bed temperature should normally be lower than 1100-1150°C to prevent slagging of the residues, but should be higher than 850°C to promote partial oxidation of the feed material. Oxidation and gasification simultaneously occur quite rapidly.

The capacity of the fluidized bed is generally dependent upon the volume of the fluid bed. Two main types of FB gasifier are:

- Circulating fluidized bed
- Bubbling bed

Major characteristics of fluidized-bed gasifiers are:

- ☆ Load flexibility having high volumetric capacity and high heat transfer rates
- ☆ Fuel flexibility, can gasify a wide range of feedstock
- ☆ Moderate oxidant and steam requirement
- ☆ Uniform, moderately high temperature throughout the gasifier
- ☆ Higher cold gas efficiency than entrained-bed gasifiers, but lower carbon conversion

Nonetheless, these advantages are offset by the complexity of the system with large blowers for blowing air and augers for feeding biomass. Besides, fluidized bed systems produce more dust and tar as compared to downdraft gasifiers. This puts a heavy load on the cooling and cleaning train.

A third type of FB gasifier is currently being developed. It is called fast internally circulating gasifier. It combines the design features of the two gasifier types.

Circulating Fluidized Bed Gasification

In a fluidized-bed gasifier, the flow of gaseous oxidant (air, oxygen, or oxygen-enriched air) is directed upwards through a distributor plate so that it permeates the bed of inert material located at the bottom, which also contains the biomass. The superficial gas velocity (the ratio between gas volumetric flow rate and the cross-sectional area) is several times larger than the minimum fluidization velocity that causes the drag forces on the particles equal to the weight of the particles in the bed. It gives the particles a fluid like behavior. The bed material is circulated between the reaction vessel and a cyclone separator, where the ash is removed and the bed material and char returned to the reaction vessel (Figure 5.6 left). This fluid like state produces intense mixing and gas-solid contact resulting in very high heat and mass transfer. The synthesis gas formed in the process moves upwards along the vertical space above the bed height, called the freeboard, and leaves the reactor. Circulating FB gasifier is able to cope with high capacity throughputs. It is commonly used in the paper industry for the gasification of bark and other forestry residues. Gasifier can be operated at elevated pressures. This characteristic is advantageous for those end uses where the gas needs to be compressed as in a gas turbine.

Bubbling Bed Gasification

Bubbling bed gasifier consists of a vessel in which gasifying agent is introduced at a velocity fast enough (0.5–1.0 m/s) to agitate the bed material which sits at the bottom part of the gasifier on a grate. Prepared biomass feed is fed from the side into the hot bed, where devolatilization occurs (Figure 5.6 right). The bed temperature is regulated in the range of 700 to 900°C by controlling the air/biomass ratio. Char particles and volatiles (tar precursors) as they come in contact with the hot fluidized bed are gasified and cracked to form char with gaseous compounds. High molecular weight compounds are cracked giving a product gas with low tar content, typically < 1-3 g/Nm^3 (gram per normal cubic meter). Additional gasifying agent can be supplied in a second zone located above the bed with the objective to convert entrained unconverted volatiles and char particles into fuel gas. Two stages *i.e.* the emulsion and the bubbles can be identified in a cross section of the bed. The emulsion contains solid particles and gas, while the bubbles contain only gas. The bubbles increase in size as they move upwards and drag high amounts of particulates. Besides, there is a change of speed, which affects the residence time of the particles in the bed. Thus, the passage of bubbles through the bed assists in mixing of solid particles and enhances heat and mass transfer rates in the gasifier. Final product is flue gas with low to medium tar content. Ash is separated from the gas in gas-solid separation units located downstream.

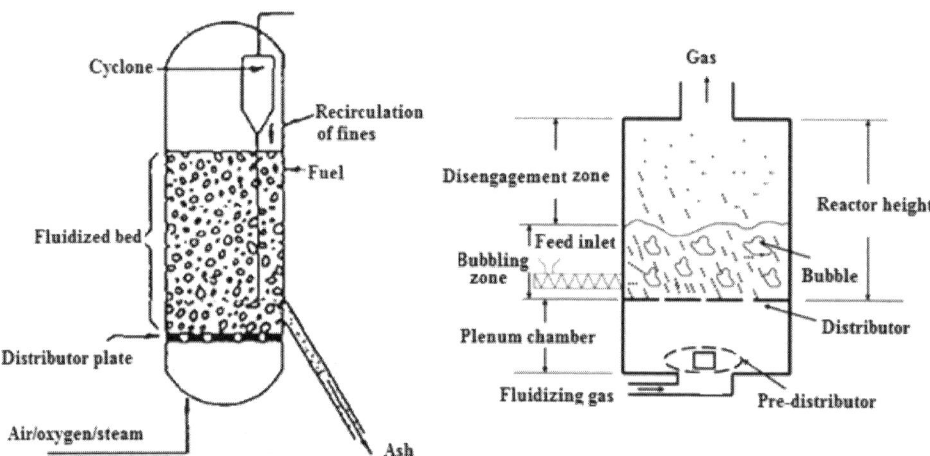

Figure 5.6. Schematic diagram of fluidized bed gasifier: circulating (left) and bubbling bed (right).

Special Boilers for Heating Applications

The producer gas from the gasifier can be used for heating (in a furnace or a boiler) or for power generation through its use in gas engines. A special kind of boiler used to produce steam is shown in Figure 5.7. It can be used to produce steam using producer gas of a gasifier.

Sugarcane Leaf-Bagasse Gasifier

Sugarcane leaves and bagasse are major residues in sugarcane crop growing areas. In order to dispose of the same, a full-fledged commercial scale gasifier system with a thermal output of 1080 MJ/h has been designed. It comprises a reactor, a gas conditioning system, a biomass feeding system and

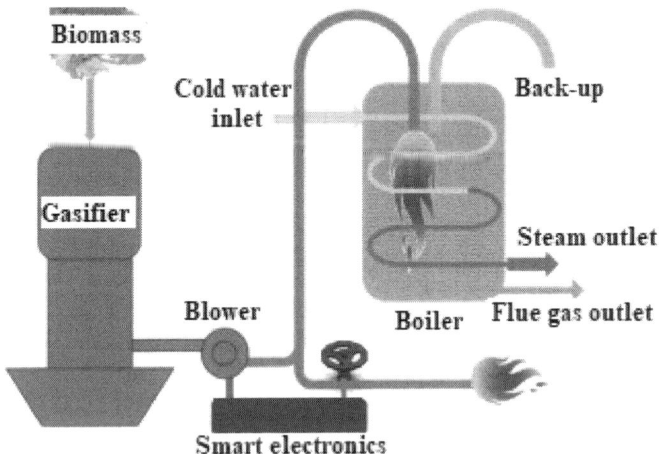

Figure 5.7. A boiler retrofitted with a gasifier for application in heating.

the instrumentation and controls. The reactor is a downdraft, throat less and open-top reactor with an internal diameter of 75 cm and an active bed height of 1.25 m. High temperature resisting firebricks conforming to IS 8 grade are used on the hot face followed by a cold face insulation. A scraper drag-out conveyor and hopper is used to convey the biomass fuel from the storage pile to the reactor. The gas conditioning system consists of a dry dust collection system to eliminate the problem of wastewater. This consists of a high temperature char/ash coarse settler and a high efficiency cyclone separator. A specifically designed high temperature resisting induced-draft fan is used to ensure that the entire system is under negative pressure. Such an arrangement ensures that outside air is sucked into the system to avoid the leaking of combustible gases in any event of leaks. The char-ash from the coarse settler and the cyclone are collected in barrels and emptied in an ash pit once every 45 minutes. The char-ash having a gross calorific value of 18.9 MJ/kg can be briquetted for fuel, or can be used as a soil conditioner. A Programmable Logic Controller (PLC)-based control system under certain critical conditions initiates automatic corrective actions. The biomass feeding and ash removal rates are fully controlled by this system. The gasification system is extremely simple to operate. A cold start takes about 10-15 min whereas a hot start can be affected in less than 5 min. Two operators per shift of eight hours are enough to operate the system including the fuel and ash handling operations.

CHARACTERISTICS OF PRODUCT GAS

By weight, gas (wood gas) produced in a gasifier unit contains approximately 20 per cent hydrogen (H_2), 20 per cent carbon monoxide (CO), and small amounts of methane, all of which are combustible, plus 50 to 60 per cent nitrogen (N_2). The nitrogen is incombustible, but does occupy volume as well as dilutes the wood gas as it enters and burns in an engine. The calorific value of product gas depends on the percentage of inflammable gases (CO, H_2, C_xH_y) contained in it. On the basis of its calorific value, gas is categorized as low (4-12 MJ/m^3), medium (12-28 MJ/m^3) and high (above 28 MJ/m^3) calorie. Direct gasification of biomass mostly yields low calorie gas (Table 5.4, Rajvanshi, 1986). The products of wood gas combustion are carbon dioxide (CO_2) and water vapour (H_20). One of the by-products of wood gasification is carbon monoxide, a poisonous gas. One should avoid breathing this gas during refueling operations or prolonged idling.

Table 5.4. Composition of producer gas from various fuels

Fuel	*Gasification Method*	*Volume Percentage*					*Calorific Value (MJ/m^3)*
		CO	*H_2*	*CH_4*	*CO_2*	*N_2*	
Charcoal	Downdraft	28-31	5-10	1-2	1-2	55-60	4.60-5.65
Wood with 12-20 per cent moisture content	Downdraft	17-22	16-20	2-3	10-15	55-50	5.00-5.86
Pellets wheat straw	Downdraft	14-17	17-19	-	11-14	-	4.50
Coconut husk	Downdraft	16-20	17-19.5	-	10-15	-	5.80
Coconut shells	Downdraft	19-24	10-15	-	11-15	-	7.20
Pressed sugarcane	Downdraft	15-18	15-18	-	12-14	-	5.30
Charcoal	Updraft	30	19.7	-	3.6	46	5.98
Corn cobs	Downdraft	18.6	16.5	6.4	-	-	6.29
Pelleted rice hulls	Downdraft	16.1	9.6	0.95	-	-	3.25
Cotton stalks cubed	Downdraft	15.7	11.7	3.4	-	-	4.32

DOWNSTREAM PROCESSING AND UTILIZATION

The gas from the gasifier is used either to provide heat or to generate a fuel gas, which can be used in internal combustion engines as a petroleum oil substitute (Figure 5.8). Although, gasifier is environmental friendly equipment without any major harmful emissions, nonetheless final product may contain some gaseous, liquid and solid products of combustion. These may include particulate matter, condensable hydrocarbons (*i.e.* tars), sulphur compounds, nitrogen compounds, alkali metals (primarily potassium and sodium), hydrogen chloride (HCl), and radioactive nuclides. Some of these are not only harmful to humans but also to engines and burners. A maximum limit of 5-15 mg solids and tar per kg of gas is allowed for use of the gas in an internal combustion engine. That is why a gasifier must have a reliable and economically cheap cooling and cleaning trains, the third step in the whole system as shown in Figure 5.8. Several unit operations are used for this purpose.

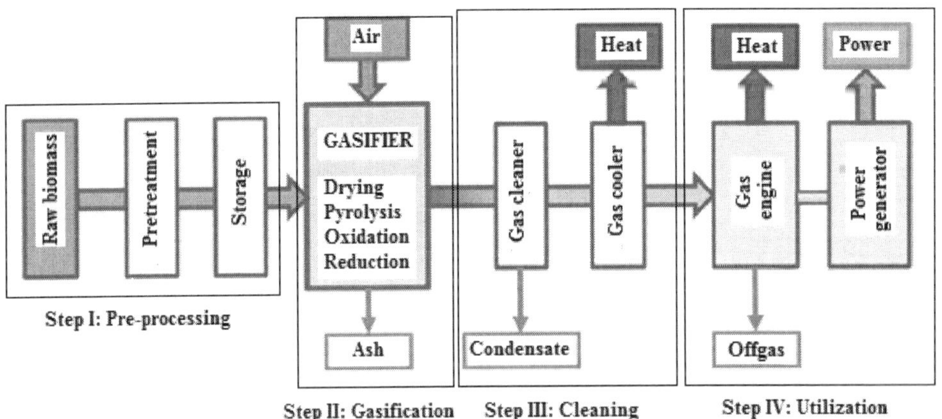

Figure 5.8. Processes involved in overall biomass gasification.

Cooling

The temperature of gas coming out of a gasifier is normally in the range of 300 to 500°C. On the other hand, the gas going to an internal combustion engine should be ideally at about ambient temperature. Thus, the gas has to be cooled not only for this purpose but to raise its energy density as well. Amongst various cooling equipment, most coolers used for this purpose are gas to air heat exchangers. The cooling is accomplished by free convection of air on the outside surface of heat exchanger. Some heat exchangers have provision of partial scrubbing of gas to take care of moisture and tar in the gas.

Tar Removal

Tar is a generic term for all organic compounds in the product gas with the exception of gaseous hydrocarbons. Tar, a part of the biomass, does not decompose completely into lighter gases. High molecular weight compounds in the gas downstream of the gasifier begin to condense at temperatures less than 450°C. The condensed material termed tar is partly deposited on the walls of the piping and partly remaining as an aerosol in the gas. Nature of the tar produced is largely determined by the type of biomass, but may also be influenced to some extent by the gasification process and the operating conditions. The primary organic compounds in tar can be grouped as mixed oxygenates, phenolic ethers, alkyl phenolics, heterocyclic ethers, poly-aromatic hydrocarbons (PAH) and larger PAH. The composition, with increased reaction severity, shifts from larger molecules (*e.g.*, larger PAH) to smaller molecules (*e.g.* mixed oxygenates). The adverse effects of excess tar in the gas are:

- ☆ The tar content hinders the removal of particulates from the gas

- It causes problems with the subsequent utilization of the product gas
- Tar is detrimental to catalysts used in conversion of syngas and fuel cell applications
- It also harms the downstream equipment such as heat exchangers, combustion engine, reactors or fuel cells in applications where gas stream is cooled. Cooling results in condensation of the tar on the wall of these equipment

Tar does not create much problems, if the product gas is used for combustion since tar itself is combustible although less readily and at higher temperatures. Overall, it may add to the calorific value of the fuel.

Removal of tar is one of the major technical challenges in the way of commercialization of gasification technology. Tar removal techniques are categorized as: (a) primary removal techniques and (b) secondary removal techniques (Table 5.5). Primary removal techniques refer to techniques that reduce the tar content in the syngas and are employed inside the gasifier without the need of a secondary reactor. These methods include design and optimization of the gasification operating conditions, and addition of appropriate catalyst in the gasifier bed (called in-bed catalysts). Numerous studies have shown that the gasification operating parameters, primarily gasification temperature profile, equivalence ratio (ER), type of gasifying agent, total gasifying agent to biomass ratio (S/B), and feed rate can effectively reduce the amount of tar in the syngas. In general, higher temperature, ER, and S/B reduce tar content in the gas due to their role in destroying the tar. Two major processes through which tar can be cracked into lower molecular weight compounds are thermal and catalytic processes. While thermal cracking takes place at 900-1100°C, catalytic cracking takes place at 800-900°C. Apparently, thermal cracking requires some additional energy to heat the gas as gasification is usually in the range of 800-900°C. It can be achieved by introducing a small volume of air to enable part of gas to combust to raise its temperature. In spite of being an effective method, it reduces the overall efficiency of conversion of biomass energy to gas. Catalytic cracking makes use of catalysts such as dolomite, limestone, olivine, alkali carbonates, Ni-based catalysts, metal-oxide catalysts, zeolite or char in the gasifier bed, either with or without sand along with the biomass feed. These help reduce tar and increase the extent of reforming reactions in the gasifier resulting in overall increase in carbon conversion efficiency. Ni-based catalysts are widely applied mainly because nickel has a very good catalytic activity and has price advantage. These are especially preferred for hydrogen or syngas production. A uniform and high temperature is ideal combination to achieve the total cracking of tar. On the contrary, application of in-bed catalysts for tar reduction in primary methods may not provide optimum conditions for yield, composition, and energy content of the syngas. Therefore, the process efficiencies need to be considered in gasifier design and in determining the best operating conditions to obtain desired product composition.

Secondary tar removal techniques comprise wet cleaning and dry gas cleaning (Table 5.5). Herein, a separate reactor is used to destruct and reform the tar content of the product gas below acceptable levels (Table 5.5). A combination of the two strategies can also be applied to remove the tar.

Table 5.5. Post gasification tar removal techniques

Method	*Techniques Used*	*Details*
	Primary techniques	
Thermal cracking	Application of high temperature with long residence time	Maximum tar destruction was found at 1250°C for 0.5 s
Catalytic cracking	Use of appropriate catalyst	Ni-based, non Ni-based, alkali metal based, acid catalysts, basic catalysts and activated carbon based catalysts
	Secondary techniques	
Wet cleaning	Use of mechanical devices	Wet cyclone, wet scrubber, electrostatic precipitator
Dry gas cleaning	Use of mechanical devices	Cyclone filter, rotary partial separator, fabric filter, ceramic filters, sand filter, activated carbon adsorber

Wet Cleaning

Water scrubbing and venturi scrubbing is used to condense the tar compounds in the syngas. The process simultaneously removes the particulates. It has been demonstrated that a venturi scrubbing system can achieve tar concentrations below 20–40 mg/Nm^3. The exit gas temperature in wet cleaning methods is in the range of 35 to 60°C resulting in loss of sensible heat. Electrostatic precipitator (ESP) scrubbers can remove most of the tar up to about 150°C. The only limitation is that these are more expensive. The tar collected using any of these physical techniques can be burned to produce heat or re-injected into the gasifier. A very low tar content gas is best produced by cooling the gas to 60-80°C with water and using electrostatic precipitators to capture any aerosols. Major disadvantages of wet cleaning are as follows:

- Wet cleaning decreases the net energy efficiency of the process because of reduced temperature of the gases
- Heavily contaminated waste water needs to be treated before it can be discharged. It is a capital-intensive process
- When the product gas is used for high temperature applications, there is net loss of energy in cooling and heating the product gas stream

Dry Cleaning

A number of devices such as cyclone filter, rotary separator and fabric filter *etc.* are used in dry gas cleaning (Table 5.5). Since 60-65 per cent of the producer gas contains particles above 60 μm in size, the cyclone filter, capable of removing particles of 5 μm and greater, has proven to be the excellent cleaning device. The cyclone filters are normally designed considering the rate of gas production and its dust content. But overall, dry gas cleaning methods have not been found to perform well.

Sulphur Removal

Several gas cleaning technologies such as absorption, adsorption, conversion, and biological transformations are capable of removing sulphur from the gas. Absorption and adsorption are normally employed in gasification.

DENDRO ENERGY

Biomass is referred to as Dendro, a Greek word for wood. *Thus, use of wood, preferably grown in a sustainable manner, in electrical generation or thermal energy refers to dendro energy.* Sustainable growing of wood instead of cutting down forest ensures that biomass is a renewable energy source having positive environmental impact. There are two primary ways of converting biomass to electricity.

- ☆ Gasification option where the biomass is converted to a combustible gas that is used as the fuel to run an engine coupled to a generator (Figure 5.9)

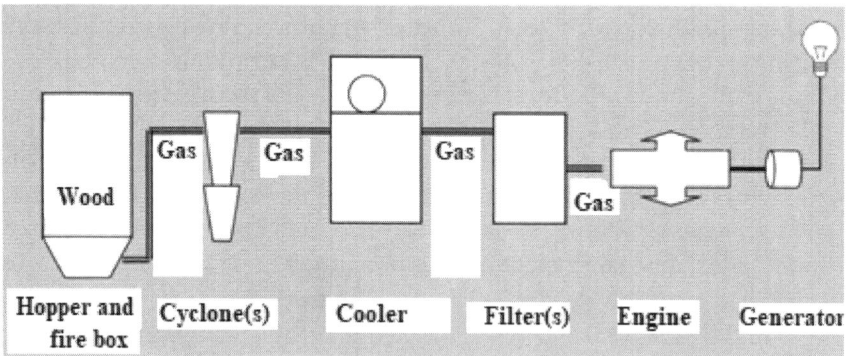

Figure 5.9. Wood to electrical energy: gasification option.

- ☆ Steam turbine option where biomass is combusted in a furnace to produce heat that in turn is used to produce high pressure steam that drives a steam turbine coupled to a generator (Figure 5.10)

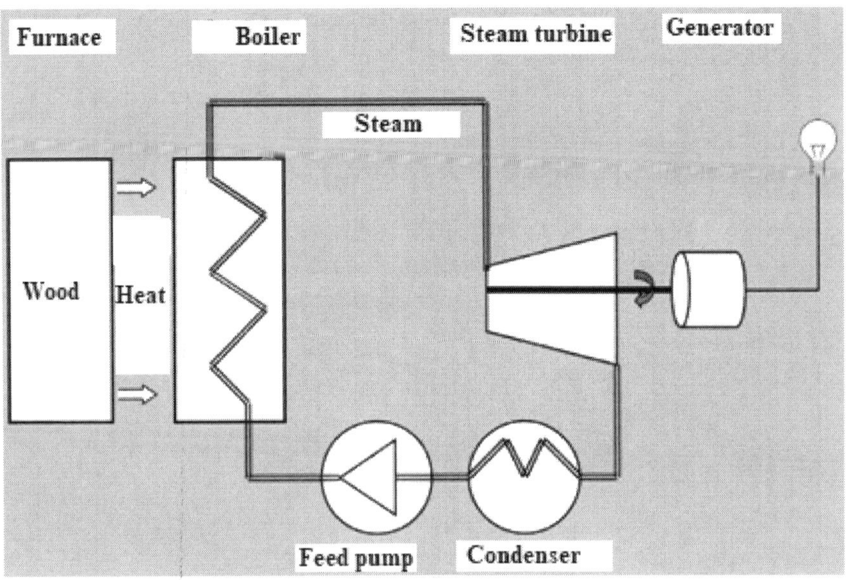

Figure 5.10. Wood to electrical energy: steam turbine option.

UTILIZATION POTENTIAL

A spark ignition engine running on producer gas derived from 1 kg of biomass on an average produces 0.55-0.75 kWh of energy. On the other hand, a compression ignition (diesel) engine cannot run completely on producer gas. In order to produce 1 kWh of energy, a diesel engine consumes 1 kg of biomass and 0.07 liters of diesel. It can result in a saving of 80-85 per cent diesel although on an average, producer gas can easily save up to 70-75 per cent diesel in a standard diesel engine.

QUESTIONS (THEORY)

1. Define gasification. Describe the fundamental gasification processes involved in gasification of biomass.

2. Name and briefly discuss the various zones encountered in a gasifier during gasification of biomass.

3. Name various types of biomass materials that can be used for gasification. Name the characteristics used to categorize the biomass as good or bad for gasification point of view.

4. Draw a neat sketch and describe a fixed bed updraft gasifier.

5. Draw a neat sketch and describe a fixed bed downdraft Imbert gasifier. How it differs from stratified downdraft gasifier?

6. Name various types of fluidized bed gasifiers. Explain the working of bubbling bed gasifier.

7. What do you understand by downstream processing? Discuss the primary and secondary tar removal processes highlighting their limitations.

8. Write short notes on pre-processing of biomass for gasification, gasification agents, cross-draft gasifier, fluidized bed gasifiers, sugarcane leaf-bagasse gasifier, dendro energy and characteristics of the gas.

Chapter 6

Biogas Production Processes

Most organic materials undergo natural anaerobic digestion in the presence of moisture and absence of oxygen to produce biogas. *Thus, biogas has been defined as a renewable energy source produced as a result of the breakdown of organic matter by certain bacteria under anaerobic conditions.* Biogas can be naturally captured from marshes, landfill or wastes such as sewage. It can also be produced artificially through anaerobic decomposition or digestion of organic materials. *Digestion is a biological process that occurs in the absence of oxygen and in the presence of anaerobic organisms (organisms that do not require oxygen for their survival or proliferation) at temperatures ranging from 35 to 70°C and atmospheric pressure.* The container in which, this process takes place is known as digester. Biogas is quite popular by the name, *Gobar Gas* mainly because cow dung has been mainly used for its production so far. The biogas normally is a mixture of methane (CH_4): 55-65 per cent and carbon dioxide (CO_2): 30-40 per cent. It has traces of hydrogen (H_2): 5-10 per cent, nitrogen (N_2): 1 to 2 per cent, water vapours (0.3 per cent) and hydrogen sulphide. Its calorific value ranges from 18.8 to 26.4 MJ/m^3 with an average of about 20 MJ/m^3. Its density is about 0.94 kg/m^3 and ignition temperature is 700°C.

RAW MATERIALS

Raw materials for biogas generation are mainly i) cow dung, ii) sewage, iii) crop residues, iv) vegetable wastes v) water hyacinth vi) poultry droppings vii) pig manure and viii) marine wastes. In fact, any easily biodegradable cellulosic organic material of animal or plant origin is a potential raw material for biogas production. A glimpse of biogas production potential of various feed stocks is given in Table 6.1. Energy equivalents of various fuels to 1 m^3 of biogas are reported in Table 6.2.

Table 6.1. Biogas production from various feedstocks

Name of the Feedstock	*Characteristics*		*Biogas Production*	
	Total Solids (TS) (per cent of wet weight)	*C:N Ratio*	*Production (L/kg)*	*Methane in Biogas (per cent)*
Cattle dung	20.8	19.7	32	50-55
Banana stem	5.2	25.3	30	65-70
Water hyacinth	7.0	18.3	53	75-80
Eucalyptus leaves	8.5	38.9	145	70-75
Composite agricultural waste	37.0	24.5	78	60-65
Grass trimmings	27.0	16.0	60	60-65

Table 6.2. Energy equivalent of various fuels to 1 m^3 of biogas

Name of the Fuel	*Kero-sene*	*Fire-wood*	*Cow Dung*	*Char-coal*	*Furnace Oil*	*Electri-city*	*LPG*
Equivalent quantity to 1 m^3 of biogas	0.60 L	3.50 kg	12.3 kg	1.50 kg	0.40 L	4.70 kWh	0.43 kg

Advantages

- Biogas is a renewable, and a clean source of energy
- Construction of biogas plant needs low initial investments
- The technology is highly suitable for rural areas
- Biogas is locally generated and can be easily distributed for domestic use
- Biogas reduces the rural poor from dependence on traditional fuel sources, which normally lead to deforestation
- The use of biogas in villages helps in improving the sanitary condition and checks environmental pollution by reducing offensive odour from overloaded or improperly managed manure storage facilities. Many volatile organic acids, the odour causing compounds, are consumed by biogas producing bacteria during biogas generation. Digester effluent with proper application reduces the likelihood of surface or ground water pollution

- The by-products like nitrogen rich manure can be used in cropped lands for high productivity. The organic nitrogen in the manure is largely converted to ammonium, which is readily available and utilized by plants
- It helps to reduce production cost as fertilizer inputs can be reduced
- Heated digesters dramatically reduce pathogen populations in a few days. Lagoon digesters isolate pathogens and allow pathogen kill and die-off prior to entering storage for land application
- Biogas reduces the drudgery of women and lowers incidences of eye and lung diseases
- Widespread use of biogas technology can create jobs related to the design, operation, and manufacture of energy recovery systems
- Biogas can be upgraded to synthetic natural gas by removing the traces of H_2, H_2S and N_2

Disadvantages

- No major technological advancement has been made in the technology and therefore, currently operational technologies are not very efficient
- Contains impurities that can corrode the metal parts of the engine. This corrosion would lead to increased maintenance costs
- Like other renewable energy sources (*e.g.* solar, wind) biogas generation is also affected by the weather. Digesters need heating under cold climatic conditions to maintain constant biogas supply
- Less suitable for dense metropolitan areas

STAGES OF ANAEROBIC DIGESTION

Anaerobic fermentation is a biochemical process in which three kinds of bacteria are involved in methane formation (Joshua *et al.*, 2014). Anaerobic digestion for methane production comprises following four concurrent stages.

- Hydrolysis
- Acidogenesis
- Acetogenesis
- Methanogenesis

Hydrolysis

Hydrolysis simply means rupture of chemical bond using water. The hydrolysis of polysaccharides to soluble sugars involves the following enzymes.

- Glycosidases or glycoside hydrolases target the glycosidic bond, which represents an oxygen atom that links the monomers together *e.g.* the bond between glucose and fructose units

- ☆ Amylases acting on starch to convert it to glucose or oligosaccharide
- ☆ Cellulases acting on cellulose convert it to glucose or other disaccharide

Large organic polymers in the biomass (carbohydrates, fats and proteins) in the process are broken down to release simple monomers (starch, fatty acid and amino acid). In the whole process, polysaccharide yield mono, di, and oligosaccharide (simple sugars), fats yield fatty acids and protein yield amino acids. The breakdown reaction of proteins is catalyzed by protease comprising a long chain of amino acids linked by a peptide bond (polypeptide). The peptide bonds are broken to release the amino acids. The amide on hydrolysis yield carboxylic acid and an amine. The carboxylic acid gains the OH^- group from the water while the amine gains the H^+ to form ammonia.

Acidogenesis

The products formed during hydrolysis are fermented during acidogenesis. It is a biological reaction where an acidogenic (fermentative) bacterium acts on the products (amino acids, sugars, and fatty acids) to produce the following:

- ☆ Volatile fatty acid (acetate, butyrate, propionate, and lactate)
- ☆ Ammonia (NH_3)
- ☆ Carbon dioxide (CO_2)
- ☆ Hydrogen sulphide (H_2S)

Bacteria involved in acetate formation like *Clostridium aceticum, Acetobacter woodii, Clostridium termoautotrophicum. Clostridium termoacetacum* and *Clostridium fermiacetacum* are able to reduce carbonic gas to acetate. They produce three molecules of acetic acids from fructose. *Thermaanaerobium brockii,* a thermophilic hydrolytic bacteria, ferment glucose to form lactate, ethanol, and hydrogen. The glucose is first glycolytically reduced to pyruvate after which reduced end products are formed enzymatically as follows.

Pyruvate → lactate (using fructose-1, 6-bisphosphate) → H_2 (by pyruvate ferrodoxin, oxidoreductase, hydroginase → ethanol (via NADH-NADPH linked alcohol dehydrogenase)

Acetobacter woodii in syntrophic association with methanogens (*methanocercina spp*) produce methane and carbon dioxide from fructose instead of three molecules of acetic acids. Together in the two processes, complex high molecular compounds of biomass are converted into simple, soluble, low molecular compounds. Organic substances such as polysaccharide, protein and lipid are converted into mono-saccharide, peptide, amino acids, and fatty acids. Then they are further converted into acetate, propionate and butyrate. The contents of the digester become slightly acidic due to the formation of these compounds.

Acetogenesis

Acetogenesis is accomplished by facultative and anaerobic group of organisms collectively called as acid formers. In this phase, monomers are further degraded into short chain acids such as: acetic acid, propionic acid, butyric acid, isobutyric acid, valeric acid, isovaleric acid, caprionic acid, alcohols, hydrogen, and carbon dioxide. The acetogens act on CO_2 and H_2 to produce acetic acid (CH_3COOH). It involves several reactions that reduce CO_2 to CO and the attachment of a methyl group ($-CH_3$) to the CO. The reduction of CO_2 to CO is catalyzed by carbon monoxide dehydrogenase while the attachment of the methyl group to CO is catalyzed by synthetase

$$2CO_2 + 4H_2 \rightarrow CH_3COOH + 2H_2O$$

The acetic acid (CH_3COOH) so produced is further utilized by the methanogens to produce methane.

Methanogenesis

Methanogenesis, also known as biomethanation, strictly involves anaerobes known as methanogenic bacteria that convert acetate or hydrogen plus carbon dioxide into gas mixture of methane (CH_4) and CO_2. These bacteria are also called methane fermentators. During methane formation, carbon is the terminal electron acceptor (instead of oxygen as in respiration) and it comes from organic compound with small molecular weight. The two methanogenic pathways involve the use of CH_3COOH and CO_2 produced in the early stages (Patel *et al.*, 2017).

$$CH_3COOH \rightarrow \quad CH_4 + CO_2$$

(Acetic acid → Methane + Carbon dioxide)

$$2CH_3CH_2OH + CO_2 \rightarrow CH_4 + 2CH_3COOH$$

(Ethanol + Carbon dioxide → Methane + Acetic acid)

$$CO_2 + 4\,H_2 \rightarrow CH_4 + 2\,H_2O$$

(Carbon dioxide + Hydrogen →Methane +Water)

Almost one-third of methane formation is due to reduction of carbon dioxide by hydrogen.

Various stages of these processes are shown in Figure 6.1. The acid formers and methane fermentators must remain in a state of dynamic equilibrium for efficient digestion. The remaining indigestible matter is referred as slurry.

FACTORS AFFECTING EFFICIENCY OF BIOGAS GENERATION

As discussed in the previous sections, different groups of micro-organisms are involved in methane production. Therefore, appropriate conditions have to

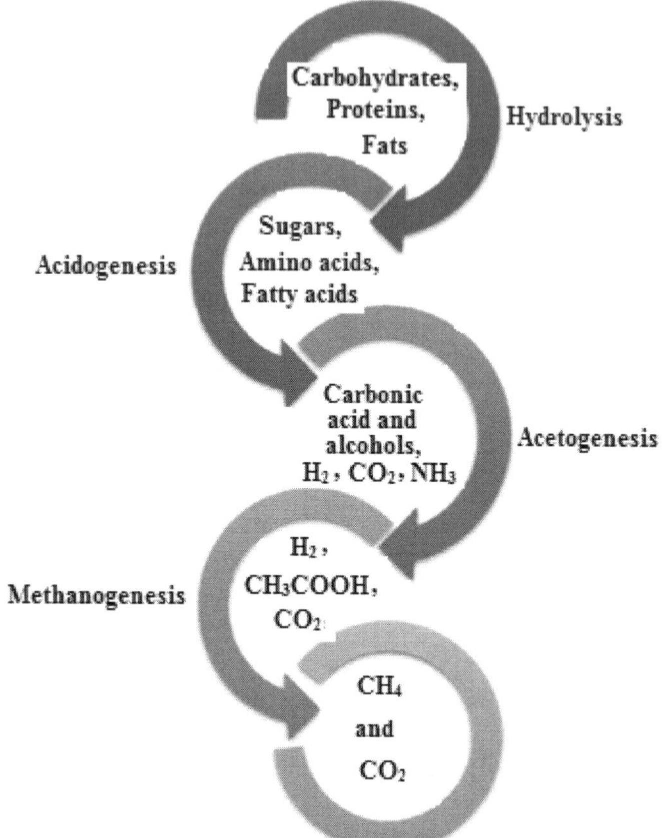

Figure 6.1. Different stages of biogas formation.

be established to keep these micro-organisms in balance. Parameters that affect the efficiency of biogas generation can be categorized in three main groups namely characteristics of the feed material, chemical/physical characteristics during digestion process and the digester characteristics. In the first category are: C:N ratio, type of feed stock, loading rate, feed rate, total solid content and particle size. Temperature, pH, nutrients and toxicity come under the chemical/ physical characteristics. Digester characteristics include retention time, diameter to depth ratio of digesters, and degree of mixing. The most appropriate conditions for efficient generation of biogas are illustrated in Table 6.3.

BIOGAS PLANT: DEFINITION AND TYPES

A biogas plant is a facility that provides oxygen-free conditions where anaerobic digestion can occur. Simply stated, it's an artificial system where it is possible to turn waste into sustainable energy and fertilizers with positive effects on the environment. The general working steps of a biogas plants are stated below.

Table 6.3. Parameters affecting efficiency of biogas generation: causal effects and guidelines

Characteristics	*Causal Effects and Guidelines*
Characteristics of feed materials	
C:N ratio	*The ratio of carbon to nitrogen present in the feed material is called C:N ratio.* Whereas micro-organisms use carbon for energy, nitrogen is used for building the cell structure. Optimum condition for anaerobic digestion ranges from 20 to 30:1. Too much carbon in the biomass will slow down and eventually stop the digestion as nitrogen is used up while carbon is still left out. If there is too much nitrogen, fermentation will stop as soon as the carbon is exhausted. The left over nitrogen will combine with hydrogen to form ammonia. This can kill or inhibit the growth of bacteria specially the methane producers. C:N ratio of various materials is given in Table 6.4 indicating a very high ratio for wheat straw.
Type of feed stocks	All plant and animal wastes can be used as feed materials. Digestion is difficult when feed stock contains high lignin as in woody materials. For efficient digestion, such feed stocks need to be combined in proper proportions. Pre-digestion and chopping may also be helpful in some materials. Animal wastes are pre-digested. Plant wastes do not need pre-digestion but excessive plant material may choke the digester.
Loading rate	*Loading rate is defined as the amount of raw material fed to the digester per day per unit volume of the digester.* An overloaded reactor accumulates excessive acid affecting gas production. On the other hand, under loading may negatively impact the designed gas production.
Feed rate	One of the pre-requisites of good digestion is the uniform feeding of the digester. It keeps the micro-organisms in a relatively constant concentration of the organic solids at all times. Therefore, the digester must be fed at the same time with qualitatively and quantitatively balanced feed every day.
Total solid content	The fresh raw cow dung contains 80-82 per cent of moisture. The balance 18-20 per cent is total solids. It is mixed usually in the proportion of 1:1 to 1:1.25 to bring down the total solid content to 8-10 per cent. This adjustment of total solid content helps in digesting the materials at a faster rate. It also helps in deciding the mixing of the various crop residues as feed stocks in biogas digester.
Particle size	Particle size is not an important factor compared to other parameters such as pH and temperature. However, the size of the particles affects the biomass degradation and ultimately the biogas production rate.
Chemical/physical characteristics	
pH	The pH of the slurry changes at various stages of the digestion. In the initial acid formation stage, the pH is around 6 or less. In the latter 2-3 weeks times, the pH increases as CH_4 is produced upon digestion of volatile acids and N_2 compounds. The digester is usually buffered if the pH is maintained between 6.5 and 7.5. In this pH range, the micro-organisms will be very active and digestion will be quite efficient. Both acidic (pH 4 to 6) and alkali pH (pH 9 to 10) are detrimental to the methanogenic (methane producing) organisms. In fact pH above 8.5 is detrimental to the survival of bacteria.

Characteristics	*Causal Effects and Guidelines*
Temperature	Most acid forming micro-organisms grow under mesophilic conditions. However, a higher temperature is favorable for methanogens. Anaerobic fermentation of raw cow dung can takes place at any temperature between 8 and 55°C. Methane bacteria work best at temperatures of 35°-38°C. The fall in gas production begins at 20°C and stops at temperatures of 8-10°C. For anaerobic digestion, temperature variation should not be more than 2 to 3°C. A 50 per cent increase in biogas production has been reported when the digester temperature was raised from 25°C to 28.3°C.
Nutrients (Seeding)	The major nutrients required by the bacteria in the digester are: C, H_2, O_2, N_2, P and S. Out of these nutrients, N_2 and P are always in short supply and therefore an extra raw material rich in phosphorus (night soil, chopped leguminous plants) should be added to the cow dung to maintain proper balance of the nutrients. It helps to obtain maximum production of gas. Addition of urea increases gas production for materials low in C:N ratio provided all other conditions remain unchanged.
Toxicity	Small quantities of mineral ions like sodium, potassium may stimulate the growth of bacteria, but high concentration of heavy metals and detergents have negative impact on gas production. Detergents like soap, antibiotics, and organic solvents are toxic to the growth of microbes inside the digester. Any addition of these substances along with the feed stock should be avoided.
Digester characteristics or physical processes employed	
Retention time	*It is defined as the theoretical time that a particular volume of feedstock remains in the digester commonly expressed in days. It means, retention time describes the length of time the material is subjected to the anaerobic reaction.* It is calculated as the volume of digester divided by the feedstock added per day. Organic substances decompose slowly under anaerobic conditions, and therefore retention time has to be more to complete the digestion. Methane-forming micro-organisms grow slowly with a doubling time of around 5–16 days. Therefore, the hydraulic retention time should be at least 10–15 days, unless these bacteria are retained by some means say entrapment.
Diameter to depth ratio	Studies have shown that gas production per unit volume of digester capacity was highest when the diameter to depth ratio was in the range of 0.66 to 1.00. It may be because of the fact that in a simple unstirred single stage digester the temperature varies with depth. The most activity of digesting sludge is in the lower half of the digester and it will be less affected by variations in night and day temperatures.
Degree of mixing	Bacteria in the digester have very limited reach to their food. It is why slurry needs proper mixing. Too much mixing stresses the micro-organisms while foaming occurs in the absence mixing. It is found that slight mixing improves the fermentation, while a violent agitation retards the digestion.

Optimum conditions for biogas production can be summarized as under.

Parameters	*Optimum Value*
Temperature	30-35 °C
pH	6.8-7.5
C : N ratio	20-30:1
Solid content	8-10 per cent
Retention time	20-50 days

Table 6.4. Carbon to nitrogen ratios of various feed stocks

Sl.No.	*Material*	*Nitrogen Content (per cent)*	*Ratio of Carbon to Nitrogen*
1.	Urine	15.18	8:1
2.	Cow dung	1.7	25:1
3.	Poultry manure	6.3	NA
4.	Night soil	5.5-6.5	8:1
5.	Grass	4.0	12:1
6.	Sheep waste	3.75	NA
7.	Mustard straw	1.5	20:1
8.	Potato tops	1.5	25:1
9.	Wheat straw	0.3	128:1

NA = Not available

- Biomass is mixed with desired quantity of water in the mixing tank to form slurry. It is fed into the digester through the inlet chamber
- During initial set-up, filling is performed both from the normal inlet and slurry outlet. The plant is left unused for about one and a half months
- During this period anaerobic bacteria present in the slurry decompose or ferment the biomass to produce biogas, which moves up in the dome of the digester
- As more and more biogas starts collecting, the pressure exerted by the biogas forces the spent slurry into the outlet chamber. From the outlet chamber, the spent slurry overflows into the overflow tank
- The spent slurry is manually removed from the overflow tank and used as manure for plants
- The gas valve connected to a system of pipelines is opened as and when biogas is required for cooking, lighting or any other use
- A functioning plant is fed on daily basis with cowdung slurry to ensure continuous supply of biogas

CLASSIFICATION OF BIOGAS PLANTS

Based on the nature of feeding, biogas plants are broadly grouped into 3 types.

Batch Type

The organic waste materials is charged only once into the digester. The feeding is between intervals. Once the process of digestion is complete, the plant is emptied. Retention time usually varies from 30 to 50 days. The gas production is intermittent. These plants are well suited for fibrous materials but needs fermented slurry to start the digestion process. Moreover, these are uneconomical to maintain.

Semi-continuous

A predetermined quantity of feed material mixed with water is charged into the digester from one side at specified interval of time; (say once a day) and the digested material (effluent) equivalent to the volume of the feed, flows out of the digester from the other side (outlet).

Continuous Type

The feed material is continuously charged to the digester with simultaneous discharge of the digested material (effluent). The main features of this type of plants are continuous gas production, requires small digestion area, lesser period for digestion, less maintenance, *etc.*

On the basis of the number of stakeholders, biogas plants have been categorized as:

- ☆ Family plants
- ☆ Community plants

The former are normally small capacity plants under the control of an individual for the use of his family. The latter are relatively bigger capacity plants. Several stakeholders pool their resources and control the operation and use of biogas from the plant. The capacity of these plants is decided on the basis of number of stakeholders involved and their gas requirement.

QUESTIONS (THEORY)

1. Define biogas. What are its constituents?
2. List the advantages and disadvantages of biogas.
3. Name and describe the various stages of anaerobic digestion to produce biogas.

4. Name various factors that affect production of biogas during anaerobic digestion. Briefly discuss the feed related parameters that affect the efficiency of gas production.

5. What do you understand by hydraulic retention time in a biogas digester? How it affects the gas production?

6. Write short notes on the effect of pH of slurry on gas production, effect of temperature on gas production, classification of biogas plants based on nature of feeding.

Chapter 7

Biogas Plants

A biogas plant as stated in the previous chapter is an artificial system where it is possible to turn waste into sustainable energy and fertilizers, with positive effects on the environment. Commonly used semi-continuous biogas plants basically are of two types (TEDIC, 1985)

Floating dome type: It is a floating-drum type plant with a cylindrical digester commonly known s KVIC (*Khadi* and Village Industries Commission) model (Figure 7.1)

Fixed dome type: The fixed-dome type plants are of two types namely Janata model with a brick reinforced moulded dome and Deenbandhu model also a fixed-dome plant with a hemispherical digester

COMPONENTS OF BIOGAS PLANTS

The biogas plants are brick, cement and steel structures. In general these plants have the following five sections (for reference see Figure 7.2).

- ☆ **Mixing tank:** It is used to mix dung/waste and water. It is located above the ground level
- ☆ **Inlet chamber:** The mixing tank opens underground into a sloping inlet chamber
- ☆ **Digester:** The inlet chamber opens from below into the digester, a huge tank with a dome like ceiling in a fixed dome biogas plant (Figure 7.2). The ceiling of the digester has an outlet with a valve to supply biogas to the point of use

- ✫ **Gas holder:** The gas accumulates in the floating drum gas holder (Figure 7.1) or dome of the digester (Figure 7.2)
- ✫ **Outlet chamber:** The digester opens from below into an outlet chamber
- ✫ **Overflow tank:** The outlet chamber opens from the top into a small over flow tank

FLOATING DOME TYPE

Floating-drum plants consist of an underground digester and a moving gas-holder. These are used chiefly for digesting animal and human feces on a continuous feed mode of operation, *i.e.* with daily input. These are most frequently used by small-to middle-sized farms (digester size: 5-15 m^3) or in institutions and in agro-industrial estates (digester size: 20-100 m^3).

KVIC Type Biogas Plant

Schematic diagram of a KVIC biogas model plant is shown in Figure 7.1. Various components of the plant are discussed in the following sections.

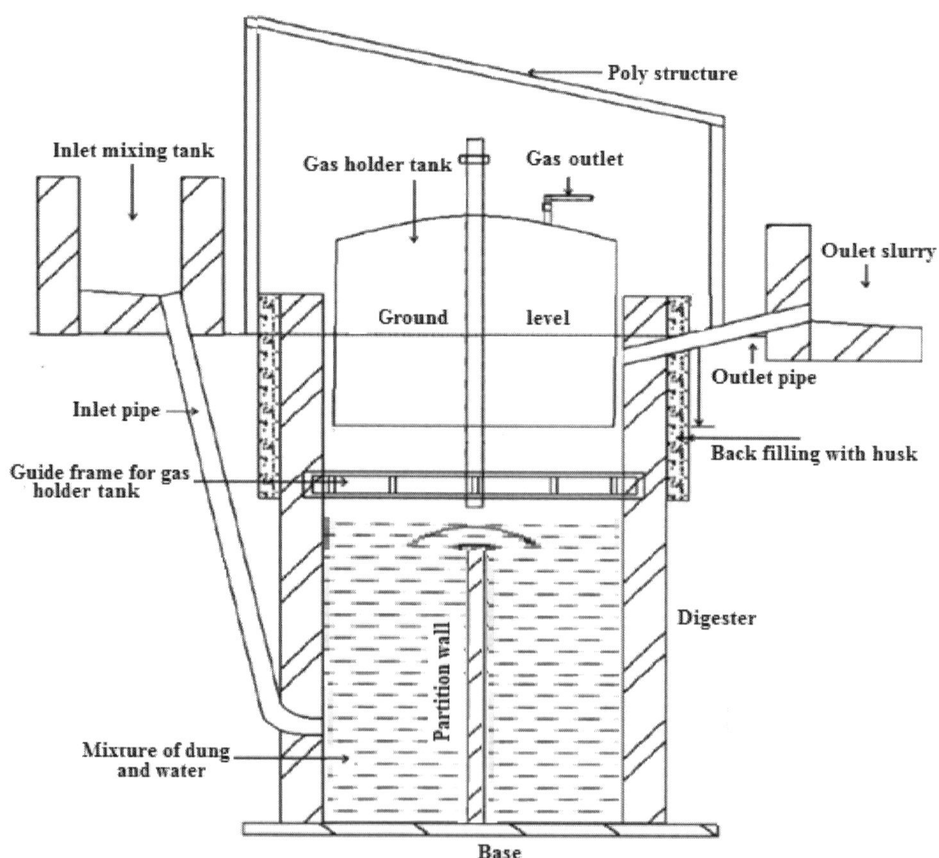

Figure 7.1. Floating dome type plant.

Digester: This consists of a cylindrical shaped pit (digester) for fermentation of the waste materials (Figure 7.1). Digester is 3.5-6.5 m deep and 1.2 to 1.6 m in diameter. The diameter of digester may be increased while depth can be proportionately reduced in areas where water table is encountered at shallow depths. A partition wall in the center vertically divides the digester in two parts. Partition wall may remain submerged in the slurry when the digester is full (Figure 7.1). The partition wall helps to increase the retention time. A pipe connects the digester to the inlet, the mixing tank. The dung mixed with water (1:1 to 1:1.25) is loaded into the digester through this pipe. The other pipe connects the digester to the outlet for slurry. The fermented material flows out through this pipe. The outlet is generally connected to a compost pit. The gas in the digester is slowly generated as discussed in the previous chapter. Firstly, complex, organic substances present in the waste are acted upon by acid formers and broken up into small-chain simple acids. Following this, acids formed in the first stage are acted upon by another kind of bacteria, called methane formers that produce methane and carbon dioxide (See chapter 6 for more details).

Gas holder: The gas holder is a cylindrical shaped drum made of mild steel sheets. The holder fits into the digester like a stopper. It sinks into the slurry from its own weight and rests upon the ring constructed for this purpose. The top is supported radially with angular iron. A central guide pipe is used to prevent the holder from tilting. As the generated gas accumulates in the holder, it rises and floats freely on the surface of slurry. The holder also acts as a seal for the gas. Depending upon the weight of the gas holder, gas pressure varies from 8 to 12 cm of water column.

Advantages and Disadvantages of Floating Dome Plants

Advantages

- Simple, easily understood operation
- Volume of stored gas is directly visible
- The gas pressure is governed by the weight of the gas holder and is nearly constant
- The construction is relatively easy. Minor mistakes during construction do not result in high material cost of the steel drum
- Scum related troubles are minimal because solids remain constantly submerged
- Higher gas production per m^3 of digester volume
- No gas leakage

Disadvantages

- High cost
- Floating drum plants have a shorter life span than fixed-dome plants because of susceptibility of steel parts to corrosion

- Regular maintenance costs for the painting of the drum
- Major problems in operation and gas yield because heat is likely to be lost through the metal gas holder. Therefore, it may pose problems during cold seasons or in colder regions

Table 7.1. List of materials required for construction of KVIC biogas plants (40 days retention time)

Materials	*Plant Size (m^3)*					
	2	*3*	*4*	*6*	*8*	*10*
Bricks- I class (No.)	2460	2765	3205	3730	4430	4650
Cement (50 kg) bags (No.)	13	17	19	23	26	28
Stone chip (12 mm/20 mm) (m^3)	0.60	0.90	0.95	1.25	1.40	1.60
River sand (m^3)	2.00	2.55	2.90	3.40	5.00	4.20
AC/PVC pipe 100 mm diameter (m)	3.90	4.60	6.30	6.50	7.70	7.10
35× 35 × 4 or 5 mm angle iron	10.90	11.90	12.70	14.25	15.00	16.50
MS pipe 40 mm diameter	1.85	2.00	2.05	2.05	2.35	2.40
Square plates 250 × 250 × 6 mm (No.)	2	2	2	2	2	2
14 mm diameter and 32 mm long bolts with nuts (No.)	16	16	16	16	16	16
35× 35 × 4 or 5 mm angle iron	16.80	19.40	21.00	30.60	35.25	43.40
M.S. pipe 50 mm internal diameter	1.15	1.15	1.25	1.25	1.45	1.50
250 mm diameter and 6 mm thick flange plate (No.)	2	2	2	2	2	2
Flats 40 × 6 mm thick	4.20	4.15	4.30	3.50	4.05	4.20
Gas outlet pipe flange 25 mm diameter (No.)	1	1	1	1	1	1
GI bend 25 mm diameter (No.)	1	1	1	1	1	1
Heavy duty gas valve 25 mm diameter (No.)	1	1	1	1	1	1
Union joint 25 mm diameter (No.)	2	2	2	2	2	2
Pipe nipple 25 mm diameter 100 mm long (No.)	1	1	1	1	1	1
Pipe nipple 25 mm diameter 150 mm long (No.)	2	2	2	2	2	2

Materials	Plant Size (m^3)					
	2	3	4	6	8	10
Clips (No.)	2	2	2	2	2	2
3 m polythene or reinforced rubber pipe 25 mm diameter (No.)	1	1	1	1	1	1
2.50 mm thick (12 gauge) mild steel sheets (2.5 m × 1.0 m) (No.)	2¼	3	3¼	4¼	4½	5½

AC = Asbestos cement, MS = Mild steel, GI= Galvanized iron.

FIXED-DOME TYPE PLANTS

Fixed-dome plants comprise closed, dome-shaped digester with immovable, rigid gas-holders and displacement pits, also named as outlet chambers (Figure 7.2). The gas is stored in the upper part of the digester. The gas pressure varies with the amount of gas stored in the holder, being low in case gas stored in the gas-holder is less. Pressure increases as the volume of gas stored increases. The gas pressure displaces the slurry into the outlet chamber and to the overflow tank. Let us have a look at two major components of the fixed dome type plant. More discussion will follow in the respective designs of the two models.

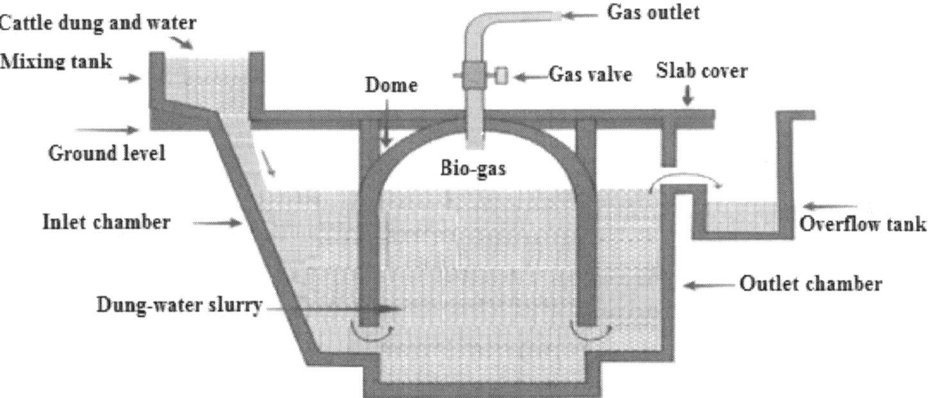

Figure 7.2. A fixed dome type biogas plant.

Digester: The digester of a fixed-dome plant is usually a masonry structure made of cement and ferro-cement. The choice of material to construct the structure is governed by the following factors.

- ☆ Technical suitability (stability, gas- and liquid tightness)
- ☆ Cost-effectiveness
- ☆ Availability in the region and transport costs

- ☆ Availability of local skills for working with the particular building material

Gas holder: It is the top portion of the fixed-dome plant where gas accumulates. It must be gas-tight. This portion is generally painted with a gas-tight layer (*e.g.* 'Water-proofer', Latex or Synthetic paints) because concrete, masonry and cement portions are not gas-tight. A flexible joint is constructed between the lower (water-proof) and the upper (gas-proof) part of the hemispherical structure to reduce the risk of cracking of the gas-holder. It prevents the cracks that might develop in the lower part due to hydrostatic pressure to move into the upper part of the gas-holder.

Advantages and Disadvantages of Fixed Dome Plants

Advantages

- ☆ Low initial cost and long useful life-span
- ☆ No moving or rusting parts
- ☆ Basic design is compact, saves space and is well insulated
- ☆ Construction generates local employment
- ☆ The underground construction saves space and minimizes the temperature variations around the digester
- ☆ Not much maintenance is required

Disadvantages

- ☆ Masonry gas-holders require special sealants and high technical skills for gas-tight construction
- ☆ Gas leaks occur quite frequently
- ☆ Variable gas pressure complicates gas utilization as burners and other simple appliances cannot be designed and operated in an optimal way. If the gas is required at constant pressure (*e.g.*, for engines), a gas pressure regulator or a floating gas-holder is necessary
- ☆ Amount of gas produced and stored in the gas holder is invisible
- ☆ Fixed dome plants need exact planning of levels
- ☆ Scum formation cannot be ruled out because there is no provision of stirring
- ☆ Excavation can be difficult and expensive especially if a bedrock is encountered

Janata Model

The original design of this plant is of Chinese origin. The model has been introduced under the name "Janata biogas plant" by *Gobar* Gas Research Station, Ajitmal in view of its low cost. The plant consists of an underground well, a sort of digester, made of bricks and cement having a dome shaped roof that remains below the ground level (Figure 7.3). At almost middle of the digester two rectangular openings facing each other come up to a little above the ground level. These openings act as an inlet and an outlet of the plant. Dome shaped roof is fitted with a pipe at the top, which serves as the gas outlet. The principle of gas production is same as that of the KVIC model. The biogas is collected in the restricted space of the fixed dome, hence the pressure of gas is much higher, and ranges from 0 to around 90 cm of water column. The plant can be constructed by village mason taking some pre-explained precautions. All the material required for the construction can be locally arranged. Good quality bricks and cement should be used to avoid structural problems like cracking of the dome and leakage of gas. Substrates other than cattle dung such as municipal waste and plant residues can also be used in the Janata type plants.

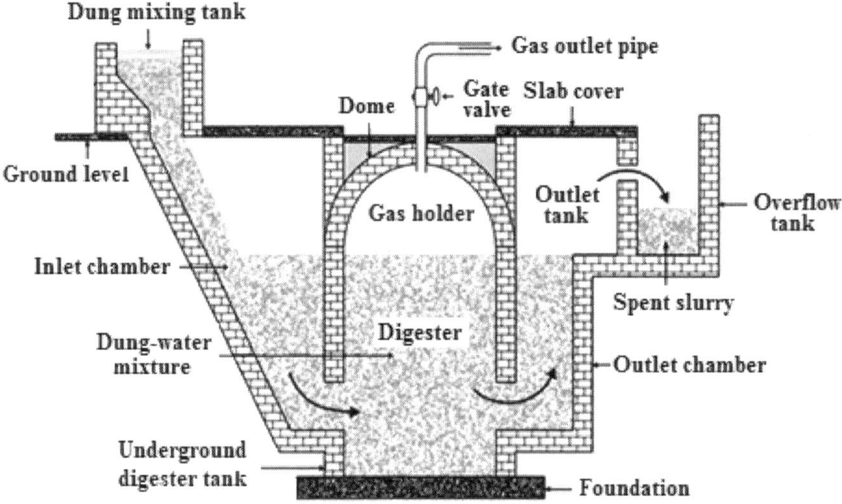

Figure 7.3. Janata type fixed dome biogas plant.

Deenbandhu Model

Deenbandhu model has been developed by AFPRO (Action for Food Production, New Delhi) in 1984. The word Deenbandhu means a friend of the poor. This plant is designed on the principle that the surface area of a biogas plant can be minimized to reduce its installation cost without sacrificing the efficiency of the plant. The design consists of segments of two spheres of different diameters joined at their bases. The structure so formed act as the digester, the fermentation chamber and the gas storage chamber. The reason to go for this kind of structure is because of high compressive strength of the brick masonry and concrete. A spherical structure loaded from the convex side is always under compression. Therefore, the internal load will not have any residual effect on the structure. The digester is connected with the inlet pipe and the outlet tank. The upper part above the normal slurry level of the outlet chamber is designed to accommodate the slurry before it is displaced out of the digester with accumulation of biogas. It is called outlet displacement chamber or simply the outlet chamber (Figure 7.4). The size of these plants is generally limited to 6 m^3 per day. The different components of Deenbandhu model biogas plant are shown in Figure 7.4. The quantities of material required for various sizes of plants are given in Table 7.2.

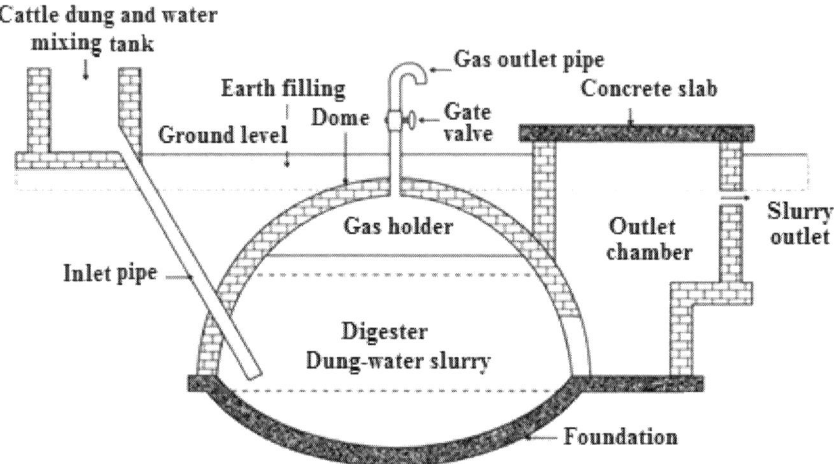

Figure 7.4. Deenbandhu fixed dome biogas plant.

Table 7.2. List of materials for Deenbandhu biogas plants of 40 days retention time

Materials	*Plant size (m^3)*				
	1	*2*	*3*	*4*	*6*
Bricks (No.)	700	1000	1300	1600	2200
Cement bags (No.)	8	14	16	22	28
Stone chip(12/20 mm)(m^3)	0.85	1.14	1.43	1.70	2.40
River sand (m^3)	0.85	1.14	1.43	1.70	2.40
GI pipe (cm)	17.78	17.78	17.78	17.78	17.78
AC/PVC pipe 6" diameter (m)	1.80	1.80	1.80	1.80	1.80
Iron bars 6 mm (kg)	6.00	7.00	10.00	12.00	15.00
Labours (man hours) for excavation	8	10	14	18	24
Masons	8	11	13	16	22
Construction labours	16	22	26	30	44

AC = Asbestos cement, GI= Galvanized iron, PVC = Poly Vinyl Chloride.

KVIC and Janata Type Biogas Plants: A Comparison

KVIC Type	*Janata Type*
Capital investment is high	Capital investment is less
Cost of maintenance is high	Cost of maintenance is less
Life span of the plant is expected to be 30 years and that of gas holder 5-8 years	Life span of the plant is comparatively more
Steel gas holder is essential, which require maintenance such as painting, repairing, and replacements of damaged parts due to corrosion	Steel gas holder is not required
Hydraulic retention time (HRT) is more as the tank is separated with a center wall	Hydraulic retention time (HRT) is quite less
Locally available materials can't be used for construction of the digester. Fabricated gas holder is to be transported from nearby towns	The entire plant can be constructed with locally available materials
The space above the movable drum can't be used for other purposes	The space above the plant can be used for parking or garden can be made above it, if constructed underground
Effect of temperature during winter is significant	Effect of temperature during winter is less
Gas pressure depends on the weight of the gasholder. The gas is released at a pressure of 8-12 cm of water column	It is engineered using hydraulic pressure balancing technique and gas pressure is about 90 cm of water column
It is suitable for processing animal dung	It is suitable for processing other waste materials along with animal dung

Plug Flow Digesters: Movable Model

A major disadvantage of the fixed dome or floating drum models is that once installed they cannot be shifted. Hence, portable models built over the ground called tubular or plug flow digesters have been developed (Figure 7.5, Rajendran *et al.*, 2012). Plug flow digesters have a constant volume in the range of 2.4 to 7.5 m^3, but produce biogas at a variable pressure. Plug-flow digesters consist of a narrow and long tank with, an average length to width ratio of 5:1. The inlet and outlet of the digester kept above ground are located at opposite ends of the tank. The remaining parts of the digester are buried below the ground in an inclined position. As the fresh substrate is added from the inlet, the digestate flows towards the outlet at the other end of the tank. The inclined position makes it possible to longitudinally separate acidogenesis and methanogenesis to produce a two-phase system. Whole system is placed under a gable or shed roof to maintain the process temperature and to minimize day/night temperature fluctuations.

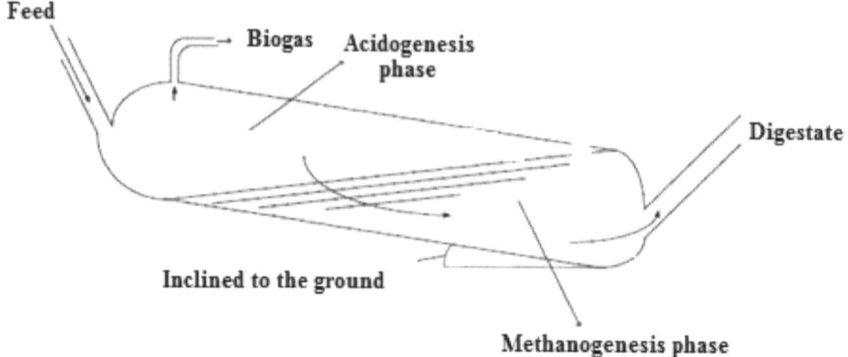

Figure 7.5. Schematic sketch of a plug flow digester.

Suitability of Various Technologies/Models

Numbers of models that have some improvement over the general types discussed in the previous sections have been designed and developed. Ministry of New and Renewable Energy has approved 6 models of floating drum type of biogas plants for installations under its various programs. Similarly, it has approved 4 different models of Deenbandhu fixed dome type biogas plants. Besides, it has approved a bag type model also called a balloon plant consisting of a UV resistant plastic or rubber digester bag (BIS, 2022), which is essentially a portable system as a plug flow digester. It has a life of about 5 years. Proper selection of the technology should be made amongst the various models approved by the Ministry as per the following recommendations.

Family size biogas plants (1 to 10 m^3): KVIC, Deenbandhu, Janta, Pragati, Flexi or bag type *etc.* (BIS, 2022)

Large scale biogas plants (10 to 140 m^3): KVIC

Large scale plants above 1000 m^3: Up-Flow Anaerobic Sludge Bed (UASB), Modified UASB, Biogas Induced Mixing Arrangement (BIMA) Digester suitable for industrial effluents, MSW, fruit and vegetable waste *etc.* (BIS, 2022)

ENERGY FROM BIOGAS PLANT

The energy (E) available from a biogas digester in MJ is given by the relation:

$$E = \eta H_b V_b \tag{7.1}$$

Here η is the combustion efficiency of burner or boiler, H_b is the heat of combustion per unit volume of biogas (20 MJ/m^3) at 10 cm water gauge pressure (0.01 atm) and V_b is the volume of biogas (m^3). It is important to note that some of the heat of combustion of methane is used to heat the CO_2 of the biogas and hence will not be available for other applications. Above mentioned equation can also be written in terms of the fraction of methane (f_m) in biogas (0.7) as

$$E = \eta H_m f_m V_b \tag{7.2}$$

Here H_m is the heat of combustion of methane (28 MJ/m^3 at standard temperature and pressure).

SELECTION OF CONSTRUCTION SITE

Selection of construction site for a biogas plant is governed by the following factors.

- The soils at the site should be easy to work with so that construction cost is minimized. At the same time, soil should have enough bearing capacity to avoid the possibility of sinking of structure
- The selected site should ensure easy operation and maintenance activities like feeding of plant, use of main gas valve, composting and use of slurry, checking of gas leakage, draining condensed water from pipeline *etc.*
- The site should guarantee plant safety
- Plant must be located as close as possible to the cattle shed. It will be then easier to operate the plant as well as minimize wastage of raw materials, especially the dung/swine manure
- The site should be at slightly higher elevation than the surroundings to avoid water logging. This also ensures free flow of slurry from overflow outlet to the composting pit
- Optimum temperature (20-35°C) has to be maintained inside the digester for its effective functioning. Therefore, avoid damp and cool places in favour of sunny sites
- Considerable quantity of water is required to prepare dung-water mixture. Thus, the plant should be near but at least 15 m away from the water source to avoid the water source pollution
- The plant should be as near to the point of application as possible because longer gas pipe unnecessarily increases the cost of gas conveyance system besides causing other problems
- The site should be at sufficient distance away from the trees to avoid damage to the bio-digester by the plant roots

DESIGN OF THE PLANT

Two main components of KVIC *gobar* gas plant are digester or pit and the gas holder or the gas collector. *Capacity of the plant is described through the quantity of gas produced in a day.* A 1 m^3 plant means it is capable of producing 1 m^3 of biogas in a day. Capacity of the plant is designed on the basis of gas requirement calculated as per guidelines reported in Table 7.3 or the availability of raw material, whichever is limiting. The capacity of the plant controls the parameters of the plant namely the digester volume and gas holder volume

(Floating). The design of these components further depends upon the daily feed and the hydraulic retention time (HRT) of the material in the plant. Following guidelines forms the basis of these parameters.

Table 7.3. Required quantity of biogas for different purposes

Sl.No.	Purpose	Size	Required Quantity of Biogas (m^3/hr)
1.	Cooking	5 cm burner	0.30
		10 cm burner	0.50
		15 cm burner	0.64
		per person/day	0.34-0.42
2.	Lighting	100 candle power lamp	0.13
		1 mantle lamp	0.07-0.08
		2 mantle lamp	0.14
		3 mantle lamp	0.17
3.	Diesel engine	1 HP	0.45-0.51
4.	Refrigerator	Per L capacity	0.0012
5.	Incubator	Per L capacity	0.0005-0.0007

Availability of Dung

Total quantity of dung available depends upon the number of cattle and other animals. The total quantity of dung is calculated as per the following guidelines.

Animal Type	Dung/day (kg)
Cow	10
Ox	12
Buffalo	15
Pig	2

When unit weight of dung is mixed with unit weight of water (to get optimum solid concentration for digestion), the density of the slurry is about 1090 kg/m^3. Thus, volume of the daily charge is calculated by the equation:

Volume = Weight of (dung + water)/Density of slurry

Digester volume can be calculated as: Volume of the daily charge × HRT

The actual volume should be about 10 per cent more to provide some empty space at the top for proper disengagement of the gas. Thus

$$\text{Digester volume} = 1.1 \times \text{Volume of the daily charge} \times \text{HRT} \tag{7.3}$$

Recommendations on hydraulic retention time for various states are given as follows. Note that regional temperature and its variations mainly govern the HRT.

Plants based on 30 days HRT

It is recommended for states of Andhra Pradesh, Goa, Kerala, Karnataka, Maharashtra, Puducherry and Tamil Nadu.

Plants based on 40 days HRT

It is recommended for states of Bihar, Gujarat, Haryana, Punjab, Jammu area of J and K union territory, Madhya Pradesh, Odisha, Rajasthan, Uttar Pradesh and West Bengal.

Plants based on 55 days HRT

It is recommended for states of Himachal Pradesh, North-eastern states, Sikkim, Kashmir area of J and K union territory, hilly districts of Uttar Pradesh and other hilly areas having long severe winters.

Digester

Digester is subjected to tensile load on inner wall surface because of slurry (hydrostatic pressure = depth of slurry × density of slurry). The part below the ground level is subjected to compressive load due to the earth pressure. Therefore, digester is designed in a circular shape that has better load bearing capacity. The height (H) to diameter (D) ratio for large capacity plants is usually kept as unity. Therefore, height of the plant is given as:

Volume of the plant = $(\pi/4)\, D^2H$

Since D = H, we have

$$H^3 = 4 \times \text{Volume}/\pi \tag{7.4}$$

Diameter of the digester, D will be same as height, H.

Gas Holder

While designing a gas holder, it is assumed that the gas produced during the night is used up for cooking breakfast and lunch in the morning. The gas produced during the day time after the morning cooking is used for cooking dinner in the evening. Thus, gas storage capacity of the plant or the volume of gas holder is kept as 50 per cent the capacity of the plant or gas produced/day.

Normally, gas produced from 1 kg of *gobar* in winter is taken as 42 liters and in summers as 55 liters. Thus, production of gas/kg of fresh dung is taken as 0.04-0.05 m^3 or 0.04 m^3/kg for a retention period of 30 days. Since only 50 per cent of it is to be stored, it works out to be 0.02 m^3/day-per kg of feed.

Total gas to be stored, m^3 = Dung fed × 0.02

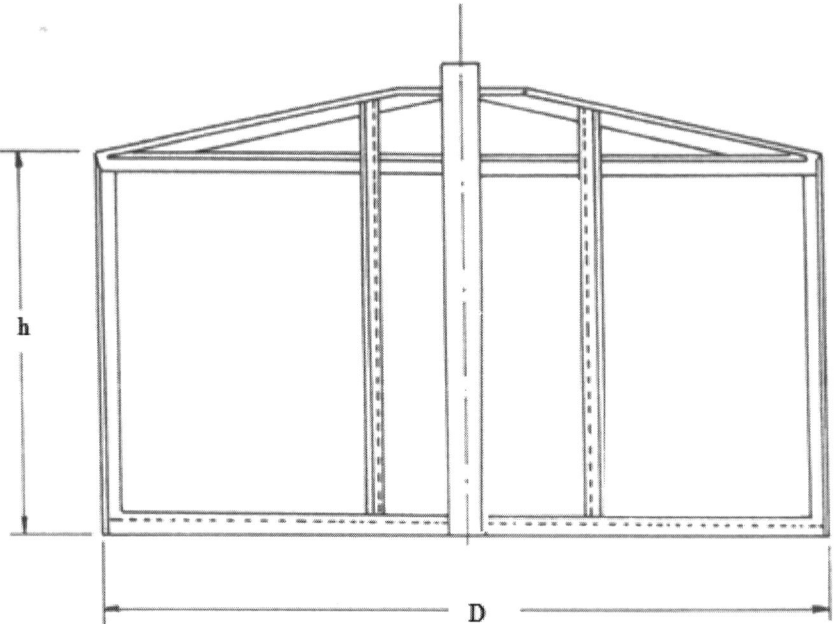

Figure 7.6. Gas holder with guides.

Diameter of the gas holder will be slightly less than the diameter of the digester so as to ensure its free movement. However to calculate the height of the gas holder, h (Figure 7.6), the diameter of the gas holder is assumed to be the same as that of the digester. Therefore

$\pi D^2 h/4$ = Volume of gas to be stored= Capacity/2

Since D is known, h can be calculated by the following equation.

$$\text{Height of the gas holder, } h = 2.0 \times \text{capacity of the plant}/(\pi D^2) \quad (7.5)$$

Field engineers can use the following approximate rules for sizing biogas plants or for estimating their performance:

1. One kg of dry cattle dung produces approximately 0.04 m^3 of biogas.
2. One kg of fresh cattle dung contains 8 per cent dry biodegradable mass.
3. One kg of fresh cattle dung has a volume of about 0.9 liters.
4. One kg of fresh cattle dung requires an equal volume of water for preparing mixture to be fed to the digester.
5. Typical retention time of slurry in a biogas plant is 40 days.

Accordingly, a general thumb rule has been framed that says that for each 1 m^3 capacity of the plant, a mixture of 25 kg of dung and 25 kg water should be sufficient.

Example 7.1:

Calculate the number of cattle required to build a 1 m^3 capacity biogas plant.

One cow yields on an average of 10 kg of cow dung in a day. Since 1 kg of cow dung will yield 0.04 m^3 of biogas.

So 10 kg will produce $10 \times 0.04 = 0.4$ m^3 of gas in a day.

Therefore, cow dung required for 1 m^3 capacity biogas plant = $1/0.4 = 2.5$. It means 2 to 3 cows will be required to operate the plant at the rated capacity.

Example 7.2

Determine the dimensions of the digester and the gas holder for a 4 m^3 capacity biogas plant. Assume appropriate values of parameters, if needed.

Assuming only cows, number of cattle required for the plant = $4/(0.04 \times 10) = 10$

The size of the digester (digester volume) is determined by the length of the retention time and by the amount of daily feed material (*e.g.*, cattle dung + water) or fermentation slurry. The volume of daily charge is

Volume of daily charge = $(100+100)/1090 = 0.18$ m^3

Assuming a retention period of 35 days, the total charge is

Total charge in 35 days = $35 \times 0.18 = 6.3$ m^3

Volume of the digester = $1.1 \times 6.3 = 6.93 \approx 7.0$ m^3

H (Eq. 7.4) = $(4 \times 7/\pi)^{1/3} = 2.07 \approx 2.1$ m

Diameter of the digester = 2.1 m (Since H = D assumed)

The height of the gas holder is determined as follows:

Gas to be stored in the gas holder = $4 \times 0.5 = 2.0$ m^3

Height of the gas holder, h (Eq. 7.5) = $2.0/[(\pi/4)\, 2.1 \times 2.1] = 0.58$ m

CONSTRUCTION

Plant Layout

Construction of the biogas plant begins with the process of layout. Herein dimensions of plant on the paper are marked on the ground to take up the digging work. Following steps are followed:

- ☆ Level the ground and determine the center line of the digester, outlet tank and inlet pit
- ☆ Decide the reference level. It is better to assume the leveled ground level as the reference level. The top of the dome (outer) should exactly be at this level

- ☆ Select the outer radius of the pit (digester diameter plus wall thickness plus space for a footing projection of at least 10 cm) and mark it on the ground as a chord
- ☆ Insert a stick or wooden peg in the leveled ground at the center of the proposed digester pit. With the help of this pole and chord prepared earlier, make a circle, which indicates the area to dig
- ☆ From the center point where the central line meets with the perimeter line, draw a tangent and measure a length equal to half of the breadth of outlet plus wall thickness (for outlet chamber) and half of the size of manhole (30 cm) plus wall thickness for manhole, on either side of this tangent. Mark the manhole in such a way that the inner size is 60 cm × 60 cm
- ☆ Draw horizontal parallel lines from the points in either side on the tangent, which will meet the dome. From the center point where the central line meets with the perimeter line, measure the length of outlet plus wall thickness to decide the outer dimension of outlet
- ☆ Use coloured powder to mark the dimensions
- ☆ Decide the location of slurry pit while laying out plant digester and outlet

Once the layout work is over, tools like, crow-bar, picks, spade, shovel *etc.* are used to dig the pit.

- ☆ Digging should be done as per the dimensions fixed during layout
- ☆ As far as practical the cuttings should be vertical
- ☆ Ensure that the excavated earth is deposited at least 2 m away from the pit on each side. It eases the construction work
- ☆ Make adjustments in depth and diameter (keeping the volume same) for areas having shallow water table. Alternatively, construct a deeper pit near the digester pit. Water in the digester pit is pumped to this pit to facilitate construction work of the digester
- ☆ Once the digging is as per the required size, level and ram the pit bottom so that the base is on a leveled land and does not settle
- ☆ Be careful to avoid accident while digging near the sides since soil in some cases may collapse
- ☆ Dig the foundation for the manhole (first step of outlet tank) along with the foundation for digester as per the dimensions in the drawing during the layout

Some of the construction activities are shown in Figure 7.7 (https://snscourseware.org/snsctnew/files/1580286076.pdf).

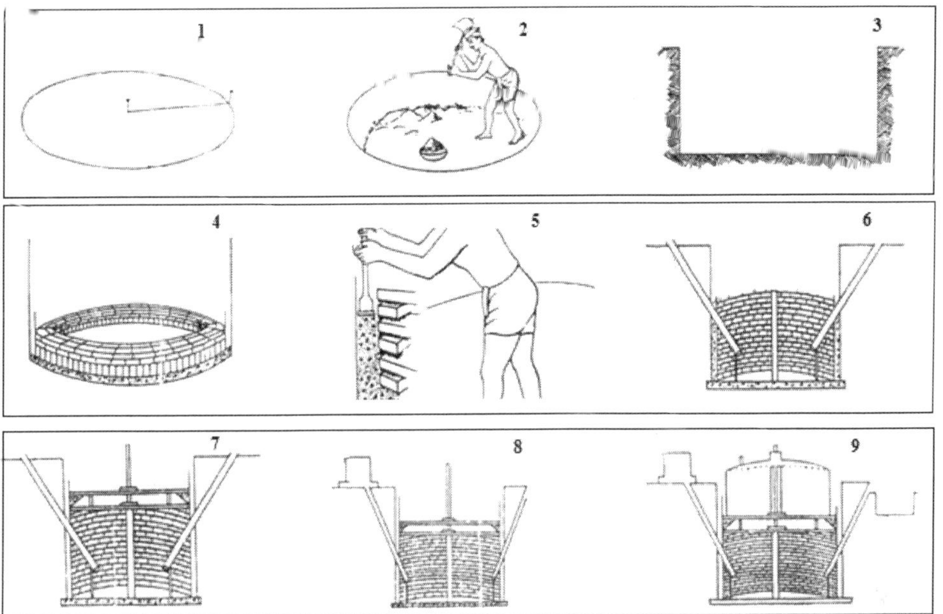

Figure 7.7. Few steps of construction activities.

1. Marking the dimensions for digging, 2. Digging. 3. Completed pit (note the vertical walls), 4. Laying the concrete foundation and brick masonry work, 5. Filling the outer layer with compost or other insulator, 6. Setting inlet and outlet pipes, 7. Fixing the guide frame, 8. Constructing the inlet tank and 9. Constructing slurry outlet tank.

The following points are considered during the construction and for maintaining the bio-gas plant.

- Soil around the well should be thoroughly compacted
- Provide slight slope to the pipe line towards the kitchen
- Lay the compost around the digester well to a depth of one meter to keep the digester warm during winter, (Figure 7.1 and Figure 7.7)
- Paint outside of the gas drum every alternate year to prevent corrosion
- Cover the gas drum with paddy straw mat especially during winters so that drum remains warm
- Mix the dung and water well prior to feeding. Feeding should preferably be done at mid-day
- Feed material should not be contaminated with soil so that sand particles do not enter into the digester
- Use PVC instead of AC pipes for inlet and outlet
- KVIC digester should be initially charged from both sides *i.e.* inlet and outlet

- ☆ Stakeholders often use molasses to get more gas during winters. Excessive addition of molasses should be avoided, as it generates more non-combustible matter
- ☆ Banana stem and water hyacinth should be chopped in 5-10 cm pieces before these are introduced into the plants
- ☆ Chopped rice straw and banana stems should be fully mixed with cow dung slurry for use in composite waste biogas plants
- ☆ *Eucalyptus* leaves should be soaked in water for five days prior to feeding into the plant
- ☆ Daily rotate the gas holder and stirrers for 50 times in the morning and 50 times in the evening to break the scum

Quality of Construction Materials

Cement, sand, aggregate *etc.* should be of good quality. Poor quality material will render substandard quality of the plant even if the plant has been properly designed and construction well executed.

Material	*Quality Considerations*
Cement	High quality Portland cement free of lumps should be used. It should be stored in a dry place
Sand	Both the coarse and granular sand for concreting and fine sand for plastering should be clean and free from soil or any other material
Gravel/Ballast	These should be neither very big nor very small. Maximum size should be ¾" or ¼ of the slab thickness. Should be clean, hard and angular in shape. If dirty, wash properly before use
Water	Use clean and fresh water, avoid water from ponds and channels, saline water of tube wells should not be used
Bricks	First quality bricks should be used. Bricks must be soaked in clean water before use followed by curing the work before and after plastering
Mild steel (MS) bars	MS bars should be as per engineering standard including weight per unit length
Mixing device	This device is used to prepare good quality water-dung solution in the inlet tank. Vertical mixing device of good quality with well galvanized and properly aligned mixing blades should be used

HANDLING THE BIOGAS

Biogas is about 20 per cent lighter than air. Its ignition temperature is in the range of 650 to 750°C. It is an odourless and colourless gas that burns with a clear blue flame quite similar to natural gas. Biogas can potentially be used for heating, combined heat and power generation and as vehicle fuel. Except for its simplest thermal use, biogas needs to be cleaned or processed prior to other uses. Gas handling equipment to transport biogas from the digester to its end-use sites comprises piping, gas pump or blower, gas meter, pressure regulator, and condensate drain(s). Biogas trapped under an airtight cover placed over the digester is removed by applying a slight vacuum on the collection pipe (*e.g.*, by connecting a gas pump/blower to the end of the pipe), which draws the collected gas under the cover. A gas meter is used to monitor the gas flow rate. Sometimes a gas scrubber is needed to clean or scrub the biogas of corrosive compounds such as hydrogen sulphide. Warm biogas as it travels through the piping is cooled resulting in condensation of water vapours. Therefore, condensate drain(s) is provided to remove the condensate.

Production of Heat or Steam

The most straightforward use of biogas is for thermal (heat) energy. In areas where petro-fuels are scarce, decentralized small biogas systems can be used to provide the heat energy for basic cooking and water heating. Although biogas burners are now available, even conventional cooking stoves can be easily adjusted for biogas by simply changing the air-to-gas ratio. Since low quality biogas can be used for this purpose, there is no need to further refine the gas for such applications. Only requirement is a gas pressure of 8 to 25 mbar and H_2S levels below 100 ppm. Biogas lighting systems can be used to illuminate rural homes.

Electricity Generation or Combined Heat and Power

Combined heat and power (CHP) systems use both the power producing ability of the fuel and the inevitable waste gas for heating. Some CHP systems are primarily designed for heating while electrical power generation is secondary. Other CHP systems are designed primarily to generate electrical power while the waste heat use is secondary. In either case, the overall (combined) efficiency is greatly enhanced than using the biogas to produce only power or heat. Since internal combustion engines are commonly used in CHP applications, use of biogas in such applications requires removal of both H_2S (to below 100 ppm) and water vapours. Gas turbines (micro-turbines, 25-100 kW; large turbines, >100 kW) with comparable efficiencies to spark-ignition engines are now available for production of both heat and power.

Vehicle Fuel

Gasoline vehicles that have been modified for natural gas can also be run on biogas provided biogas is upgraded to natural gas quality. Most vehicles in this category have been retro-fitted with a gas tank and a gas supply system in addition to the normal petrol fuel system. Contrarily, dedicated vehicles (using only biogas) are more efficient than these retro-fits.

A complete list of equipment and systems where biogas can be used is as follows.

- Internal combustion engines – electrical power generation, shaft power
- Gas turbine engine (large) – electrical power generation, shaft power
- Micro-turbine engine (small) – electrical power generation
- Stirling heat engine – electrical power generation
- Boiler (steam) systems
- Hot water systems
- Process heaters (furnaces)
- Space or air heaters
- Gas fired chiller - refrigeration
- Absorption chiller - refrigeration
- Combined heat and power (CHP) - large and small scale – electrical power and heat
- Fuel cells – electrical power, heat

One cubic meter of biogas can do the following operations:

- It can illuminate a mantle lamp (60 W) for a period of 7 hours
- It can cook three meals for a family of five
- It can run 2 HP engine for one hour
- It can run 100 L capacity refrigerator for 9 hr
- It can generate 1.25 KWh of electrical energy

HANDLING THE MANURE

A by-product of the anaerobic digestion is effluent commonly known as bio-slurry, a wash out of the digester. This effluent is a stabilized organic solution having high value as a manure/fertilizer besides having other potential uses. Thus, management of slurry is quite important for its effective utilization. Currently, biogas slurry is just released in depressions around the biogas plants. It results in loss of its main nutrient, the nitrogen. Manure management system besides reducing nutrients losses is needed to collect and store manure in a systematic manner for sanitary, environmental, and farm operational considerations. It essentially includes some storage facilities because the effluent cannot be applied to land and crops every day around the year. The size of the storage facility and storage period must be adequate to store the effluent during the non-growing season. Moreover, it should also take into account whether the manure is collected and stored in the form of liquid, slurry, semi-solid, or solid.

Raw Manure

Depending upon animal type, raw manure has a solid content in the range of 8 to 25 per cent. It can either be diluted by water or thickened by air drying or by adding bedding materials.

Liquid Manure

Liquid manure is prepared by lowering the solids content to less than 5 per cent. It can be flushed using fresh or recycled water right from the outlet of the plant. The manure and flush water can be pumped to treatment and storage tanks, ponds, lagoons, or other suitable structures before its application to cropped lands. Liquid manure systems can also be used for biogas production and energy recovery. In colder climates, biogas recovery is usually limited to gas flaring to minimize odour problems.

Slurry Manure

Solid content of the manure handled as slurry is about 5 to 10 per cent. Thus, the by-product from the plant is diluted to that extent. Slurry manure is usually collected by a mechanical scraper system. This can be pumped, and is often treated or stored in tanks, ponds, or lagoons prior to land application. Manure managed in this manner may be used for biogas recovery and energy production, depending on climate and dilution factors.

Semi-solid Manure

Solid content of manure handled as a semi-solid is 10 to 20 per cent. This manure is typically scraped. No additional water is added to the manure. It is stored until it is spread on local fields. Fresh scraped manure (less than one week old) can be used for biogas and energy production in all climates, because it can be heated to promote bacterial growth.

Solid Manure

Aged solid manure or manure left unmanaged is allowed to dry until its solid content is more than 20 per cent. It is handled as a solid by a scoop loader. It is not suitable for biogas recovery.

Advantages of Biogas Slurry

☆ Biogas slurry contains major plant nutrients especially its nitrogen content is high compared to dry and semi-dry slurry. It nourishes the soil and accelerates the growth of plants, especially root growth. Its application enhances crops yield in a sustainable manner. Following table indicates how biogas slurry is more effective than other organic slurries in relation to NPK (Sharma and Samar, 2016).

Slurry	*N_2 (per cent)*	*P_2O_5 (per cent)*	*K_2O (per cent)*
Fresh cattle dung	0.3-0.4	0.1-0.2	0.1-0.2
Farmyard slurry	0.4-1.5	0.3-0.9	0.3-1.9
Compost	0.5-1.5	0.3-0.9	08.-1.2
Biogas slurry	**1.5-2.5**	**1.0-1.5**	**0.8-1.2**
Poultry slurry	1.0-1.8	1.4-1.8	0.8-0.9
Cattle urine	0.9-1.2	Trace	0.5-1.0
Paddy straw	0.3-0.4	0.8-1.0	0.7-0.9

- It contains several micro-nutrients such as Mn 188 ppm, Zn 144 ppm, Fe 3500 ppm, Cu 28 ppm. Thus, it increases the availability of these nutrients to plants
- Biogas slurry is free from weed seeds
- It is practically odourless without any foul smell and pathogen
- Well digested slurry is easier to spread and does not attract insects/ flies
- It acts as a soil conditioner and improves soil physical condition having organic matter content of about 65 per cent
- It enhances aeration and water holding capacity of soil for root penetration resulting in better growth. It also improves internal drainage and soil structure
- Application of biogas slurry improves the quality of produce. Qualitative improvement in taste, smell, size, and colour *etc.* of the produce has been reported
- Its dark colour absorbs more sunlight resulting in warming up of the soil
- The effluent repels termites
- Improves microbial properties of the soil by enhancing bacteria and fungi growth crucial for high yield agriculture
- Good for acidic soils as it reduces harmful elements like aluminium and minimizes its toxicity
- Positively impacts membrane permeability of root hairs that enhances nutrient uptake
- Biogas slurry can also be used as feed for fish culture, which acts as a supplementary feed. Bio-slurry stimulates the primary production of algae and micro-organisms in a fish pond, which serve as feed for the fishes. On an average 15-25 L wet slurry can be applied per day in a 110 m^2 fish pond. Biogas slurry mixed with oil cake or rice bran in 2:1 ratio remarkably increases fish production
- Biogas slurry can also be used for the production of bio fertilizers like Azolla and aquatic biomass Spirulina

Disadvantages

- Bio-slurry being a liquid is difficult to transport and apply to fields located far away from the digester
- It cannot completely replace the inorganic fertilizers
- Bio-slurry is regularly produced. But many times there is no crop or crops are at a stage that the application of nutrients may not be usefully exploited by the plants. Thus, it has to be stored or needs to

be converted to compost. Improper storage may result in loss of NH_4 through volatilization of NH_3. Bio-slurry has to be properly covered to minimize these losses

- Another pathway for nutrient losses is leaching of the liquid fraction of bio-slurry. This is especially the case for storages without a bottom liner
- Water soluble P decreases by anaerobic digestion, probably caused by a pH increase, leading to a shift of HPO_4^{2-} towards PO_4^{3-}, and subsequent precipitation of calcium or magnesium phosphates and/or struvite (magnesium ammonium phosphate)
- Bio-slurry is often dried or composted to get solid organic fertilizer. From theoretical perspective both drying and composting can lead to N losses through volatilization of NH_3, while other nutrients are normally retained in the slurry
- Stricter precautions for product safety might be required when human excreta and/or municipal solid wastes are included in the feedstock. Such feed stocks may contain pathogens and heavy metals
- Both farmer and biogas technicians need training on bio-slurry management especially on timing of the application of bio-slurry to the field and crops and identification of crops to which bio-slurry can be applied best including the best application strategy

COST

According to Verma and Sahu (2019), the cost of KVIC and Deenbandhu model were on an average Rs. 45841 and Rs. 25618 for a plant capacity of 3-4 m^3/day (Table 7.4). The estimate also reveals that the cost of drum in a KVIC model constitutes about 20 per cent of the total cost of the plant. The cost estimate is based on 2015-16 prices in Sehore district of Madhya Pradesh.

MAINTENANCES AND GENERAL CARE

Some operational problems commonly encountered in biogas plants can be sorted out through general maintenance and care. These activities must begin immediately after the biogas plant is installed.

Daily

- Add dung slurry to the plant with dung and water ratio as 1:1
- Make sure that no stones and sand makes its way into the plant during feeding
- Clean the gas burner
- The water traps should always contain water otherwise the gas will leak out through the gas trap

Table 7.4. Cost of installation of KVIC and Deenbandhu model of biogas (Rs./plant)

Sl.No.	*Components*	*KVIC Model*	*Deenbandhu Model*
A.	**Civil construction**		
1	Bricks	16900	7611
2	Cement	6000	5087
3	Sand	1250	1292
4	*Gittee*	875	808
5	Boulder	0	366
6	Iron	256	1138
7	AC/PVC Pipe	1220	300
8	Gate valve	250	274
9	Cost of supply line and pipe fitting	2270	612
10	Burner	1100	1027
11	Steel gas holder with guide frame	7800	0
12	Paint	740	367
	Sub-total (A)	38661 (84.4 per cent)	18882 (73.7 per cent)
B.	**Labour cost**		
1	Digging the pit	1980	2216
2	Construction	2340	902
3	Mason	2860	3618
	Sub-total (B)	7180 (15.6 per cent)	6736 (26.3 per cent)
C.	**Total cost (A+B)**	45841	25618
D	**Subsidy**	11500 (25 per cent)	11500 (45 per cent)
E.	**Net cost of installation**	34341	14118

Monthly

- ☆ Check gas pipeline for leaks with a soap solution

Annual/Long-term

- ☆ Check for gas and water leaks and repair them
- ☆ Check gas pipelines for leakages
- ☆ Check for solid sediment at the bottom of the digester plant every 5-6 years. It can be accomplished by inserting a long stick in the plant to determine any change in depth of the digester. If necessary, completely empty the digester to remove the solids. Plastering of the inside portion of the plant can also be done at this stage. Take necessary safety precautions when performing these tasks

PROBLEMS: CAUSES AND REMEDIES

A biogas plant may stop generating gas or the gas may not be reaching the utility site. It can be ascribed to several factors that affect the biogas production or its availability. Look for the following.

Biogas leaks: Leaks somewhere in the system may leave very little biogas for use even when the plant is operational. Check the gas holder and biogas pipeline for leakages. Simple soap solution can be used to detect the leakages.

Temperature problems: A drastic reduction in biogas production can be anticipated when the ambient temperatures reach below 20°C. In such cases, adopt some kind of heating system to heat the contents of the bio-digester. Following tips may be useful.

- ☆ Use warm water for diluting the dung
- ☆ Diluted dung slurry can be prepared in a mixing tank and kept all day to warm up. The digester may be loaded in the evening when the mixture is still warm
- ☆ Addition of organic matter containing high percentage of nitrogen like urine, night soil *etc.* may prove helpful
- ☆ The gas holder may be covered with plastic sheet during day time to increase the digester temperature. During night, the same should be covered with gunny bags to minimize heat losses
- ☆ A fraction of the digester material may be recycled along with fresh slurry to improve the bacterial population in the digester

pH: The most common pH problem is one of acidity as acids are produced during anaerobic processes in a digester. The pH of the contents in the digester tank should be as close to neutral as possible. A simple litmus test on the digester contents can be used to identify the problem. If the test results show pH below 7, addition of a small amount of lime or grounded lime stone can help to normalize the pH. Note that excessive amounts of lime may not dissolve in the mixture causing harm to the bacteria. Therefore, lime concentration should never exceed 500 mg/L of mixture in the digester tank.

Other problems: A number of other problems can arise during the life time of a digester. To investigate such problems, recall the basics of what makes a digester work (organic material, strong seals, and warmth). Eliminate anything that could possibly affect its functioning. For example, never introduce unnecessary chemicals into the tank. Never use the dung of the livestock that have recently been given antibiotics or other medications. The chemicals present in the dung may harm the bacteria in the digester. Also, make sure to use non-corrosive materials for handling the gas and water. Cement and plastic cause no harm to the mixture in the tank, but metals should be avoided in the tank, and for tubing through which the biogas travels. Colour of the flame at the burner may also indicate problems in the functioning of the digester. A blue flame means that the digester is producing quality methane. If the flame is yellow, the level of carbon dioxide in the biogas is high. Too much carbon dioxide in the digester could hint at a digestion process problem. It could also cause malfunction for the equipment fueled with digester gas. Some possible causes/defects and remedies to improve biogas production are reported in Table 7.5 (Biogas World, 2022).

Table 7.5. Possible problems/defects, their causes and possible remedies

Problems	*Causes*	*Remedial Measures*
No gas after the first filling of the plant	Lack of time	Wait for 3- 4 weeks
The feeding material does not enter into the digester	Blocked inlet pipe	Take out the inlet pipe for inspection or open the inlet pipe
	The position of the inlet pipe is below the overflow point	Ensure right vertical dimensions are used
Reduced gas yield	The position of the inlet pipe is below the overflow point	Ensure right vertical dimensions are used
	Reduction in feedstock quality	Check the level of potential inhibitor compounds
	Temperature drops in the digester	Check the heating system or take precautions as listed in a previous section
	Inhibition by foreign compounds	Mix bio-slurry from another digester to improve methanogenic bacteria which might have dropped
	Non-homogenous substrates	Use homogeneous substrate
Methane concentration drops	Reduced feedstock quality	Assure substrates mixing and quality
	Temperature drop	Check the heating system or take precautions as listed in a previous section
	Inhibition by foreign compounds	Check the composition of the inhibitor compounds in the feedstock
Foaming problem	A new substrate with high protein content has been added	Reduce or stop feeding the digester with such kind of substrate
	Air is introduced in the digestion chamber	Reduce the introduction of air in the digester
pH drop in the digester	Feeding rate is too high or variable	Reduce the input in the biogas plant until the system returns to normal
	Introduction of foreign compounds like soap	Check thoroughly what is entering the digester

Problems	Causes	Remedial Measures
	Inhibition by foreign compounds	Analyze the chemical composition of the feedstock and the input entering the digester
	Agitation system is not working	Ensure proper agitation
Gas yield has dropped (Insufficient gas pressure or maximum pressure not reached)	Underfeeding the plant	Follow the feeding instructions
	Gas leaks	Check for gas leaks
	Scum formation in the digester	Check the digester climate/ constituents
	Improper mixture ratio of water and dung	Follow normal mixing ratio preferably 1:1. or 1:0.75. Do not make the mixture too thin or too compact
	Siltation in the digester	Clean the digester every 5-6 years
Bio-effluent smelling at the expansion chamber and the channel	Overfeeding of the digester	Follow feeding instructions
	Leakage in the digester	Check for gas leaks
The feeding material does not enter into the digester	Blocked inlet pipe	Take out the inlet pipe for inspection or open the inlet pipe
Thick bio-slurry overflow	The position of the inlet pipe is below the overflow point	Ensure right vertical dimensions are used
	Incorrect water/dung mixing ratio	Operate with proper mixing ratio
	No hydraulic movement in the digester	Ensure daily use of the gas to allow hydraulic movement
Slurry entering the gas pipe	Water leakage in the digester	Check for leakages
	Water leakage in the expansion chamber	Check for cracks in the digester and the mixing and expansion chamber
	Gas outlet pipe placed below the overflow point	Check slurry overflow point
		Reduce slurry overflow point to lower level
		Put a filter to allow only gas to pass through, blocking the slurry in the process
Low gas quantity in the digester	Underfeeding/irregular feeding of the digester	Follow feeding instructions
	Scum formation in the digester	Proper mixing and removal of impurities
		Remove the scum, agitate the mixture

Problems	Causes	Remedial Measures
No gas reaching the appliances	Blocked water passage in the pipeline system	Check the presence of water
	Leakages in the system/pipeline	Check for leaks
	Disconnected gas pipeline	Check for pipe and connect it back
	Closed main gas valve	Check the main gas valve
Flame dies quickly	Insufficient pressure	Check quantity of gas
		Increase gas pressure by breaking the scum by stirring the slurry
Gas does not burn	Wrong kind of gas	Add properly mixed slurry
Flame far from burner	Pressure too high or deposition of carbon on the nozzle	Adjust gas outlet valve and clean nozzle

COOK STOVE

The most useful application of biogas is for heating and lighting in homes. An efficient cook stove or a lamp is essentially required for these activities. A scientifically designed burner will provide efficient combustion, improve heat transfer leading to reduced fuel demand, improve health of women and children resulting from reduced emissions and lower cost of cooking. Initially tin burners were used for cooking purposes. In these burners gas was supplied from base pipeline. Perforations were provided at upper portion of the tin to enable the gas to spurt out where it used to burn. A number of small sized stones were kept outside to facilitate improved air and gas mixing. Tin burner could also be easily embedded inside the earthen *'chulah'* to suit flat and round bottomed cooking vessels. These burners were quite inefficient. Currently, either LPG stoves or KVIC designed burners (utility type and deluxe biogas burners, Figure 7.8 left) made of cast iron with injectors made of gun metal are commonly used. While LPG stoves for biogas are less efficient, KVIC burners are designed keeping in view the characteristics of the biogas. As a result, these cook stoves offer high efficiency, and clean cooking energy solutions. A detailed categorization of cook stoves is given in Table 7.6. The categorization reported in this table is based on several factors such as technology, fuel type, feeding, and design type *etc.* For the purpose of this section, we are only concerned about biogas cook stoves.

Combustion of Gas in the Burner

The combustion of gas in a burner involves proper mixing of air with fuel gas, and burning of the resultant air-gas mixture. The chemical reaction of combustion of biogas (containing 60 per cent methane and 40 per cent carbon dioxide) and air mixture is given as:

$$0.6\,CH_4 + 0.4\,CO_2 + 1.2\,O_2 + 4.5\,N_2 \rightarrow CO_2 + 1.2\,H_2O + 4.5\,N_2 + \text{Energy}$$

Table 7.6. General categorization of cook stoves for biomass fuels

Basis	*Categorization of Traditional/Improved Stove*
Technology and evolution	Traditional
	Improved
	Advanced
Fuel type	Wood
	Charcoal
	Agro-residue
	Dung cake
	Pellets/Briquette
	Miscellaneous and multi-fuel
Feeding	Batch feeding
	Continuous feeding
	Mouth feeding
	Side feeding
Design type	Natural draft
	Force draft
	Direct combustion cook stoves
	Gasifier based cook stoves
	Cook stoves with chimney
	Cook stoves without chimney
Material	Mud
	Stone/brick
	Metallic
	Non-metallic
	Hybrid
Pots	Single
	Double
	Triple and miscellaneous
Demography	Domestic
	Community size
Functions	Single function say cooking
	Multi-function say cooking, heating *etc.*
Portability	Fixed
	Portable

Thus, one volume (0.6+0.4) of biogas requires 5.7 volumes of air (1.2+4.5). It means that the stoichiometric requirement is 1/(1+5.7) = 0.149, *i.e.*, 14.9 per cent biogas in air. The biogas in different kinds of burners burns over a narrow range of mixtures from approximately 9 per cent to 17 per cent of biogas in air. If the flame is 'too rich', *i.e.*, has too much fuel, then it will burn badly and incompletely, giving poisonous carbon monoxide and soot particles. Therefore, the design of appliances aims at to maximize the conversion of methane into carbon dioxide in order to minimize the release of unburned methane and products of incomplete combustion. Stoves usually run slightly lean with a small excess of air to avoid the danger of the flame becoming rich. The amount of energy released on complete combustion of biogas is about 20-24 MJ/m^3 as compared to pure methane which releases 36 MJ/m^3 of energy.

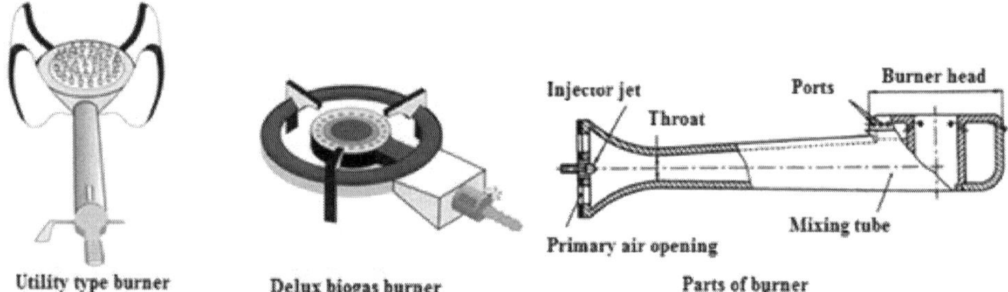

Figure 7.8. KVIC utility burner (left), deluxe burner (middle) and parts of burner (right).

KVIC Burner

A typical biogas stove consists of gas supply tube, gas tap/valve, gas injector jet, primary air opening(s) or regulator, throat, gas mixing tube/manifold, burner head, burner ports (orifices), pot supports and body frame (Figure 7.8). KVIC biogas stoves are single or double burner types, varying in capacity and consume 0.22 to 0.44 m^3 of gas per hour or more. These two types of burners are adequate for 1-2 and 5-6 member family, respectively. Although gas pressure in gas holder and burners vary from one plant to another, the burners are mostly designed for average conditions assuming initial gas pressure to be 8.8 cm water column and in burner from 7.6 cm and 8.3 cm water column. KVIC designed biogas burners are also available in sizes suitable for commercial use.

A scientifically designed stove should meet the following criteria.

- ☆ Gas inlet pipe should be smooth to minimize resistance to flow of gas and air
- ☆ Spacing and size of air holes should match with the requirement of gas combustion
- ☆ Volume of burner manifold should be large enough to allow complete mixing of gas with air
- ☆ Size, shape and number of burner port holes should allow easy passage of the gas-air mixture, formation of stabilized flame and complete

combustion of gas, without causing lifting up of flame, off the burner port or flame back flash from burner port to gas mixing tube and injector jet

- ☆ The flame should be self stabilizing *i.e.*, flameless zones must re-ignite automatically within 2 to 3 seconds
- ☆ Under ideal conditions, the pot should be cupped by the outer cone of the flame without being touched by the inner cone
- ☆ Size and shape of the burner in terms of diameter of jet, length of intake holes measured from the end of jet, the length of mixing pipe and its diameter and number and diameter of flame port holes should result in high efficiencies

For quality certification purposes National Standard Biogas Stove Specification, 2002 (revised edition) of the Bureau of Indian Standards (BIS) is in force in India at present (BIS, 2002). It provides information on construction, operation and safety requirements and tests for stoves intended for use with biogas up to 708 liters per hour flow rate. Following three shortcomings have been reported by the consumers who are using various types of biogas stoves.

- ☆ Air intake is small and it is not adjustable
- ☆ High heat flow with low heat efficiency
- ☆ Big diameter of injector jet

DESIGN CONSIDERATIONS FOR BIOGAS BURNERS

The following rules should be observed while designing a biogas stove:

- ☆ Air must get thoroughly mixed-up with gas before it reaches the flame ports and should flow near the gas jet ideally with a venturi (In the absence of sufficient air there will be some smell from burning gas)
- ☆ Total area of the flame ports should be between 8-200 times the area of the gas jet. It may be noted that biogas stoves have large flame port diameters than LPG stoves
- ☆ Distance from the flame ports to the surface of the cooking pot should be between 2.5 cm to 3.0 cm
- ☆ Supports for cooking pots should not prevent air from getting to the flame
- ☆ To allow cross-lighting from one flame port to the next, the distance between flame ports should not be more than 2.0 cm
- ☆ Cast iron is preferred for its corrosion-resistance properties

The orifice area per unit flow rate (A_0/Q) should be about 0.01 mm^2 hr/L for gas pressures in the range of 5-10 cm WG. For higher pressures orifice diameter is calculated with the help of the following equation.

$$Q = 3.6\, C_d A_0\, [2g\, (\rho_w/\rho_g) h_g]^{1/2} \tag{7.6}$$

Here Q is the gas flow rate, L/hr, C_d is the coefficient of discharge (about 0.6), A_0 is the area of the orifice, cm^2, g is the acceleration due to gravity, 981 cm/s^2, ρ_w is the density of water, 1000 kg/m^3, ρ_g is the density of gas, 1.128 kg/m^3 and h_g is the pressure of the gas, cm WG (cm of water gauge). Two 8 mm diameter holes with an adjustable check-nut over them should be provided immediately after the orifice for air entry. The mixing chamber should be about 15-20 cm long. For high pressures, the length can be increased. The total area of the flame ports should be about 300 times the orifice area for proper flame propagation. Thus, the diameter of each hole can be decided once the total number of holes is fixed. A total of 12 holes should be reasonable. The distance of the flame ports to the base of the utensil should be about 2.5 cm. The thermal efficiency shall be at least 55 per cent for each burner when the pan is supported correctly on the burner (BIS, 2002).

Sustainability of flame is another important test to adjust the air to fuel ratio. It shall be possible to operate the stove with taps open at gas pressure of 747 N/m^2 (7.62 gf/cm^2) without flame either blowing off or striking back. If the flame does not sustain but is put off as soon as the match stick is removed, it can be concluded that either air supply or fuel supply is in excess. Some specifications for a 0.45 m^3 per hour biogas stove for domestic use are as follows:

Jet size	2.25 mm diameter
Area of jet	3.98 mm^2
Flame port size	6.0 mm diameter
Number of ports	20
Total area of ports	565 mm^2
Ratio of jet area to flame port area	1:142
Length of gas mixing pipe	20 mm
Diameter of gas mixing pipe	20 mm

Performance

Performance of cook stoves varies greatly across different models and designs. Below are few parameters that give an idea of the performance of cook stoves either qualitatively or with a measurable target.

- Cooks 1 kg of rice in no more than one hour
- Weighs no more than 5 kg
- Achieves 30 per cent thermal efficiency
- Uses materials available within the country
- Additional cost of cook stove is recovered through fuel savings in less than one year

- ☆ Parts of the stove that are most likely to fail can be easily replaced
- ☆ Aesthetically pleasing to potential users
- ☆ Achieves a biomass stove safety score of at least 88

These parameters can be grouped under the heads namely performance, affordability and usability as listed below.

Performance	*Affordability*	*Usability*	
Energy efficiency	Sale price	Time saving	Portability
Minimum pollution	Low unit cost	Weight	Easy maintenance and servicing
Minimum greenhouse gas emissions	Long service life	User interface	Cleanliness
Safety	Low fuel consumption	Ease of ignition	Attractiveness
Durability		Tending requirement	
Time for the operation			

GAS LAMP

The principal of a gas lamp is similar to that of a gas stove. In a lamp, the burning gas heats a mantle until it glows brightly. The maximum light flux values of 400-500 lumens can be obtained with biogas. The values correspond to luminous efficiency in the range of 1.2 to 2.0 lm/W. It is equivalent to a normal 25-75 W incandescent light bulb. The key factors that determine the luminous efficiency are the type and size of mantle, the inlet gas pressure and the fuel-air mixture. The hottest inner core of the flame should match exactly with the form of the mantle (Figure 7.9). If the mantle body is too large, it will show

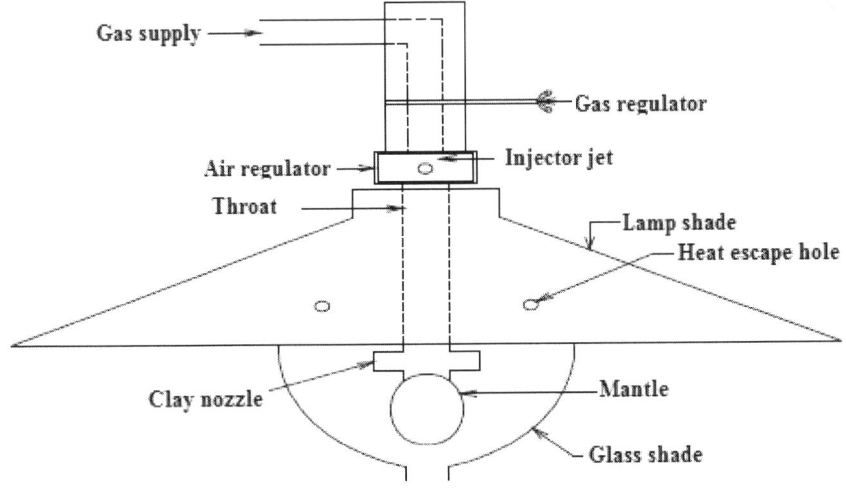

Figure 7.9. Schematic diagram of a biogas lamp.

dark spots. If the flame is too large, then the gas consumption will be too high for the same light flux yield. The performance of the lamp is also dependent upon gas pressure. Lighting is poor for the inlet gas pressure below 75 mm WG (Nijaguna, 2006). The lighting is satisfactory if the gas pressure is above 100 mm WG and excellent at 150 mm WG.

QUESTIONS (THEORY)

1. Define biogas plant. Name and discuss its various sections.
2. Draw a neat diagram of KVIC biogas plant. Discuss its various components and working.
3. List the advantages and disadvantages of floating dome biogas plants.
4. List the advantages and disadvantages of fixed dome biogas plants.
5. Draw a neat diagram of fixed dome Janta type biogas plant. Discuss its various components and working.
6. Draw a neat diagram of fixed dome Deenbandhu biogas plant. Discuss its various components and working.
7. How will you proceed to select a site for the location of a biogas plant on the farm?
8. A farmer has 10 cows. Proceed to design the volume, diameter and height of the digester. Assume that the normal guidelines are applicable.
9. How will you proceed to design the volume, diameter and height of the gas holder of a 4 m^3 capacity biogas plant? Assume that the normal guidelines are applicable.
10. List the advantages and disadvantages of biogas slurry.
11. List the daily, monthly and annual maintenance and care steps for the upkeep of a biogas plant.
12. List the general characteristics of a good cook stove for biogas.
13. Write short notes on plug flow digester, applications of biogas, handling of manure, resolving the temperature problems, and gas lamp.

Chapter 8

Vegetable Oils

A question often arises whether we can use vegetable oils (VO) or leftover oils from cooking and other processes as fuels in diesel vehicles without any processing? Such questions had often come up in the past and crop up every now and then. Thought of using vegetable oils as fuel came up as early as in the year 1900. The inventor of diesel engine, Rudolf Diesel tried to use peanut oil to energize the engine at the Paris exposition. Besides the technical feasibility, idea of using vegetable oils as fuel could not get acceptance, as the oils remained normally expensive than petroleum fuels. This retarded the scientific efforts to investigate further acceptability of vegetable oils as fuel. Later numerous factors such as OPEC oil embargo of the '70's, fuel shortages and rapidly escalating fuel prices created renewed interest in researches on vegetable oil as substitute fuel for diesel engines. Certain advantages associated with vegetable oils as fuels are listed as follows.

- Vegetables oils are available everywhere in the world
- They are renewable because the crops which produce oil seeds can be planted year after year
- They can be prepared locally on-farm
- They are greener to the environment, as they seldom contain sulphur element in them. Moreover, vegetable oils derived from crop and forest resources are carbon neutral, making them as future diesel alternative
- Liquid nature with easy portability
- Heat content (80 per cent of diesel fuel)

The disadvantages associated with vegetable oils are:

- Higher viscosity
- Lower volatility
- The reactivity of unsaturated hydrocarbon chains
- Gum formation due to oxidation and polymerization during storage and combustion

It is now emerging that certain vegetable oils up to 15 per cent and certain commodity oils up to 100 per cent can be used in industrial applications. It may be noted that efforts to use vegetable oils in tractors have also been made in the past.

RAW MATERIALS

Biomass-derived feedstock for producing liquid biofuels as per source has been classified into three categories, *i.e.* triglyceride based biomass, starch-and-sugar derived biomass, and cellulosic biomass. A variety of liquid biofuels can be produced from triglycerides based biomass such as vegetable oils, animal fats, waste cooking oils and microalgae oils. Over 350 oil bearing crops and 70 species of oilseed crops have been listed in the literature having reasonably good oil production potential. While seed yields ranged from 200 to 14000 kg/ha, oil yield has been reported to vary from 8 to 73 per cent. The oils have further been categorized as edible and non-edible oils (Table 8.1). The most predominant edible oils such as sunflower, safflower, soybean, cotton, winter rape, canola, and peanut have been used as fuel substitutes. In a developing country like India edible vegetable oils are in short supply and have applications in day to day life. Therefore, the emphasis has to be on non-edible vegetable oils as alternative fuel. Attempts have been made to use oils like *Karanja, Neem, Mahua, Jatropha, etc.* either as partial or complete replacement of diesel in compression ignition (CI) engines. Another classification of vegetable oils is virgin vegetable oils and recycled vegetable oils. Virgin oils also termed as pure plant oils or straight vegetable oils are extracted from plants solely for use as fuel. Production of vegetable oils for use as fuels is theoretically limited only by the agricultural capacity of a given region. But it may impact the supply of pure vegetable oils from other uses resulting in their price escalation. Virgin vegetable oils possess almost the same heat values as diesel fuel, but limitations arise on account of their inherent high viscosity. High viscosity may lead to poor atomization of the fuel, incomplete combustion, coking of the fuel injectors, ring carbonization, and accumulation of fuel in the lubricating fuels. Note that fuel-injection systems in modern diesel engines are quite sensitive to viscosity changes. Recycled vegetable oils, also termed used vegetable oils (UVO), waste vegetable oils (WVO), used cooking oils (UCO), or yellow greases are recovered from businesses and industry that use oils for cooking. Prospects of used vegetable oil for use as fuel are limited by the capacities of such industries and logistic problems of collection *etc.* Most discussion in this chapter centers on virgin vegetable oils except one separate section that deals with used vegetable oils

Table 8.1. Edible and non-edible vegetable oils having potential as fuel

Edible Vegetable Oils	*Non-edible Vegetable Oils*
Peanut	*Jatropha curcas*
Safflower	Karanjia (*Pongamia glabra*)
Palm	*Pongamia pinnata*
Soybean	*Mahua*
Sesame	*Neem*
Rapeseed/Canola	Pine seeds
Sunflower	Castor
Mustard	*Nagchampa*
Linseed	*Kusum*
Coconut	*Ark* (*Calotropis gigantia*)

OILS AND FATS

Oils and fats are water insoluble, hydrophobic substance primarily composed of fatty esters of glycerol (triacylglycerol). Chemically, triacylglycerol (TGs) are the triacylglyceryl esters of various fatty acids with glycerol. These are mainly hydrocarbons made up of glycerol (backbone) with three fatty acids chemically bonded to it - triglycerides (Figure 8.1). R_1, R_2, and R_3 in this Figure represent the aliphatic chains of the fatty acids. There is no difference when we say triacylglycerol, a correct chemical name for an ester derived from glycerol bound to three fatty acids, and triglyceride, a common name for the substance.

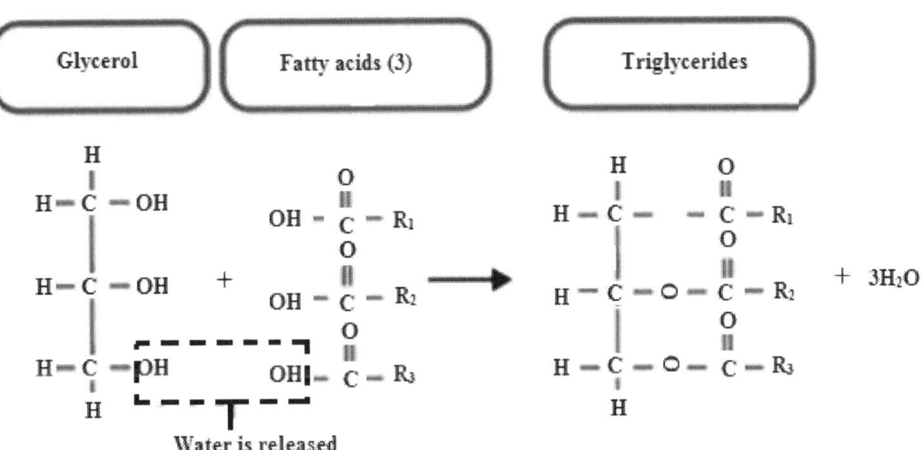

Figure 8.1. Structural notation of vegetable oil.

Triglycerides are the main constituent of the vegetable and animal fats such as peanut oil, sunflower oil, corn oil, butter, lard and tallow. Triacylglycerol that are liquids at room temperature are generally called as oils and those that are solids are called as fats. Triacylglycerols can be simple triacylglycerols (all three acyl groups are the same) or mixed- triacylglycerol (acyl groups are different). Triacylglycerols molecules have molecular weights between 800 and 900 and are thus nearly four times higher than typical diesel ($C_{16}H_{34}$) fuel. Due to their higher molecular weight, VOs have low volatility, but are inherently more reactive than diesel fuels because of their unsaturation. As a result they are more susceptible to oxidation and thermal polymerization reaction.

CHARACTERISTICS OF FUELS

Acid value: The total acid number is an indication of the presence of free fatty acids formed due to oil degradation and combustion. It can also result from improper manufacturing as a result of remaining catalyst or excessive neutralization.

Ash percentage: Ash is a measure of the amount of metals contained in the fuel. Ash constituents, such as vanadium, nickel, sodium, aluminum and silicon are harmful for the engine, and their presence increases engine wear and corrosion.

Calorific value: Also known as heating value or heat of combustion is *the amount of heating energy released by the combustion of a unit value of fuels.*

Carbon residue: An important indicator of the quality of biofuel is the carbon residue, which corresponds to the content of glycerides, free fatty acids, soaps, polymers and remaining catalyst.

This indicates the tendency of fuel to form carbon deposits in the engine.

Cetane number (CN): *A relative measure of the interval between the beginning of injection and auto ignition of the fuel.* Long-chain, unbranched, saturated hydrocarbons (alkanes) have high CNs and good ignition quality while branched hydrocarbons (and other materials such as aromatics) have low CNs and consequently poor ignition quality. High CNs can cause operational problems such as combustion occurs before the fuel and air are properly mixed, resulting in incomplete combustion and smoke. Too low CNs may cause engine roughness, misfiring, high air temperatures, slower engine warm-up and incomplete combustion. Most engine manufacturers designate a range of required CNs for their engines. In most cases, CN range is around 45-50 while vegetable oil has 35 to 45.

Cloud point (CP): *The temperature at which oil starts to solidify*. Heating will be necessary to avoid waxing of the fuel for operating an engine at temperatures below oil's cloud point.

Density: Density is the weight per unit volume. Denser oils contain more energy. For example, gasoline and diesel fuels give comparable energy by weight, but diesel is denser and hence gives more energy per liter. Biodiesel is generally denser than diesel fuel ranging from 877 kg/m^3 to 884 kg/m^3 compared with diesel at 835 kg/m^3 (Verma and Pandey, undated).

Flash point (FP): *The flash point temperature of the fuel is the minimum temperature at which the fuel will ignite (flash) on application of an ignition source.* Flash point varies inversely with the fuel's volatility. Minimum flash point temperatures are required for proper safety and handling of fuel. Neat biodiesel has a flash point (150°C) well above the flash point of petroleum based diesel fuel (±70°C).

Heating value: *Heating value or heat of combustion is the amount of heating energy released by the combustion of a unit value of fuel.* One of the most important determinants of heating value is the moisture content, heating value decreasing with increasing moisture content. Whereas air dried biomass typically has about 15-20 per cent moisture, the moisture content for oven-dried biomass is negligible.

Iodine value (IV): *The iodine value is a measure of the total unsaturation of vegetable oils. It is also an indicator of their susceptibility to oxidation.* Iodine value is measured by the amount of iodine in grams that reacts with 100 mL of given oil. On the basis of their iodine value vegetable oils have been divided into four categories: saturated oils (iodine value between 5 and 50), mono-unsaturated oils (50 and 100), di-unsaturated oils, also called semi-siccative (100 and 150) and tri-unsaturated oils called siccative (over 150).

Melt point or pour point: *Melt or pour point refers to the temperature at which oil in the solid form starts to melt or pour.* If the temperature falls below the melt point, the entire fuel system including all fuel lines and fuel tank will need heating.

Sulphur percentage: *It is the percentage of sulphur in the fuel by weight.* First use vegetable oil and animal fat based biodiesel has less than 15 ppm sulphur. Many researchers claim that pure biodiesel is essentially sulphur free and is considered to be an ultra-low sulphur fuel. Sulphur content in the diesel fuel used in on-road applications is limited by law to very small percentages.

Viscosity: *Viscosity represents flow characteristics and the tendency of fluids to deform with stress.* It is determined by measuring the amount of time taken for a given measure of oil to pass through an orifice of a specified size. Viscosity affects fuel atomization and injector lubrication. Diesel fuels with high viscosity tend to form larger droplets upon injection which can cause poor combustion, increased exhaust smoke and emissions. Fuels with low viscosity may not provide sufficient lubrication for the precision fit of fuel injection pumps, resulting in leakage or increased wear.

The characteristics of some edible and non-edible oils are reported in Table 8.2 and Table 8.3 respectively along with diesel (Bettis *et al.*, 1982). Note some variations in the values of diesel in the two Tables because they may represent different categories and different manufacturing processes. Perusal of these Tables reveal that vegetable oils compared with diesel in general have acceptable cetane number (37-58), high viscosity, high carbon residue, high flash point (174-293), cloud point and pour points, acceptable calorific values (88-94 per cent of diesel), and low sulphur content (< 0.02 per cent).

Table 8.2. Selected physical properties of vegetable oils as they relate to their use as fuel

Oil or Fat	*Iodine Value*	*CN*	*HG (kJ/kg)*	*Viscosity (mm^2/s) at 38°C*	*CP (°C)*	*PP (°C)*	*FP (°C)*
Castor	82-88	-	39500	29.7	-	-31.7	260
Corn	103-140	37.6	39500	34.9	-1.1	-40.0	277
Cotton seed	90-119	41.8	39468	33.5	1.7	-15.0	234
Linseed	168-204	34.6	39307	27.2	1.7	-15.0	241
Peanut	80-106	41.8	39782	39.6	12.8	-6.7	271
Rapeseed	94-120	37.6	39709	37.0	-3.9	-31.7	246
Safflower	126-152	41.3	39519	31.3	18.3	-6.7	260
High-oleic safflower	90-100	49.1	39516	41.2	-12.2	-20.6	293
Sesame	104-120	40.2	39349	35.5	-3.9	-9.4	260
Soybean	117-143	37.9	39623	32.6	-3.9	-12.2	254
Sunflower	110-143	37.1	39575	37.1	7.2	-15.0	274
No. 2 DF	-	47	45343	2.7	-15.0	-33.0	52

CN = Cetane number, HG= Total enthalpy, CP = Cloud point, PP = Pour point, FP = Flash point.

Table 8.3. Characteristic and composition of few non-edible oils and diesel

Properties	*Jatropha*	*Karanja*	*Polanga*	*Rubber*	*Mahua*	*Neem*	*Diesel*
Saturated fatty acid							
Palmitic ($C_{16:0}$)	16.0	11.7	12.0	10.2	17.8	14.9	–
Stearic ($C_{18:0}$)	6.5	7.5	12.9	8.7	14.0	14.4	–
Total	22.5	19.2	24.9	18.9	31.8	29.3	–
Unsaturated fatty acid							
Oleic ($C_{18:1}$)	43.5	51.6	34.1	24.6	46.3	61.9	–
Linoleic ($C_{18:2}$)	34.4	16.5	38.3	39.6	17.9	7.5	–
Linolenic ($C_{18:3}$)	0.80	2.7	0.30	16.3	–	–	–
Total	78.7	70.8	72.4	80.5	64.2	69.4	–
Cetane number	52.3	55.8	57.3	–	40	57.8	46
Oilseed content	55	33	65	40–50	50	44.5	–

Properties	*Jatropha*	*Karanja*	*Polanga*	*Rubber*	*Mahua*	*Neem*	*Diesel*
FFA, per cent weight	14	2.5	22	17	20	–	–
Specific gravity	0.92	0.91	0.90	0.91	–	–	0.84
Viscosity 40°C, mm²/s	18.2	27.8	72.0	76.4	24.6	–	7.50
Flash point	174	205	221	198	232	–	50
Calorific value, Mj/kg	38.2	34.0	39.3	37.5	36	–	42.25
Iodine value	93.0	80.9	93.8	135.3	74.2	69.3	38.3

FFA = Free fatty acid.

Raw oils are refined as per the following processes mainly to remove some impurities and to improve their self life *etc.*

- Removal of free fatty acids
- Degumming - removes mucilaginous material
- Bleaching - removal of pigments
- Deodorized - steam heating
- Winterize - prevents clouding by chilling oil and filtering out particles
- Hydrogenation - yields vegetable lards, margarine and cheese substitutes

COMBUSTION CHEMISTRY, EMISSIONS AND DEPOSITS

Pre-combustion of Triglycerides

When a diesel fuel is injected into the combustion chamber of a diesel engine, it takes some time before it is ignited. *The brief period after injection of the fuel into the combustion chamber but before the onset of ignition is termed as pre-combustion phase.* The fuel during this phase is subjected to chemical reactions while it passes through the pre-combustion temperature and pressure gradients until the conditions for ignition are attained. During the pre-combustion phase, numerous classes of compounds, including alkanes, alkenes, cycloaliphatics, and aromatics, and oxygenated species, such as aldehydes, shorter-chain esters, and furans are formed from the fatty compounds of the fuel. Some of these species formed during this period greatly influence the subsequent combustion phase. The formation of low-CN compounds during pre-combustion may contribute to the lower CN of some fatty compounds, especially more highly unsaturated ones. This has been demonstrated most conspicuously for aromatics. Therefore, enhanced understanding of pre-combustion of fatty compounds can aid in the development of techniques such as cetane improvement for reducing emissions, especially NO_x. In general, decreasing the ignition delay time (pre-combustion phase) of a fuel can raise the CN because less CN-decreasing intermediates are formed.

Combustion

Ideally, hydrocarbons on complete combustion produce carbon dioxide (CO_2) and water as per equation shown for alkanes or saturated hydrocarbons.

$$C_nH_{2n+2} + (1.5n + 0.5)O_2 \rightarrow nCO_2 + (n + 1)H_2O$$

But combustion process in a diesel engine is always incomplete because it mainly occurs through a diffusion flame. This causes the formation of partially oxidized materials such as carbon monoxide (CO), other oxygenated species such as aldehydes, and hydrocarbons. In general, exhaust emissions of neat vegetable oils have similar constituents in some emission categories, compared with conventional DF. For example, exhaust emissions observed in the combustion of conventional DF and biodiesel are smoke, particulates (particulate matter), polyaromatic hydrocarbons (PAHs), hydrocarbons, CO, and oxides of nitrogen (NO_x; also referred to as nitrous oxides, or nitrogen oxides). An important difference is lack of sulphur-containing emissions from biodiesel emissions since it lacks sulphur. Some additional problems associated with emissions from neat vegetable oils are listed as under.

- PAH emissions from neat vegetable oils are lower than conventional DF emissions, especially very little amounts of alkylated PAHs. PAHs are compounds composed of fused aromatic rings that may carry alkyl substituent such as a methyl group. They are known carcinogens
- Emissions from neat vegetable oils have higher NO_x levels arising from reaction of nitrogen and oxygen from air at early stages of the combustion process. It is often difficult to control NO_x emissions because such techniques may increase other constituents in the emissions or fuel consumption
- There is dramatic increase in total aldehydes with vegetable oils as fuel. Formaldehyde formation is consistently higher than with DF2
- Component TGs in vegetable oils can lead to formation of aromatics via acrolein (CH_2=CH-CHO) from the glycerol moiety
- Significantly lower emissions of C3 aldehydes (for example, acrolein) has been reported for methyl esters of rapeseed oil than for the oil itself
- Hydrocarbons represent a broad category of compounds including hydrocarbons and oxygenated species such as aldehydes, ketones, ethers, *etc.* Increased emissions of aldehydes and ketones with vegetable oils as fuels are attributed to the formation of acidic water during decomposition of the oils. This acidic water is an indication of the formation of short-chain oxygenates, which are likely to ignite poorly compared to the long-chain carbon-rich fatty compounds

☆ Particulate matter in emissions with rapeseed methyl ester as fuel in DI engines showed large amounts of volatile and extractable compounds adsorbed on the soot, which caused the particulate emissions to be higher than with conventional DF. The composition of particulate matter from conventional DF on the other hand has a high carbon to hydrogen ratio of approximately 10:1. Thus, particulates are mainly carbon in the form of crystallites.

In conclusion, it can be said that understanding both the pre-combustion and combustion processes of the fuel is essential to suppress undesirable emissions and deposits. Modification in any of the two phases can help to improve the combustion properties and emissions from VOs/biodiesel.

PROBLEMS WITH NEAT VEGETABLE OIL AND THEIR EASTERS AS FUEL

Most studies have shown that neat VOs give satisfactory engine performance and power output. In short-term trials, performance often equaled or was even slightly better than conventional DF. However, several kinds of problems encountered in long-term use resulted in reduced engine life. Some such problems are of general nature while others are specific to particular oil. Vegetable oils cause serious damage to combustion systems due to formation of deposits, and downgrade the performance and emission quality, especially in direct injection engines. The molecules cracks due to high molecular weights and low volatility of the oils during fuel vaporization and combustion processes resulting in the formation of deposits. The high viscosity of the oils radically modifies the phenomena associated with spraying of the fuel and the combustion timing, which is also disturbed by the low cetane number of these oils. Note that cetane number is about 30 to 40, which is well below the 51 limit imposed by Diesel standard EN 590. Long-term use of vegetable oils often results in coking (chemically degraded components of the fuel and combustion products adhere to the internal surfaces of an injector) and trumpet formation on the injectors. Besides, other problems encountered are: carbon deposits around the nozzle orifice, upper piston ring grooves and on piston rings, oil ring sticking, fuel filter blockages, crankcase oil dilution, and thickening and gelling of the lubricating oil as a result of contamination by the vegetable oils. Some of the problems mentioned above will be further amplified with improvements in diesel engine technologies (smaller injector nozzles with more injectors). Such modifications may result in higher risk of clogging. Some of these problems are discussed in the following sections.

Polymerization

Polymerization is any process in which relatively small molecules, called monomers, combine chemically to produce a very large chainlike or network molecule, called a polymer. Polymerization of vegetable oils can occur both during storage as well as during its use as a fuel. Most problems have been traced to the polymerization of TGs via their double bonds that leads to formation of engine deposits. An oxidative free-radical mechanism was observed to govern TG polymerization. The polymerization of the vegetable oil in the ring belt area results in the rings to seize. It is often associated with increase in blowby, a corresponding increase in the viscosity of the lubricating oil and resulting catastrophic failure of the engine. Coking of injector nozzles severely limits engine performance by limiting the amount of fuel delivered to the combustion chamber and by altering the spray pattern. Following remedial measures have been suggested.

- If the vegetable oil is used in direct injection diesel engines without modifications, its use should be limited to short-term operations. Long-term operation results in operational and durability problems
- Blend the VO with diesel fuel. The blend should not contain more than 25 per cent VO
- Use transesterified vegetable oil that may pose lesser problems and may even completely replace diesel fuel

Problems due to High Viscosity

Most engine problems with VOs are traced to higher viscosity as listed below.

- High viscosity and low volatility of VOs is responsible for poor atomization, which results in incomplete combustion and carbon deposits
- It interferes in the fuel injection and contributes to combustion nozzle clogging
- High viscosity causes excessive engine deposits and ring sticking
- Contaminations of lubricating oil
- High fuel viscosity causes premature wear of the fuel pumps and injectors
- It dramatically alters the structure of the fuel spray coming out of the injectors by increasing droplet size, decreasing spray angle, and increasing spray penetration. These effects tend to increase wetting of the engine's internal surfaces, thereby diluting the engine lubricant and increasing the tendency for coking
- High viscosity of oils causes higher fuel line pressure compared with conventional fuel. This reduces the required seating pressure of the injector needle valve creating a situation where post injection dribble may occur

Problems in Association with other Properties

Few problems occur due to high viscosity in association with other related properties.

- Diesel indices of the vegetable oils are poor for some of the oils *e.g.* sesame (24.9), coconut (24.5)
- In the presence of phosphide in oils it can absorb atmospheric moisture to form gum in the fuel system and the cylinder. It is separately discussed in the following section
- Published literature strongly indicates that use of VOs reduce diesel engine life because of the build-up of carbon deposits inside the engine and the accumulation of VOs in the engine lubricant. These issues are attributable to VO's high viscosity and high boiling point relative to the required boiling range for diesel fuel
- Most researchers have reported fuel filter plugging with vegetable oils requiring appropriate precautions or frequent replacement of filters
- Problems also arise as VOs are much more reactive to oxygen, and have higher cloud point and pour point temperatures
- Vegetable oils contain significant amounts of oxygen impacting ignition characteristics such as poor cold engine start-up, misfire, and ignition delay. Latter results in incomplete combustion, *e.g.* deposit formation, carbonization of injector tip, ring sticking, lubricating oil dilution and degradation, polymerization during storage

Gumming of Engines

Crude vegetable oils are obtained by pressing and/or extracting oil from the seeds that contain several compounds other than triglycerides. Some of these compounds such as diglycerides, tocopherols, sterols and sterol esters need not necessarily be removed during refining. Other compounds such as phosphatide, free fatty acids, odours, colouring matter, waxes and metal compounds must be removed because they affect taste, smell, appearance and storage ability of the refined oil. When the vegetable oils are used as fuel in the diesel engine, gum present in these oils causes gumming of engines comprising plugging of the engine filters, lines and injectors. Besides, gumming may result in decline in engine power. The use of crude vegetable oils without degumming in transesterification for biodiesel production decreases conversion rate and poses problems while separating glycerol from biodiesel following the transesterification reaction.

VOs, their Easters and Engine Emissions

Vegetable oils are biodegradable, far less toxic than other fuels such as gasoline, petroleum-based diesel, ethanol, or methanol having a potential to significantly reduce pollution. The VOs have much higher flash point that reduces the risk of accidental ignition. Use of vegetable oils as fuel in diesel engines generally produces high concentrations of CO, HC and PM because of the need to adapt to the combustion time. In contrast, NO_x emissions are low as poor volatility and lower heating value of vegetable oil gives lower premixed combustion resulting in lower combustion temperatures as compared to diesel fuel. Vegetable oils and their derivatives in diesel engines help to reduce the emissions of sulphur oxides, carbon monoxide (CO), poly aromatic hydrocarbons (PAH), smoke, particulate matter (PM) and noise. The results may vary from one kind of VO to other and/or whether it is a methyl or ethyl ester. Few examples are listed to illustrate the issue. Emissions of CO and hydrocarbons were lower with methyl and ethyl soyate, a product prepared from soy oil and methanol or ethanol as fuel, but NO_x were consistently high than the conventional DF. NO_x emissions were slightly higher, but hydrocarbon, CO, particulate and PAH emissions for rapeseed oil methyl ester were in ranges similar to the DF reference. The differences in exhaust gas temperatures corresponded with the differences in NO_x levels. The rapeseed esters emitted less aldehydes than the corresponding neat rapeseed oil. Unrefined rapeseed methyl ester emitted slightly more aldehydes than the refined ester, while the opposite was true for PAH emissions. A 31 per cent increase in aldehyde and ketone emissions has been reported when rapeseed methyl ester was used as fuel, mainly due to increased acrolein and formaldehyde, while hydrocarbons and PAHs were significantly reduced, NO_x increased slightly, and CO was nearly unchanged. PAH emissions of sunflower ethyl ester were somewhere in between DF and the corresponding neat vegetable oil. Reduced PAH emissions may correlate with the reduced carcinogenetic nature of particulates with rapeseed methyl ester as fuel. The general trend on reduced emissions except NO_x was confirmed by later studies, although some studies report little changes in NO_x. In a DI engine, sunflower methyl ester produced equal hydrocarbon emissions but less smoke than a 75:25 blend of sunflower oil with DF.

IMPROVING THE ENGINE PERFORMANCE

Engine performance with VOs as fuel can be improved by several means. These means are grouped in two categories namely improvement in oil characteristics or modifications in the engine. In the latter category redesigning the engine or its major components is not desirable being a costly alternative. Therefore, only minor modifications that help to overcome the problems are included in this category.

Vegetable Oil Improvement

Choice of oils: Vegetable oils vary widely in characteristics. Therefore, choice of particular oil can be made to minimize the problems in engines. Palm and peanut oils are more relevant for regions with hot climates. Oils with fewer double and triple bonds in their fatty acids, *i.e.* less unsaturated vegetable oils have fewer tendencies to polymerize.

Additive to minimize fuel filter plugging: Addition of certain additives to the vegetable oil can improve certain characteristics of the fuel to resolve several problems. Hydrogen peroxide or kerosene is used to improve the physico-chemical properties of vegetable oils. Fuel filter plugging can be reduced by addition of DuPont FOA 2. It has been observed that there was fourfold reduction in filter plugging with the addition of this material.

Cetane enhancers: Cetane enhancers reduce the ignition delay, and lower the premixed fuel combustion temperature resulting in lower NO_x in emissions, which are mainly formed at high temperatures in the combustion chamber. In a study conducted in a turbulence combustion chamber at an intake air temperature of 105°C, 8 per cent hexyl-nitrate in vegetable oils (cotton seed, rape, and palm) exhibited the same ignition delay as conventional DF. Various compounds such as alkyl nitrates have been used as cetane-enhancing additives in conventional DF. Only few studies on such compounds in biodiesel exist.

Changes in feedstock properties: Changes in the feedstock properties give a more engine-friendly fuel. It can be achieved by applying different methods such as: preheating, blending with diesel fuel, transesterification, cracking/ pyrolysis or emulsification. As an example, lowering the viscosity of vegetable oils can be achieved to a greater extent by adopting techniques such as heating, dilution, emulsification, *etc.*

Pre-heating: Viscosity of oils is a function of the temperature. Thus, pre-heating of the fuel is commonly used to reduce the viscosity. A heating device warms up the fuel and decreases its viscosity to levels close to diesel fuel before the fuel is injected into the system. Among all the methods, pre-heating is most favourable, least time consuming, very feasible and quite cheap. Change in kinematic viscosity of few oils before and after heating is shown below.

Vegetable oil	*Kinematic Viscosity (mm²/s)*	
	Before Heating	*After Heating*
Coconut oil	31.11	5.44
Mustard oil	36.59	6.58
Palm oil	49.56	6.71
Sesame oil	40.35	7.09
Kusum oil	42.08	5.9 (Heated to 110°C)
Kusum oil	42.08	3.6 (Heated to 120°C)

Dilution: The principle of dilution was experimented in 1980 in Brazil to solve some of the major problems related to use of vegetable oils as engine fuel. Vegetable oils are diluted in varying proportions with diesel, solvents or ethanol with or without the addition of other additives. Dilution reduces the viscosity and the density of vegetable oils. Depending upon the mixing ratio, the viscosity and density of the blended fuel will be somewhere in between the viscosities and densities of two neat components. For example, viscosity of kerosene is 2.25 times lower than that of peanut oil at 20°C and 1.55 times at 100°C. Thus, the viscosity of the mixture of peanut oil and kerosene will be lower than the pure peanut oil. A mixture of 30 per cent diesel fuel, 30 per cent kerosene, 40 per cent peanut oil has the density of about 850 kg/m^3 at 20C°, which is very similar to the diesel fuel. This mixture can be used as a fuel without making additional adjustments in the engine fuel system. Initially, a mixture of 10 per cent vegetable oil with conventional fuel was used without any alterations or modifications in the engine. Later on, it emerged that a blend of 20 per cent vegetable oil and 80 per cent diesel can be effectively used. Some short-term experiments even used a 50-50 ratio of VO and conventional fuel. A model study showed that blends of DF2 with 0 to 34 per cent vegetable oil can result in proper atomization. Addition of 4 per cent ethanol to diesel fuel has been found to increase the brake thermal efficiency; brake torque and brake power, but decreased the brake specific fuel consumption. Different oils may have to be blended with different blending percentages to achieve comparatively similar performance. If the final mixture still has high viscosity, then it may not be possible to use it in unmodified engines. Research on the direct use of straight vegetable oils (SVOs) and several blends such as vegetable oil-diesel blend, vegetable oil-alcohol blend and vegetable oil-alcohol-diesel blend on diesel engine reveal that brake specific fuel consumption (BSFC) for all blends fall between that of SVO and diesel. SVO-diesel blends containing less than 50 per cent SVO have comparable brake thermal efficiency (BTE) to that of diesel fuel. Generally, NO_x emitted by SVO, SVO-diesel and SVO-alcohol blends are lower or comparable to that of diesel fuel. Blends of SVO-diesel-alcohol are found to reduce smoke, especially at high loads. It is found that CO and unburnt hydrocarbon emissions vary depending on engine specification, dynamometer operating condition and types of vegetable oil used. Overall, results from this technology have been mixed and not much change in engine problems has been noticed compared with neat vegetable oils. Data reported in Table 8.4 give some overview to arrive at some conclusion on the utility of dilution.

Table 8.4. Various blends, their major properties and recommendations

Sl.No.	*Blend*	*Property*	*Remarks*
1.	75:25 (volume per cent) petro-diesel/sunflower oil blend	Viscosity 4.88 mm^2/s at 40°C	Not recommended for long-term use in the diesel engine
2.	75:25 (volume per cent) petro-diesel/high-oleic safflower oil blend	Viscosity 4.92 mm^2/s at 40°C	Passed 200 hr engine manufacturer's association test May lead to lubricant thickening
3.	70:30 winter rapeseed oil/DF1 mixture or blends of ≤ 15 per cent rapeseed oil with DF2, and an 80: 20 DF2/safflower oil blend	-	Reduced CO and hydrocarbon emissions

Sl.No.	Blend	Property	Remarks
4.	75:25 DF/crude sunflower oil blend	-	Produced greatest solids contamination in the lubricating oil
5.	80:20 DF/sunflower oil blends	-	Exhaust smoke increased after prolonged use due to carbon build-up, which also resulted in power loss as well
6.	50:50 blend of Stoddard solvent*with soybean oil	Low CP (-18.9°C) and PP (-31.7°C)	Performance was not as good as a diesel engine with DF2

*A dry-cleaning fluid, viscosity 0.95 mm²/s, estimated CN 50, heat of combustion 46,800 kJ/kg, CP < -16°C, PP < -35°C, flash point 42.2°C

Several other studies have revealed that the best blending is B10 for *Kusum*, B40 for *Karanja* oil (Acharya *et al.*, 2009), B20 Mustard and *Neem* oils, B80 for *Jatropha* oil, and B20 for Linseed oil. Properties of these blends are found to be closest to properties of diesel fuel. Soybean oil having less carbon residue, better DI (diesel index) and good flash point has been suggested as the best for blending with diesel.

Emulsification: Vegetable oils do not form uniform air/fuel mixture because of their low atomization potential. They are also less volatile and burn at a higher temperature than diesel. The average flame temperature of the most commonly used VO fuels (sunflower, cotton seed, soybean, corn, opium poppy, and rapeseed) are over four times greater than that of diesel. The concept of diluting vegetable oil with ethanol, another agricultural based energy source, may be possible for on-farm preparation of fuel. However, mixing of vegetable oil and ethanol is confronted with the difficulty of phase separation. *Emulsification is the action, process or method for obtaining emulsions. Two non-miscible phases are blended using a substance known as surfactant, a surface active component that under certain experimental conditions leads to a stable system that acts as a single phase.* For example, 1-butanol is used as a non-ionic surfactant to disperse the water and alcohol in vegetable oil to form a micro-emulsion. Note that preparation of an emulsion produces no by-products but consumes surfactant and improves fuel properties. Data in Table 8.5 reveal that all the fuel properties of the emulsion changed with emulsification compared to the neat honge oil.

Table 8.5. Properties of diesel, honge oil and micro-emulsions

Properties	Diesel	Neat Honge Oil	Emulsion of Honge Oil and Diesel
Viscosity in Cst	4.25	40.25	25.78
Flash point, °C	79	190	42
Fire point, °C	85	210	50
Calorific value, kJ/kg	42700	37258	34633
Specific gravity (at 25°C)	0.830	0.925	0.900

Waste Oils and Emulsification: The waste cooking oils and fatty acid distillates are available around the world. These are produced on a daily basis but have low commercial value and little use. Therefore, these are managed in different ways depending on the kind of the waste oil. Two most important challenges for the vegetable refining oil, and food industries are to reduce generation of the waste products and to find solutions for the use of the wastes. Emulsification is one of the techniques to treat these waste products to make them suitable for use in engines (Senthil and Jaikumar, 2014). Summarized results from the review of literature reveal significant decreases in exhaust emissions of nitrogen oxides, cylinder pressure, brake specific fuel consumption, hydrocarbon, smoke opacity, carbon monoxide, particulate matters from emulsified waste cooking oils and fatty acid distillates compared with diesel fuel. The brake thermal efficiency with emulsified vegetable oil improved being very close to diesel. Emission percentage of various components improved with water/oil emulsions as compared to pure oil (Table 8.6). Based on engine performance and exhaust emissions, it is concluded that fatty acid distillates and waste cooking oils can be used as engine biofuels through emulsification.

Table 8.6. Comparison summary between water/oil emulsions and oil only fuels

Emissions	*Emission Percentage Improvement using Water/ Oil Emulsions as Compared to Pure Oil*
NO_x	0-10 per cent less than oil
CO	44 per cent less than oil
Unburned hydrocarbons	60 per cent less than oil
Particulate	50-90 per cent less than oil
Smoke	60-80 per cent less than oil

Degumming of vegetable oils: *Degumming is the process of removing phospholipids or gums from the crude vegetable oils.* The conventional degumming processes are: hydratable or water degumming and non-hydratable or acid degumming. Water degumming process comprises transfer of the feedstock into the degumming vessel where the temperature of oil is slowly raised. Water or steam (*e.g.* 3 per cent water for soybean oil) is added to hot crude oil (*e.g.* 70°C). The resulting gum layer formed after a contact time of about 5 min is separated from the oil by centrifuging and processed into commercial lecithin. The resulting degummed oil has considerably lower phosphorus content than the original crude oil yet it still contains non-hydratable phosphatides (NHPs). These NHPs are commonly removed during alkali refining. This unit operation comprises the dispersion of an acid, *e.g.* phosphoric acid in water degummed oil (or even the crude oil). The amount of phosphoric acid required depends on the amount of gums present as well as the type of oil being refined. Thereafter, slight excess of caustic soda liquor is added. It is followed by the separation of the soaps. The soap stock so obtained contains free fatty acids originally present in the crude or water degummed oil, some triglyceride oil, the NHPs and other mucilaginous compounds such as sucrolipids and lipoproteins. Fatty acids contained in the soap can be recovered before its disposal. The pollution potential of the soap stock is considerably reduced with this step.

The acid degumming process is an alternative to the water degumming process. Crude oil, either water degummed or not, is treated by a degumming acid, usually phosphoric acid, citric acid or malic acid in the presence of water. For oils containing relatively low amounts of NHP (*e.g.* sunflower oil) this process can lead to degummed oil with a lower residual phosphorus content (5 to 30 ppm) than water degumming. The acid degummed oil is then dry degummed and physically refined. The process is usually carried out at elevated temperature of around 90°C. Ultra-filtration or dry gumming is another potential technology for degumming, although it is a difficult process since most oil constituents have nearly similar molecular weight.

Esterification and transesterification: Cracking and refining effectively upgrade the quality of vegetable oils, but with a substantial increase in the cost of fuel. Moreover, the step cannot be practiced for direct on-farm utilization of the harvested vegetable oil. *Esterification is any reaction typically between a fatty acid and an alcohol that results in the production of an ester.* If oil feedstock contains an excess of free fatty acids (FFA) they need to go through the process of esterification to convert FFA to esters (usually with strong mineral acids such as H_2SO_4) as per the following reaction.

$$ROH + R'\text{-}COOH \rightarrow R'COOR$$

Here ROH is an alcohol, R'COOH is a fatty acid and R'COOR is an ester. The key to effective esterification process is to make sure that the feedstock has been sufficiently filtered to remove all contaminants. The feedstock is then fed to the acid or resin esterification process. The catalyst sulphuric acid dissolved in methanol is then mixed with the pretreated oil. The final step of esterification is to dewater and feed the product to the transesterification process to produce biodiesel.

Transesterification is the reaction of an ester with an alcohol to replace the alkoxy group or simply the conversion of component TGs to simple alkyl esters using alcohol. A generic transesterification reaction in the presence of a catalyst is given as:

$$ROH + R'COOR'' \rightarrow R'OH + RCOOR''$$

Here R'COOR'' indicates an ester, ROH an alcohol, R'OH another alcohol (glycerol), and RCOOR''an ester mixture. Base catalysts such as sodium hydroxide are preferred over acid catalysts in the transesterification process because of the rapidity of the reaction. The main difference between esterification and transesterification is that esterification includes an ester as the end product whereas transesterification includes an ester as a reactant. When methanol is used in the transesterification process, the product of the reaction is mixture of methyl esters. If ethanol is used in the transesterification process, the reaction product will be mixture of ethyl esters. In both cases, glycerin will be the co-product of the reaction. These two processes are discussed in greater details in chapter 9. Readers may refer this chapter for more information.

Easters of vegetable oils improve the injection process and ensure better atomization of the fuel in the combustion chamber because of their significantly lower viscosities than the parent oils and fats. The polymerization reaction is also decreased. A possibly more benign advantage is avoidance of formation of undesirable acrolein. Transesterification to reduce viscosity-related problems of vegetable oils is more popular because of its ease and rapidity of the reaction. Besides, methyl esters are the most popular esters for several reasons. One reason is the low price of methanol compared to other alcohols. The popularity of methyl esters has contributed to the term biodiesel that refers to vegetable oil esters. The results from biodiesel as a fuel are much better than pre-heating of VOs. Few studies on transesterification of *Jatropha* oil into its methyl ester show that kinematic viscosity decreased from 35.98 mm^2/s to 4.84 mm^2/s, and from 50 mm^2/s to 5.51 mm^2/s. On an average, viscosity ranged in between 5 mm^2/s to 7 mm^2/s. In general, most emissions parameters except the NO_x are reduced with esters compared to the conventional DF

Vegetable oil refining: Vegetable oils can be used as feed stocks in an oil refinery. There it can be transformed into fuel either by hydrocracking (which breaks big molecules into smaller ones using hydrogen) or hydrogenation (which adds hydrogen to molecules). These methods are capable of producing gasoline, diesel, or propane. Some commercial examples of vegetable oil refining are Neste NExBTL, Topsoe HydroFlex technology, Axens Vegan technology, H-Bio, ConocoPhilips process, and the UOP/Eni Ecofining process. Diesel fuel so produced is known as green diesel or renewable diesel. The processing technologies and chemical make-up of the fuel is quite different than biodiesel. The green diesels can be introduced into any diesel engine or infrastructure without many mechanical modifications at any ratio with petroleum-based diesels.

Engine Modifications

To use vegetable oils as fuel in the engines, several investigators explored the possibility of modifying the engines rather than attempting to modify the characteristics of VOs. Some of these modifications are an offshoot of altering the fuel characteristics. For example, heaters are provided on the fuel line of the engine to preheat the VO prior to its injection into the engine. The overall performance of this system depends on the heating system, the engine, the type of vegetable oil (thinner is easier), and the climate (warmer is easier). Similarly, modification has been made for dual fuel operation. A valve system is used to prepare a blend of two fuels before their introduction into the diesel engine. These modifications might mitigate the problems to some degree but don't eliminate them entirely as discussed in the previous sections of this chapter.

Modification has been made to increase the injector valve opening pressure (IVOP) to overcome the problem of larger droplet diameters and lower injection velocities. The Increase in IVOP decreases fuel spray droplet diameter, increases velocity and penetration distance resulting in improved engine performance besides improving the emission characteristic. Note that initial average fuel spray droplet diameter is inversely related to the square of the velocity. Reports have emerged to show that this modification improved brake thermal efficiency, ignition, and combustion stability with various VOs. But it has its own problems. Increasing the IVOP beyond a point becomes counterproductive because of increased spray penetration resulting in wall impingement. For this reason, a particular engine's IVOP needs to be tuned for the specific VO used.

Advancing the injection timing is another minor modification employed to help the performance and emissions of VO fuelled diesel engines. Now engine modification kits are available at least for slow speed, stationary, indirect-injection diesel engines. The kit allows use of waste vegetable oil as fuel with its performance similar to the diesel fuel. The modification kit includes the tuning of the injector pressure and timing to better atomize given the unique properties of VOs.

Some attempts have also been made to design engines that can better handle the VOs. The Elsbett engine is designed to run on straight vegetable oils. However, even by 2007, no vehicle has been produced that allow straight burning of vegetable oils. In Germany, some large haulage vehicle fleets are adopting the VOs as fuel, but it is largely for economic reasons. Deutz and John Deere are both trying to develop engine that can run on VOs. A German standard DIN 51605 has been developed for grape seed oil fuel.

ON-FARM PRODUCTION OF VEGETABLE OILS

Vegetable oils can be produced on small scale for on-farm utilization to run tractors, pumps and small engines for power generation, and irrigation purpose *etc.* Two major issues that need attention from this angle are:

- On-farm oil recovery from the oil seeds should be as inexpensive as possible to compete with the fuel prices
- To identify optimum parameters of press operation and processing associated with transesterification

Commonly used methods of oil expression from the oil seeds are as follows.

1. Grinding with stones
2. Cold pressing – high quality
3. Steam driven stone press - hot pressing
4. Screw press - continuous feed
5. Solvent extraction
6. Combination of 4 and 5

Amongst all, the last three appears to be most promising methods for on-farm oil extraction. Screw pressing using a mechanical screw press is quite simple and appears to be the most appropriate technique for on-farm use. The second method is solvent extraction. In this method, oil seeds are crushed and soaked in a solvent such as hexane. The mixture is filtered and hexane is allowed to evaporate leaving the oil in the container. The complexity and hazards associated with solvent extraction limit the application of this technique for small on-farm processing plants. The last option combines the solvent extraction with a light screw press operation. This process is most commonly used in large commercial oil seed extraction plants.

Filtration of extracted oil is second most important step. In this process, extracted oil is allowed to settle in a holding tank for a minimum of about 2 days. It is then filtered through a series of elements ending with 4 to 5 micron filtration. The process can be simplified by using a simple filtration system consisting of throw away filters and a final fuel filter of a diesel engine. Small amounts of phosphatides called gums and free fatty acids from vegetable oils can be removed by hydrating with steam or hot water. Even the last step may not be required in some cases such as rape seed oil. Storage, the least difficult task, is the last step. It is possible to maintain the vegetable oil quality for long periods with reasonable care during storage.

QUESTIONS (THEORY)

1. List the advantages and disadvantages of vegetable oils as fuel for diesel engines.
2. Name various characteristics of fuels and explain at least 5 of them.
3. Discuss the role of oil viscosity while using oils as fuel in engines.
4. What do you understand by gumming of engines? Briefly discuss the degumming operations of vegetable oils.
5. Name various options to improve the engine performance through modification in oil characteristics.
6. Discuss various options of engine modification to improve their performance with vegetable oils as fuel.
7. Write a short note on on-farm preparation of vegetable oil for small engines.
8. Differentiate between edible and non-edible oils, oil and fat, cloud point and flash point, esterification and transesterification.

Chapter 9

Biofuels

Any fuel derived from biomass that is plant or algae material or animal waste is known as biofuel. In its most general sense, biofuels are all types of solid, gaseous and liquid fuels that can be derived from such biomass. Examples of solid biofuels include wood, charcoal and bagasse. Examples of gaseous biofuels include methane gas and producer gas. Examples of liquid biofuels include methanol, ethanol, plant oils and the methyl esters produced from these oils commonly known as biodiesel. Two major commercially available liquid biofuels are ethanol and biodiesel. The focus of this chapter is on liquid biofuels that are used primarily as transport fuel. Although it mainly covers biofuels from plant oils, yet the production of ethanol from sugar crops, starchy crops, and cellulosic biomass is briefly discussed to provide a complete overview.

FEEDSTOCK FOR BIOFUELS

The feedstock used for producing biofuels has been classified into three groups namely cellulosic biomass, sugar and starchy crops, and oil-containing or oil-producing plants.

Cellulosic Biomass

Cellulosic biomass comprises a wide range of heterogeneous solid materials including, but not limited to: (a) agricultural residues comprising leftover material from crops such as rice straw, or coconut coir, or the stalks, leaves and husks of corn plants; (b) forestry wastes such as chips and sawdust from lumber mills, dead trees, and tree branches; (c) municipal solid wastes; (d) food processing and other industrial wastes such as slops from alcohol distilleries and black liquor from pulp and paper manufacturing; and(e) energy crops grown for fuel purposes such as fast growing trees and grasses. The main components of cellulosic biomass are cellulose, hemicelluloses, and lignin. Both cellulose and hemicellulose can be converted to simple monomeric sugars, largely glucose and xylose, for the production of ethanol via fermentation. Although the name sugar is most often used to refer to sucrose or table sugar, in general, sugars are water-soluble carbohydrates that have relatively low molecular weight and are characterized to have sweet taste.

Depending on the source of biomass, cellulose may account for 40 per cent to 60 per cent by weight of the biomass and is thus the most common form of carbon in the biomass. It is a complex sugar polymer, or a polysaccharide, and is made from six-carbon sugar called glucose. Because of its crystalline structure, it is resistant to hydrolysis, the chemical reaction that enables the production of simple, fermentable sugars from a polysaccharide. Hemicellulose is also a major source of carbon in the biomass and may account for 20 per cent to 40 per cent by weight of the biomass. It is also a complex polysaccharide that is made from a variety of five-carbon and six-carbon sugars. Although it is relatively easier to hydrolyze it into simple sugars compared to cellulose, the sugars so produced are not easily fermented to ethanol. Lignin comprises from 10 per cent to 24 per cent by weight of biomass and provides structural integrity and strength in plants. It remains as the residual material after the sugars in the biomass have been converted to ethanol. As a highly complex carbon-containing polymer, it contains a lot of energy and can be burned to produce steam and electricity for use in the biomass-to-ethanol manufacturing process. The cellulosic biomass is increasingly becoming important for producing biofuels. Its attractiveness as a feedstock results from the fact that it is made up of very complex sugar polymers that are not usually used as a source of human food.

Biomass from Sugar and Starchy Crops

The second group of feedstock is sugar and starchy crops. These are plants such as sugar cane and sugar beets that can store the energy from the sun by converting it into simple sugars through the photosynthesis process. Carbohydrates, on the other hand, are a group of organic compounds that include sugars, starches, celluloses and gums. These compounds are also produced through photosynthesis by plants and contain only molecules of carbon, hydrogen and oxygen, usually in the ratio 1:2:1. Whereas simple sugars are called monosaccharides, more complex sugars comprise between two and ten monosaccharides linked together. As examples, dissacharides are those that contain two monosaccharides, trisaccharides contain three and so on. These biomass products such as corn, cassava and sweet potato are mainly used as human or animal food, being the major sources of energy in the diet. Notwithstanding this statement, these products are increasingly being diverted for the production of biofuels, particularly ethanol as gasoline substitute or for blending with gasoline. It is in consequence of the need to find alternative sources of energy to replace fossil fuels.

Oil-Containing or Oil-Producing Plants

The third group of feedstock is composed of oil-containing or oil-producing plants. A large number of plants are capable of producing oils, in particular fixed oils. *Fixed or vegetable oils are esters of fatty acids, dense and non-volatile.* Examples of vegetable oils are sunflower oil, sesame oil, olive oil, almond oil, coconut oil and many more. The common characteristic of these oils is that they may be decomposed into glycerin and one or more acids of the type called fatty acids. The physical properties commonly found in these plant oils are the following:

- They float on water but are not soluble in it
- They are greasy and have lubricating properties
- They are not readily volatile
- They may be burned without leaving any significant residue

Most of these oils such as soybean oil, coconut oil and palm oil so far have been mainly used for human or animal food or production of various types of cosmetics and pharmaceuticals. Increasing amounts of these and many more oils are now being processed to produce biofuels. The biofuels so produced are used either as petrol/diesel substitute or for blending in the petrol/diesel. Ethanol for example has become one of the most widely used biofuel as it can be blended with gasoline. In the same manner, biodiesel can be blended with or used as a substitute for diesel. It is why; their production and use have multiplied. The additional advantages of using biofuels lie in reducing emissions of air pollutants, reduced build-up of greenhouse gases and reduced dependence on imported oil. Besides, the cultivation and harvesting of this kind of feedstock will have positive impacts on the agricultural and rural economies. Microalgae are likely to provide a very lucrative alternative for future biodiesel production due to their capability of giving high oil yield. Nonetheless only some species of microalgae are useful for biofuel production.

BIODIESEL

Direct use of vegetable oils and/or their blends have generally been considered unsatisfactory and impractical for both direct and indirect diesel engines (See chapter 8). Therefore, attempts have been made to produce various kinds of fuel oils such as biodiesel, ethanol, and methanol *etc.* (https://www.canr.msu.edu). The term biodiesel has no unambiguous definition. It stands for neat vegetable oils used as DF as well as neat methyl esters prepared from vegetable oils or animal fats and blends of conventional diesel fuel with vegetable oils or methyl esters. With increasing emphasis on the use of esters as DF, the term biodiesel increasingly refers to alkyl esters of vegetable oils and animal fats and not the oils or fats themselves. Thus, *biodiesel is a generic name for fuels obtained by transesterification of a vegetable oil. Since, it is derived from natural (biological) renewable resources, it is named as biodiesel.* The term biodiesel appears to have been first used in a Chinese paper published in 1988. The next paper using the term appeared in 1991 and since then the use of the term biodiesel in the literature has expanded exponentially. *As per ASTM standards, biodiesel has been defined as "the mono alkyl esters (methyl or ethyl) of long chain fatty acids derived from renewable lipid feedstock, such as vegetable oils or animal fats, for use in compression ignition (diesel) engines designated as B100"*. Small amounts of hydrocarbon or non-hydrocarbon additives may be incorporated to improve ignition or other characteristics. The material shall be free from grit, suspended matter and other visible impurities including any residual oil. Dyes or markers are allowed for the purpose of identification. Unlike fossil diesel, pure biodiesel is biodegradable, non-toxic, and essentially free of sulphur and aromatics and releases less emissions during combustion. Blends with diesel fuel are indicated as "Bx", where "x" is the percentage of biodiesel in the blend. For instance, "B5" indicates a blend with 5 per cent biodiesel and 95 per cent diesel fuel; in consequence, B100 indicates pure biodiesel. Fuel blends consisting of 6 to 20 volume per cent of biodiesel with remainder being a light middle or middle distillate grade diesel fuel can be used without any major modifications in the engines.

Biodiesel is sometimes called FAME (fatty acid methyl ester) or FAEE (fatty acid ethyl ester). It can be produced from edible oils such as palm oil, soybean oil, rape seed oil, sunflower oil and some other vegetable oils; animal fats and non-edible oils like *jatropha*, castor beans, and *pongamia pinnata*. In Europe, rapeseed oil is the major feedstock used to make biodiesel, although sunflower oil is also used to some extent. Soybean oil is the major feedstock for biodiesel production in the USA. On the contrary, use of edible oils to produce biodiesel is not encouraged in China because China imports more than 400 million tonnes of edible oils annually to satisfy its consumption needs. Indian situation is no different. India imports about 65 per cent of its requirement of edible oils consumption. Attempts are being made to make biodiesel from recycled waste oils but the scale of production is limited due to high operating costs.

Advantages of Biodiesel

- ☆ Produced from sustainable, renewable and domestic biological sources
- ☆ Eco-friendly biodegradable with minimum adverse environmental consequences of biofuel spills
- ☆ Emissions are sulphur free with less carbon-monoxide (CO), hydrocarbons (HC), particulate matter and aromatic compounds compared to diesel as reveled by the data reported in Table 9.1
- ☆ Reduced tail pipe emissions and global warming
- ☆ Income generation for rural communities
- ☆ Fuel properties similar to the conventional fuel
- ☆ Higher flash point (100°C minimum)
- ☆ Can be used in existing unmodified diesel engines

- Reduced expenditure on oil imports
- Non toxic and safe to handle
- It is energy efficient (The energy efficiency of biodiesel is 320 per cent versus 80 per cent of petroleum diesel)
- It is a good solvent and may clean out fuel line and tank sediments (this may result in fuel filter clogging during initial replacement period from diesel to biodiesel)
- Used cooking oils and fat residues from meat processing may be used as raw materials

Table 9.1. Emission constituents of B100 and B20 compared to petro diesel

Emission	*B100*	*B20*
Carbon monoxide	-48	-12
Unburned hydrocarbon	- 67	-20
Fines	-47	-12
NO_x	+10	+2
Sulphates	-100	-20
Air toxics	-60 to -90	-12 to -20

Limitations of Biodiesel

- The purity of biodiesel changes during storage due to oxidative and hydrolytic reactions
- Availability of feedstock (raw material) for production may pose problems
- Biodiesel contains approximately 8 per cent less energy per gallon
- It has a higher cloud and pour point (will freeze at a higher temp) than conventional diesel
- It is not compatible with some hose and gasket materials, which may cause them to soften, degrade, and rupture
- It may increase nitrogen oxide emissions
- It is less stable than diesel fuel, and therefore long-term storage (more than six months) of biodiesel may pose problems
- It dissolves the deposits of sediments and other contaminants from diesel fuel in storage tanks and fuel lines, which can enter the engine through the biofuel

The disadvantages are significantly reduced when biodiesel is blended with diesel fuel.

METHODS OF PRODUCTION

Biodiesel can be produced from edible vegetable oils, non-edible vegetable oils or animal fat and various other biofuels such as methanol, ethanol *etc.* Step-wise procedure of producing biodiesel includes the following.

- ☆ Treatment of raw materials
- ☆ Alcohol-catalyst mixing
- ☆ Chemical reaction
- ☆ Separation of the reaction products
- ☆ Purification of the reaction products

Following three methods are commonly used to convert vegetable oils to CI engine fuel.

- ☆ Pyrolysis
- ☆ Micro-emulsification
- ☆ Transesterification

Pyrolysis

Pyrolysis is the conversion of one substance into another by means of heat with or without the aid of a catalyst. Pyrolytic chemistry is difficult to define because of the variety of reaction paths and the variety of reaction products formed during the reactions. The liquid fractions of the thermally decomposed vegetable oils are likely to approach diesel fuels. The material used in this process can be vegetable oils, animal fats, natural fatty acids and methyl esters of fatty acids. It involves heating in the absence of air or oxygen so as to crack the chemical bonds to yield small molecules. In 1947, tung oil saponified with lime was thermally cracked to yield crude oil, which was refined to produce diesel fuel and small amounts of gasoline and kerosene. The pyrolyzate has a lower viscosity, flash point, and pour point than diesel fuel and equivalent calorific values. The cetane number of the pyrolyzate is lower than diesel fuel. The pyrolyzed vegetable oils contain acceptable amounts of sulphur, water and sediments and give acceptable copper corrosion values. It has unacceptable ash, carbon residual and pour point. For more details of this process readers may refer chapter 4.

Micro-emulsification

Vegetable oils have high viscosity which should be reduced before they can be used in standard engines (See chapter 8). The micro-emulsification is one solution to solve the problem of high viscosity. *Micro-emulsion is defined as a mixture of two immiscible liquids to form homogenous and stable solution stabilized by surfactant(s) or combination of surfactant(s)* (the substance when added to a liquid reduces its surface tension, thereby increasing it's spreading and wetting properties). In practice, micro-emulsion is made from vegetable oils with an ester and dispersant (co-solvent), or from vegetable oils, alcohol, a surfactant and a cetane improver. All micro-emulsions with butanol, hexanol and octanol meet the maximum viscosity requirement for diesel fuel.

Transesterification

Transesterification is the process in which fat or oil reacts with an alcohol to form esters and glycerol. Biodiesel is produced by transesterification of large, branched triglycerides into smaller straight chain molecules of methyl esters using an alkali or acid or enzyme as catalyst. Although the transesterification reaction can be catalyzed by either acids or bases, the base-catalyzed reaction is more common. A triglyceride has a glycerin molecule as its base with three long chain fatty acids attached (Figure 9.1). The characteristics of the oil/fat are governed by the nature of these fatty acids attached to the glycerin. The nature of the fatty acids in turn also affects the characteristics of the produced biodiesel. The reaction besides Figure 9.1 is also explained through Figure 9.2. Key parameters to achieve high conversion efficiency in the transesterification reaction are the content of free fatty acids, water and non-saponificable substances in raw materials. Amongst the various kinds of alcohols (methanol, ethanol, propanol, butanol and amyl alcohol) methanol (CH_3OH) and ethanol (C_2H_5OH) are frequently used in the transesterification process. Methanol because of its physical and chemical characters can quickly react with triglycerides and sodium hydroxide as both are easily dissolvable. Moreover, the emulsions formed have higher stability with ethanol than methanol. Therefore, it is much easier and quicker to destabilize the emulsion with methanol during phase separation and purification of biodiesel. The most common form uses methanol to produce methyl esters, FAME and ethanol to produce an ethyl ester, FAEE. Higher alcohols such as isopropanol and butanol have also been used but are less common. Transesterification can be carried out by several ways including common batch process, supercritical processes, ultrasonic methods, and microwave methods. A by-product of the transesterification process is the glycerol. For every tonne of biodiesel manufactured, about 100 kg of glycerol is produced. The by-product glycerol

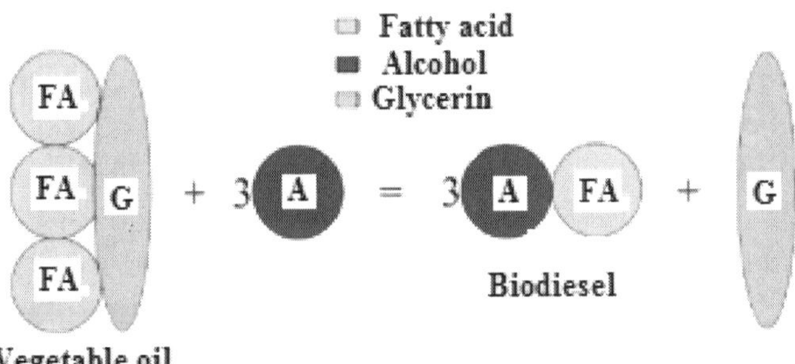

Figure 9.1. Schematic view of the chemical reaction during formation of biodiesel.

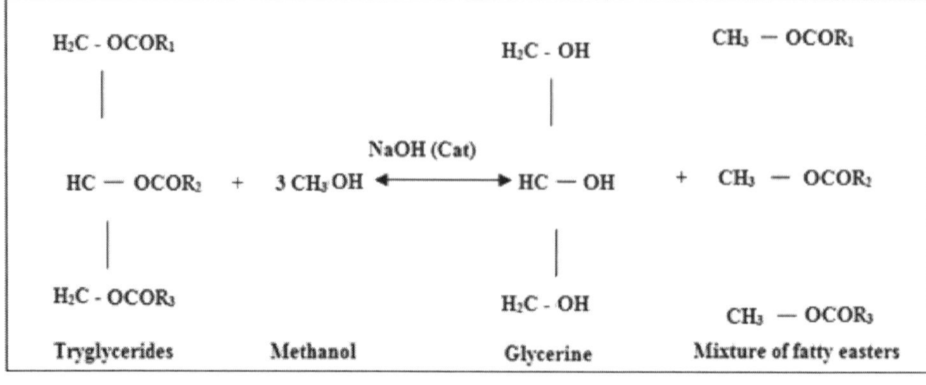

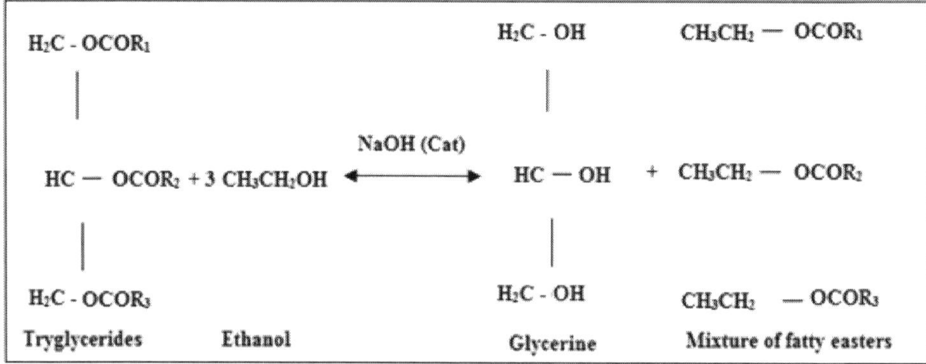

Figure 9.2. Basic transesterification reactions: with methanol (top) and with ethanol (bottom).

is used in the pharmaceutical, food and cosmetics industries. Some important steps of the transesterification process are as follows.

Alcohol-catalyst Mixing

The water-free (anhydrous) alcohol is used in the transesterification process. Due precautions must be exercised, and all applicable safety regulations must be followed when working with methanol, hydroxides and methoxides. Catalyst is first mixed with the alcohol. The mixture is stirred until the catalyst is completely dissolved in the alcohol. Thereafter, oil is added to this mixture, the alcohol-to-oil volume ratio being the key variable of the transesterification process.

The stoichiometric ratio requires 1 mol of oil to react with 3 mol of alcohol to produce 3 mol of fatty acids methyl esters (FAME) and 1 mol of glycerin. In practice, an alcohol excess of 100 per cent *i.e.* 6 mol of alcohol per mole of oil is used. Since the reaction is reversible, excess alcohol shifts the equilibrium to the products side. This corresponds to 1:4 alcohol-to-oil volume ratio ($R = 0.25$). Besides alcohol-to-oil volume ratio, several other factors influence the chemical reaction. Some of these factors are: molar ratio of alcohol to vegetable oil, type of catalyst (alkaline or acidic), temperature, reaction time, degree of refinement of the vegetable oil, presence of moisture and free fatty acids. Although crude oils can also be transesterified, but that usually gives lesser ester because crude oils contains gums and other extraneous materials. Altering certain conditions such as reaction time, temperature and stirring can improve the yield to some extent.

Since alcohols and oils do not mix at room temperature, the chemical reaction is carried out at a higher temperature with continuous stirring. It increases the mass transfer between the phases. For the same reaction time, conversion is higher at high temperature. As an example, transesterification was 99 per cent complete in 4 hr at 32°C with an alkaline catalyst (NaOH or NaOMe). The same reaction can be completed in 1 hr at 60°C with alcohol:oil molar ratio of at least 6:1 with fully refined oils. Under normal conditions the reaction rate will double with every 10°C increase in reaction temperature. But there is a limit beyond which temperature cannot be increased. The temperature during transesterification is usually kept in the range of 50 to 70°C at atmospheric pressure. It is because the boiling point of methanol is approximately 68°C.

Catalytic Transesterification

Catalytic transesterification (also called alcoholysis) is the reaction of a triglyceride (fat/oil) with an alcohol in the presence of some catalyst that improves the reaction rate and the yield (Figure 9.3). The catalysts used in transesterification are classified as basic, acid or enzymatic. Basic catalysts include sodium hydroxide (NaOH), potassium hydroxide (KOH), carbonates and their corresponding alcoxides (for instance, sodium methoxide or ethoxide). Acid catalysts include sulphuric acid, sulfonic acids and hydrochloric acid. Heterogeneous catalysts used in biodiesel production include enzymes, titanium silicates, and compounds from alkaline earth metals, anion exchange resins and guanidines in organic polymers. Lipases are the most frequently used enzymes for biodiesel production.

The mixture of fatty acids methyl esters (FAME) and glycerin forms two distinct phases having different densities. Whereas fatty acid alcohol esters have density of about 0.88 gm/cc, the glycerol phase has a density of the order of 1.05 gm/cc, or higher depending upon the amount of methanol, water, and catalyst in the glycerol. The two phases begin to form immediately after the stirring is stopped. Most of the catalyst and excess alcohol will concentrate in the lower phase (glycerin), while most of the mono-, di-, and triglycerides will concentrate in the upper phase (FAME). Once the phases are clearly and completely defined, these can be physically separated by gravity, the density difference being sufficient for the use of simple gravity separation techniques. But, separation under the action of gravity alone takes several hours to complete proving to be a major bottleneck in the production process. Therefore, other equipments *viz.*, decanter system, centrifuge system or hydro cyclones are

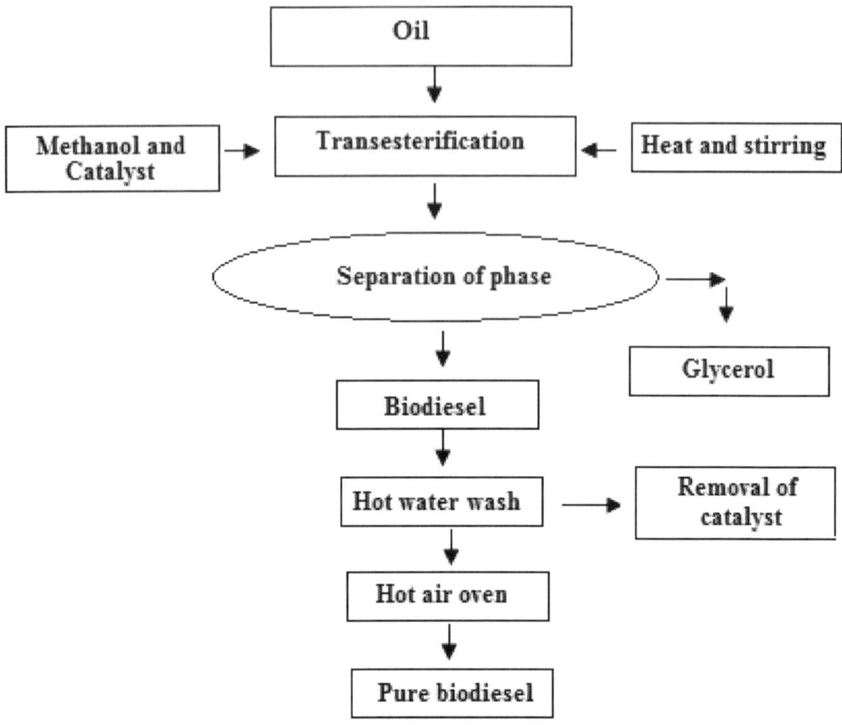

Figure 9.3. Flow diagram of catalytic transesterification process of biodiesel production.

used to separate the two phases. Centrifugation is faster albeit more expensive. Since the FAME mixture contains several impurities notably remnants of alcohol, catalyst and mono-, di-, and triglycerides, the product will have certain undesirable characteristics such as higher cloud point, higher pour point, and lower flash point, *etc.* Therefore, a purification process is used to make the final product that complies with prescribed standards. This process is discussed in a forthcoming section of this chapter.

Batch Process

The flow diagram of the batch process to manufacture biodiesel is shown in Figure 9.3. Catalyst (sodium hydroxide or potassium hydroxide or some other similar salts) is dissolved in the alcohol using a standard agitator or mixer. The alcohol/catalyst mixture is then charged into a closed reaction vessel and the biolipid (vegetable or animal oil or fat) is added. The system from here on is totally closed to the atmosphere to prevent any loss of alcohol. The reaction mix is kept just above the boiling point of the alcohol (around 70°C) to speed up the reaction. Some manufacturers recommend that the reaction for safety reasons should take place anywhere from room temperature to 55°C. Recommended reaction time varies from 1 to 8 hr. Excess alcohol is used to ensure total conversion of the fat or oil to its esters. The system is agitated during the reaction time and thereafter agitation is stopped. In some processes, the reaction mixture is allowed to settle in the reactor to give an initial separation of the esters and glycerol. In other processes, reaction mixture is pumped into a settling vessel, or the phases are separated using a centrifuge. Once the glycerin (the component that gums up the engine when vegetable oil is used directly to run the engine) and biodiesel phases have been separated, the excess alcohol in each phase is removed with flash evaporation process or by distillation. Sometimes biodiesel is purified by gently washing with warm water that removes residual catalyst or soaps. It is thereafter dried, and sent to storage. The glycerol stream is neutralized, washed with soft water and then sent to the glycerol refining section. Enough care has to be exercised to monitor the amount of water and free fatty acids in the incoming oil or fat. Too high levels of either free fatty acid or water may cause problems that may result in soap formation or causing hardships in separation of the glycerin by-product downstream.

Supercritical Transesterification

Simple transesterification process is time consuming besides needing separations of the catalyst and saponified impurities from the biodiesel. To overcome these problems, supercritical methanol at high temperature and pressure is used in the supercritical transesterification process (Demirbag, 2002). In the supercritical state (at 340°C and 43 MPa) oil and methanol are in a single phase. Therefore, reaction occurs spontaneously and rapidly within 2-4 min. Purification of biodiesel is much easier, trouble free and environment friendly because the step to remove the catalyst is eliminated as no catalyst is required in this process. The process can be used even if the feedstock contains some water. Although, high temperatures and pressures are used in this process, yet the overall energy cost of production is similar or even less than the catalytic production process.

Ultra- and High-Shear in-line or Batch Reactors

Ultra- and high shear in-line or batch reactors are quite flexible as biodiesel can be produced in continuous, semi-continuous, or in batch-modes. This method drastically reduces production time as well as increases production volume. The reaction takes place in the high-energetic shear zone of the ultra- and high shear mixer that reduces the droplet size of the immiscible liquids such as oil or fats and methanol. As a result, catalyst reaction is faster because of the higher surface area resulting from smaller droplet size.

Ultrasonic Reactor Method

Ultrasonic waves are used in the ultrasonic reactor method to produce as well as to collapse the bubbles in the reaction mixture in a continuous manner. This provides simultaneous mixing and heating needs of the transesterification process. The ultrasonic reactor method reduces reaction time, reaction temperatures, and energy input of the biodiesel production process. Industrial scale ultrasonic reactors are capable of producing several thousand barrels of biodiesel per day.

Microwave Method

The microwaves can provide intense localized heating much higher than the recorded temperature of the reaction vessel. Thus, it consumes only one-fourth of the energy required in the batch process. Although still in the lab-scale, current research is directed to develop and use commercial microwave ovens to provide heat needed for the transesterification process. It appears that the microwave method holds great potential and can prove to be an efficient and cost competitive method for commercial-scale biodiesel production. A continuous flow process capable of producing 6 L/min at 99 per cent conversion rate has already been developed.

Purification of Biodiesel/Glycerin

Mix of methyl esters obtained after transesterification contain several impurities such as remnants of catalyst, water and methanol. The mix must be purified to reduce the concentration of these impurities through washing, neutralization, drying, other ester treatments and additions *etc.* Purification is necessary to ensure that the properties of biodiesel must comply with the requirements set by national and international standards irrespective of the kinds of oils and fats used as raw materials.

Successive washings of the end product with water are used to neutralize residual catalyst, and to remove soaps, residual free glycerol and methanol. First washing is carried out with acidified water, which is followed by two additional washings with pure water. It must be ensured that no emulsions are formed during washing as it will reduce the efficiency of the process. An alternative to washing process is the use of ion exchange resins or silicates.

There are stringent limits on the amount of water present in the final biodiesel product. Therefore, any traces of water in the biodiesel must be eliminated by drying. Although purified product is now ready to be labeled as biodiesel, but other treatment processes may be needed as per requirement. These are mainly needed to reduce colour bodies in the fuel and remove sulphur, phosphorus, and glycerides. Next step is the addition of certain materials having specific functionalities that modify one or more than one fuel properties. Examples include cloud point/pour point additives, antioxidants, or other stability enhancing agents.

Raw glycerin produced in the process is of low quality having little commercial value. The purified glycerin on the other hand has commercial value as it is used in many industrial applications. Many novel uses of glycerin to produce high-value-added products are in the pipeline. Therefore, refining the by-product glycerin makes the process of biodiesel production economically attractive. Glycerin purification assumes greater significance in large-scale production plants, where besides getting high quality glycerin; part of the methanol is also recovered. Reutilization of recovered methanol in the transesterification reaction can considerably reduce the biodiesel production cost.

BIODIESEL FROM *JATROPHA* OIL

Edible oils are in high demand and expensive than diesel fuel. Therefore, attempts have been made to use non-edible oils for biodiesel production. Amongst many such sources, *Jatropha curcas* has been identified as one of the potential biodiesel sources. *Jatropha curcas* is a poisonous, semi-evergreen shrub or small tree that can attain a height of up to 6 m. A native of the Mexican tropics, it has now dispersed and naturalized throughout the tropics of the whole world.

Why *Jatropha curcas*

- The plant is suitable for tropical and sub-tropical regions of the world
- Rapid growth with high seed productivity
- Plant is resistant to drought and pests
- It is not a part of the food chain
- Seed contain 27-40 per cent oil, oil : seed ratio by weight is about 1:3
- Easy to propagate with low plantation cost
- Detoxified by-product after oil extraction can be used as manure and animal feed
- Offers business opportunities and employment to rural people

Production Process

The full cycle of biodiesel production comprises the following steps.

- Plantation of *Jatropha* and extraction of oil
- Transesterification of oil

Harvested *Jatropha* seeds are first dehulled, and then pressed to extract oil using a hydraulic press or a solvent extractor. The oil produced from the hydraulic press is generally turbid and contains significant amount of water, metals and dust that makes it less suitable for biodiesel production. In the solvent extraction method, the grounded seeds are placed inside a soxhlet and brought into contact with a condensed solvent. The solvent dissolves the oil, which is later separated using a rotor vapour. The oil yield from the solvent extraction is higher than hydraulic press extraction and is quite suitable for biodiesel production.

Transesterification

The process flowchart of biodiesel production from *Jatropha curcas* seeds including use of by-products is shown in Figure 9.4. Note that transesterification process is similar to what has been discussed in the previous section. Briefly, the *Jatropha* oil is blended with alcohol and catalyst mixture in appropriate proportions at appropriate temperature. This mixture is continuously stirred for two hours. Glycerol in the solution is separated out through settling or by other means discussed in the previous section. Liquid biodiesel is transferred to a washing tank to obtain purified biodiesel.

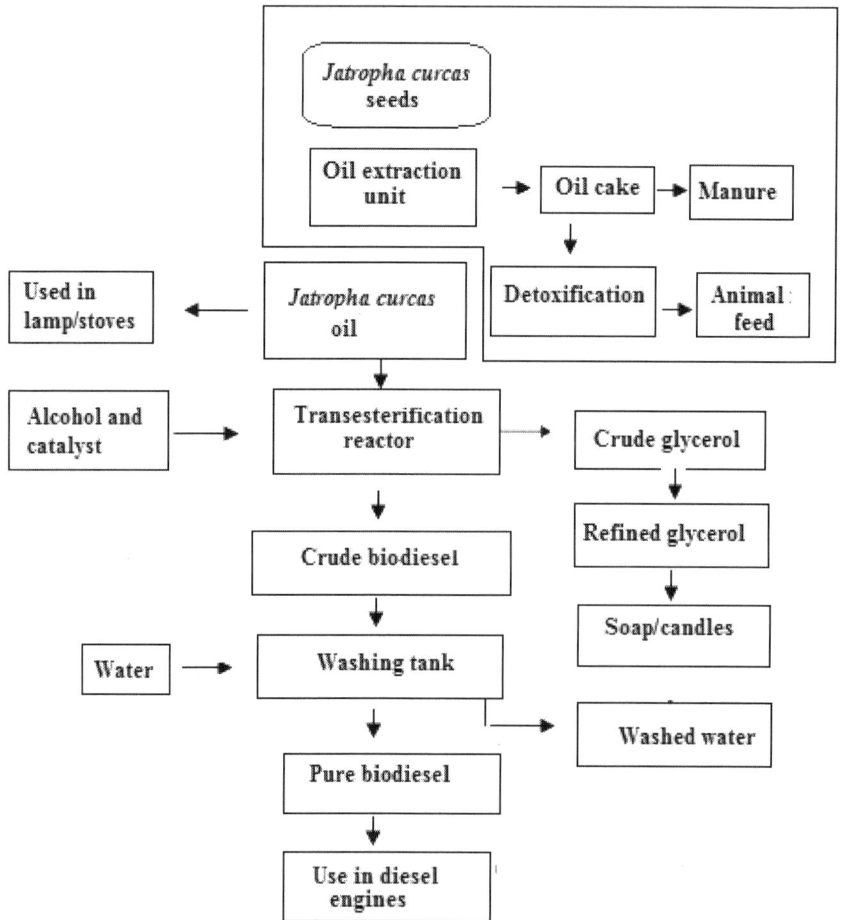

Figure 9.4. Process flowchart of biodiesel production from *Jatropha curcas* seeds.

CHARACTERISTICS AND APPLICATIONS

Biodiesel, depending upon the feedstock used in its production, is golden to dark brown in colour. It is immiscible in water, has a high boiling point and low vapour pressure. The flash point of biodiesel (>130°C) is significantly higher than diesel (> 52°C) or gasoline (–43°C). Biodiesel has a density of ~ 0.87 g/cm^3 (Table 9.2). Its calorific value is about 39.2 MJ/kg (Table 9.2). Biodiesel has better lubricating properties and much higher cetane ratings than any of low sulphur diesel fuels. Fuel properties of some of the methyl esters (biodiesel) from various plant-based oils are reported in Table 9.3. Specific gravity of biodiesel methyl esters of six non-edible oils ranging from 0.86 to 0.89 meets the international standards related to this property. The biodiesel specifications should meet the standard established by BIS (BIS, 2016, Table 9.4). Biodiesel is the only alternative fuel that can be used directly in any existing unmodified diesel engine or may be blended with petroleum diesel in any ratio. Addition of biodiesel reduces fuel system wear, and increases the life of the fuel injection equipment that relies on the fuel for its lubrication. Since biodiesel is virtually free from sulphur, it is

Table 9.2. Fuel properties of *Jatropha* oil and its biodiesel

Properties	*Jatropha Oil*	*Jatropha Biodiesel*	*Petro-diesel*
Density, g/mL	0.920	0.87	0.84
Viscosity, mm^2/s at 40° C	3.5	5.2	4.5
Calorific value, MJ/kg	39.7	39.2	42.0
Flash point, °C	240	175	50

Table 9.3. Select properties of methyl esters of 6 non-edible oils

Properties	*JME*	*KME*	*PME*	*MME*	*NME*	*RME*	*Diesel*	*Biodiesel Standard*	
								EN 14214	*ASTM 6751–09*
Specific gravity, g/mL	0.86-0.88	0.88-0.89	–	–	–	0.87	0.84	0.86–0.90	0.87–0.90
Calorific value, MJ/kg	42.7	42.1	41.4	37.0	40.1	36.5	42.5	–	–
Viscosity, mm^2/s at 40°C	4.2	4.4	4.0	4.0	8.8	5.8	3.8	3.5–5.0	1.9–6.0
Flash point, °C	148	163	140	208	–	130	45	≥120	≥130
Cloud point, °C	10.2	14.6	13.2	–	–	4.0	–1.0	–	–
Pour point, °C	4.2	5.1	4.3	6.0	–	–8.0	–16.0	–	–
Cetane number	60.7-63.3	59.7-60.9	52.5	–	–	–	–	–	47 minimum

JME= Jatropha methyl ester, K = Karanja, P = Palm oil, M = *Mahua*, N = *Neem* and R= Rubber.

often used as an additive to ultra-low sulphur diesel (ULSD) fuels. The solvent properties of biodiesel being different than petro-diesel, it can easily degrade natural rubber gaskets and hoses in the vehicles. It can also break down deposits of residue in the fuel lines of engines that have previously been running with petro-diesel. Thus, fuel filters are likely to get clogged with particulates if a quick transition is made from petro-diesel to pure biodiesel. Therefore, fuel filters of engines should be replaced shortly after switching is made from pure biodiesel to biodiesel or a biodiesel blend.

Table 9.4. BIS standards for biodiesel

Sl.No.	*Characteristic*	*Value*
1.	Density at 15°C, kg/m^3	860-900
2.	Kinematic viscosity at 40°C, cSt	2.5-6.0
3.	Flash point (PMCC) 20°C, Min	120
4.	Sulphur, mg/kg, Max	50
5.	Carbon residue, per cent by mass, Max	0.05
6.	Sulphated ash, per cent by mass, Max	0.02
7.	Water content, mg/kg, Max	500
8.	Total contamination, mg/kg Max	24
9.	Copper corrosion, 3 h at 50°C, Max	1
10.	Cetane No., Min	51
11.	Acid value, mg KOH/g, Max	0.50
12.	Methanol, per cent by mass, Max	0.20
13.	Ethanol, per cent by mass, Max	0.20
14.	Ester content, per cent by mass, Min	96.5
15.	Free glycerol, per cent by mass, Max	0.02
16.	Total glycerol, per cent by mass, Max	0.25
17.	Phosphorous, mg/kg, Max	10.0
18.	Sodium and potassium, mg/kg, Max	To be reported
19.	Calcium and magnesium, mg/kg, Max	To be reported
20.	Iodine value	To be reported
21.	Oxidation stability at 110°C, hr, Min	6
22.	Mono-glycerides content, per cent by mass, Max	0.7
23.	Di-glyceride content, per cent by mass, Max	0.2
24.	Triglyceride content, per cent by mass, Max	0.2
25.	Cold filter plugging point (CFPP), °C, Max (Summer, winter)	18,6
26.	Linolenic acid methyl ester, per cent m/m, max	12
27.	Polyunsaturated (> 4 double bonds) methyl ester, per cent m/m, Max	1

Max means maximum and Min means minimum.

ETHANOL FROM AGRICULTURAL PRODUCE

Ethanol (C_2H_5OH) is basically an alcohol fuel produced through the process of fermentation of starches and sugars with the help of enzymes and micro organisms. It can be used as an alternative biofuel to petrol. Ethanol with less than 1 per cent water called anhydrous ethanol can also be blended with petrol in varying quantities to serve as fuel. Currently, all spark ignition petrol engines can operate with mixtures of up to 5 per cent bioethanol (E5). On the other hand many engine manufacturers do not discourage and actually suggest higher blends of bioethanol, as it offers excellent performance and reduced air pollution compared to conventional fuels. The other type of alcohols is methanol (CH_3OH). Between the two types, ethanol has attracted relatively more attention. A significant advantage of bioethanol is that it can be blended with petrol as both an oxygenator and octane enhancer. Bioethanol, as an octane enhancer, is an ideal solution for substituting MTBE (methyl tert-butyl ether), having many environmental concerns and a possible carcinogen. Ethanol is also suitable as a feedstock for producing other octane enhancers such as ETBE (ethyl tertiary butyl ether) or TAEE (tertiary amyl ethyl ether), considered safer than MTBE.

Advantages of Bioethanol

- Ethanol as against petrol and diesel is a pure compound having fixed set of physical as well as chemical properties
- Bioethanol is an environment friendly fuel because of its carbon neutrality on a lifecycle basis – meaning the carbon dioxide (CO_2) emitted during its use is offset by the carbon dioxide absorbed during the growth of products from which ethanol is produced
- If emissions of CO_2 and nitrous oxide are taken together, lifecycle greenhouse gas emissions with bioethanol can be reduced by 90 per cent compared to petrol. Emission reductions even by current European standards are significant – around 35-65 per cent depending upon the processing method
- Taxes by governments are lower on bioethanol because of its industrial and environmental benefits
- Additional revenue streams are available from the spent products that result from ethanol production. These products are usually dried to produce distiller's dried grains with solubles marketed as a protein and nutrient rich animal feed for cattle, pigs and poultry. Similarly sugarcane can be used for ethanol production and large amounts of bagasse (the remaining wood fibers after the juice is extracted) for heat energy
- The biomass can be utilized as a renewable fuel source, producing electrical power and heat for the process, or for conversion into methane for use as a fuel. These energy conversion technologies

enhance the overall production energy balance and at the same time earn carbon credits

Limitations of Bioethanol

Few concerns of bioethanol as a long-term alternative fuel for the engines are as follows.

- Biodiversity – a large area under arable farming is required to grow crops of interest. It might led to destruction of some natural habitats including rainforests
- The food versus fuel debate – The lucrative prices of bioethanol may lure some farmers to sacrifice food crops for biofuel production, which might increase food prices around the world
- Carbon emissions – The carbon neutrality of bioethanol is debatable when all elements are taken into consideration including the cost of changing land use of an area, transportation and the burning of the crop
- Other concerns – Many old engines may not be equipped to handle even 10 per cent ethanol. Moreover, using 100 per cent ethanol may decrease fuel economy by around 15-30 per cent compared with 100 per cent petroleum

Irrespective of what is stated in previous paragraphs, ethanol is a strong candidate for biofuel since it can be produced from the agricultural resources. Earlier, it was not prepared because of its low energy density, high production cost and fear of corrosion of various engine parts. The current shortage of gasoline has necessitated the substitution of ethanol as fuel in SI engines. As per statistics of 2017, USA is the largest producer of ethanol to the tune of 15250 million gallons followed by Brazil with 7295 million gallons. Source wise, major production of ethanol in USA is from corn. Brazil, on the other hand, produces ethanol on a large scale from sugarcane and happens to be the producer of cheapest bioethanol in the world. It is widely used fuel in cars in Brazil. In the Indian context, ethanol production from sugarcane is hugely attractive because the productive capacity can be quite high, of the order of 1345 L/ha.

Methanol versus Ethanol

Methanol and ethanol are two variants of alcohol. These have quite different properties and uses. Methanol, also known as methyl alcohol, is composed of only one carbon atom that combines with a methyl group (CH_3–) and a hydroxyl group (-OH) represented as CH_3OH. Methanol is also known as wood alcohol because it was originally produced by the distillation of wood. It is now derived through synthetic processes and biomass and is poisonous in nature. It is generally produced by steam-reforming natural gas to synthesis gas, which is then fed into a reactor with a catalyst to produce methanol and water vapours.

It can also be produced from a broad group of biomass materials, as well as from coal and natural gas. Methanol is cheaper than ethanol. It upon burning produces a bright white flame. It is the safest motor fuel, being much less flammable than gasoline, a fact that has led to its widespread use in racing cars. On the contrary, it is aggressive to some materials like plastic components and some of the metals in the fuel system. Therefore, necessary precautions have to be exercised while using it as a fuel. Moreover, it is not used in alcoholic beverages being highly poisonous. Methanol is generally used in the manufacturing of products such as formaldehyde *etc.*

Ethanol, also known as ethyl alcohol, has two carbon atoms combining an ethyl group (CH_3CH_2–) to a hydroxyl group (-OH). Its formula is given as CH_3CH_2OH (C_2H_5OH). Thus, it is cheaper to produce methanol than ethanol with a given biomass. Commercial ethanol is produced by factory fermentation of food crops. Ethanol has a heavy burning smell and emits bright blue flame. It is superior to methanol in the sense that it has wider ignition limit (3.5 -17.0) than methanol (2.15 -12.8). Calorific value of ethanol (26880 kJ/kg) is considerably higher than methanol (19740 kJ/kg). Ethanol is much more superior fuel for diesel engines as its cetane number is 8 compared to the 3 of methanol. Ethanol has even replaced methanol in racing cars due to its very high heat of vaporization. Ethanol is less corrosive and chemically toxic than methanol. It carries more energy per gallon, about 75 per cent of the energy of gasoline/L, compared to 67 per cent for methanol. Ethanol besides its fuel value is a primary ingredient of alcoholic beverages.

Both ethanol and methanol are water soluble and biodegradable in the environment. The adverse consequences of spilling of any of these are much less than that of the petroleum products. Both cover fewer km per liter than gasoline, but about as many km per dollar at current prices in USA. A comparative statement of various properties of the two kinds of alcohols along with those of gasoline is reported in Table 9.5 (Simeon, 2018).

Table 9.5. Comparative statement of various properties of gasoline, methanol and ethanol

Properties	*Gasoline*	*Methanol*	*Ethanol*
Chemical formula	C_8H_{15}	CH_3OH	C_2H_5OH
Molecular weight	111.21	32.04	46.07
Oxygen content, wt. per cent	–	49.93	34.73
Carbon content, wt. per cent	86.3	37.5	52.2
Hydrogen content, wt. per cent	24.8	12.5	13.1
Stoichiometric air-fuel ratio	14.5	6.43	8.94
Lower heating value, MJ/kg	44.3	20	27
Heat of evaporation, kJ/kg	305	1.178	840
Research octane number	96.5	112	111

Properties	Gasoline	Methanol	Ethanol
Motor octane number	87.2	91	92
Vapour pressure, psi at 37.7°C	4.5	4.6	2
Density, g/cm^3	0.737	0.792	0.785
Normal boiling point, °C	38–204	64	78
Auto-ignition temperature, °C	246–280	470	365

ETHANOL PRODUCTION TECHNOLOGIES

Three kinds of feed stocks namely sugar feed stock *i.e.*, sugarcane and sugar beet; starch feed stock *i.e.*, cereal grains and potato and cellulosic feed stock *i.e.*, forest products, agricultural residues and industrial and household wastes are used to manufacture ethanol. Ethanol produced from first two types of feed stocks is called first generation ethanol because it is produced from crops directly from the fields. The one produced from the third kind of biomass is termed as second generation ethanol as it is produced from residuals and waste products. The chemical make-up of ethanol is uniform across the board irrespective of whether it is produced from corn, sugarcane, or cellulose, but the production processes and technologies are quite different. According to the Union Ministry of Petroleum and Natural Gas, the approximate ethanol availability in India is 3000 million liters in 2018. Of this, about 1300 million liters goes into making liquor, 600 to 800 to make chemicals leaving about 1000 to 1200 million liters for blending fuels.

Ethanol from Starchy Feed Stock (Grains)

Ethanol production from cereal grains such as barley, wheat and corn is a much easier process than from cellulosic material. The process includes the following steps.

- Milling of grains
- Hydrolysis of starch to sugar
- Fermentation by yeast
- Distillation
- Removal of water from ethanol

Raw material after grinding is mixed with water and enzymes that convert starch into sugars. The yeast or bacteria is then used to ferment free sugars into ethanol and carbon dioxide. Fermentation process is controlled by the concentration of ethanol, as high alcohol concentration can kill the yeast or bacteria. The rate of fermentation is considerably reduced at alcohol concentration of about 15 per cent. Therefore, ethanol needs to be separated out from the other material at this stage. Concentration of ethanol is improved to about 95 per cent

(190 Proof) using the distillation process. Ethanol content is usually measured in terms of a unit called Proof. One per cent ethanol content equals 2 proofs. Ethanol is then dried by different drying agents raising its concentration to 99.5 per cent ethanol or absolute ethanol. Extractive distillation with benzene yields anhydrous ethanol. About 3 kg of wheat is required to produce 1 L of absolute ethanol. Ethanol used for fuel is then denatured with a small amount (2-5 per cent) of some product like petrol to make it unfit for human consumption. The by-product distiller's dried grains with solubles (DDGS) consist of the protein value of corn, yeast, fiber and water discharge from the first-stage distillation column. The distiller's grain has a high protein content of about 27 per cent. Therefore, it makes an excellent animal feedstock supplement. It is competitive with soybean meal, which is about 44 per cent protein. Another by-product carbon dioxide (CO_2) that evolves from the fermentation process can also be used in grain storage silos to control spoilage. The process flow chart for ethanol production from grains (corn) is shown in Figure 9.5. The ethanol production potential of various crops is given in Table 9.6.

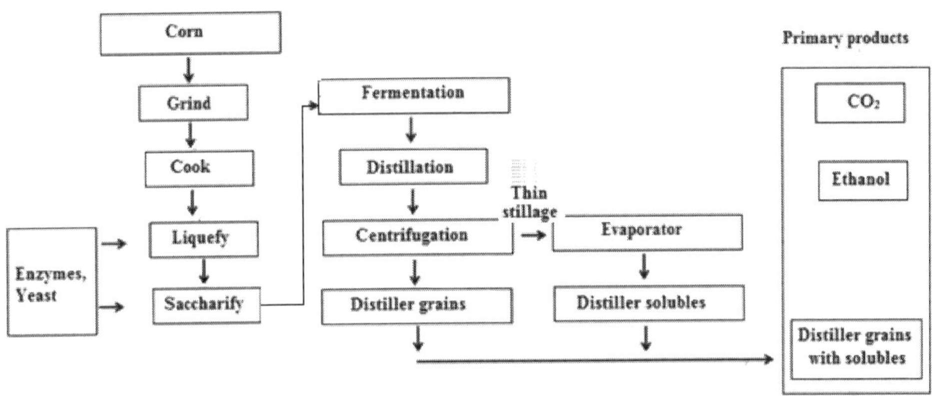

Figure 9.5. Process flow chart for the production of ethanol from corn.

Table 9.6. Ethanol yields of various crops

Crop	*Liters of Ethanol per tonne of Crop*	*Liters of Ethanol per hectare per Year*
Sugarcane	70	3500
Cassava	180	2160
Sweet sorghum	86	3010
Sweet potato	125	1875
Corn (maize)	370	2220
Wood	160	3200

Ethanol from Sugarcane

One of the easiest and most efficient processes to produce ethanol is from sugarcane. It contains about 15 per cent sucrose. The glycosidic bond in the disaccharide can be broken down into two sugar units, which are free and readily available for fermentation. The cane is harvested and its juice is extracted. Juice now takes two routes one to produce sugar and another to produce ethanol. Some by-products of juice segment also find entry into the fermenter. For the ethanol segment, the juice is clarified and concentrated by boiling. The concentrated juice is fermented with yeast to produce raw ethanol. A series of distillation steps including a final extractive distillation with benzene produces anhydrous ethanol. The normal yield is about 8.73 L of ethanol per tonne of cane. The process flow chart for production of ethanol from sugarcane is shown in Figure 9.6. Even sugarcane bagasse can be used to produce ethanol, the most promising fuel of the future. This issue will be briefly discussed in the cellulosic ethanol section.

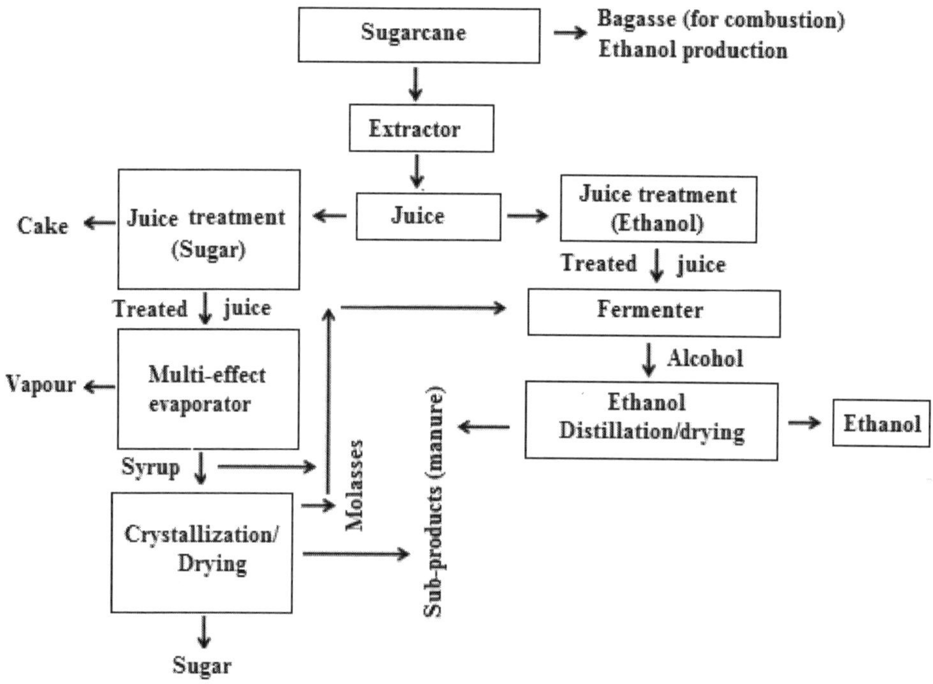

Figure 9.6. Process flow chart for the production of ethanol from sugarcane.

Cellulosic Ethanol

Cellulosic ethanol is made from stalks and stems of plants and a variety of non-food plant materials, including: agricultural residues (rice, corn stover, wheat straw) agricultural wastes (sugarcane bagasse waste, rice husks, citrus pulp), forestry wastes and wood wastes (including willow and poplar), municipal solid wastes, paper pulp, sawdust and grass/crops (switch grass, *miscanthus*). The process of producing ethanol from cellulosic materials is far more complicated than the processes employed for starch- or sugar-based ethanol production. Processes namely fermentation and thermo-chemical conversion are used to produce ethanol from cellulosic materials. Enzymes are used to break up the complex cellulose-hemicellulose-lignin structure of various biomass types before the fermentation process can begin (Table 9.7).

Table 9.7. Main components of several biomass types

Biomass Type	*Biomass Composition (per cent w/w)*			
	Cellulose	*Hemicellulose*	*Lignin*	*Others*
Softwood	41	24	28	7
Hardwood	39	35	20	7
Wheat straw	40	28	17	15
Rice straw	30	25	12	33
Bagasse	38	39	20	3
Sunflower seed hull	26.7	18.4	27	-
Poultry litter	27	17.8	11.3	20
Deciduous plant	42	25	21.5	11.5

Fermentation Method

The two main steps of this process are hydrolysis and fermentation following some pretreatment (Figure 9.7). Pretreatment of cellulosic materials is necessary to prepare it for hydrolysis, which converts the cellulose and hemicellulose into glucose. Standard pretreatments include either a chemical treatment involving an acid or a base or a physical pretreatment such as grinding, followed by biological treatment to catalyze the reaction. The latter employs enzymes secreted by certain organisms such as fungi or bacteria. The combinations of these processes can also be used. These steps make the cellulose more accessible to the cellulase enzymes that digest the cellulose to turn it into glucose. Once the sugars are derived from the cellulosic materials, they are fermented using yeast or bacteria, causing the organic substance to break down into simpler substances; especially, the anaerobic breakdown of sugar into alcohol, a process similar to one used in the corn-based ethanol production. The liquid resulting from the fermentation process contains ethanol and water. The water is removed through distillation, again similar to the corn-based ethanol process. The type of feedstock and method of pretreatment both influence the amount of ethanol produced. This description of the processes illustrates as to why ethanol production from cellulose takes more time, is expensive, and complex relative to the procedures used to convert corn or sugarcane into ethanol. The co-product namely lignin resulting during pretreatment and hydrolysis processes can be burned to generate heat or electricity. One of the main areas of research in the development of cellulosic ethanol has been to find the most effective and low-cost enzymes for the pretreatment and organisms for the fermentation processes.

Ethanol from Bagasse

As per statistics of 2017 (RFA, 2017), India is one of the second largest producers of bagasse followed by Brazil (Table 9.8). Bagasse, a waste material from sugarcane is an energy source and can also be used to produce second generation ethanol (Figure 9.6 and 9.7). Sugarcane bagasse is mainly composed of 33-36 per cent cellulose, 28-30 per cent hemicellulose, and 17-24 per cent lignin. Sugarcane bagasse hemicellulose is composed of heteroxylans with a predominance of xylose. Hence it can be easily hydrolyzed chemically than other cellulose types. Ethanol from bagasse can be produced like any other cellulosic material in a four step process of pretreatment, enzymatic hydrolysis, fermentation and distillation (Figure 9.7).

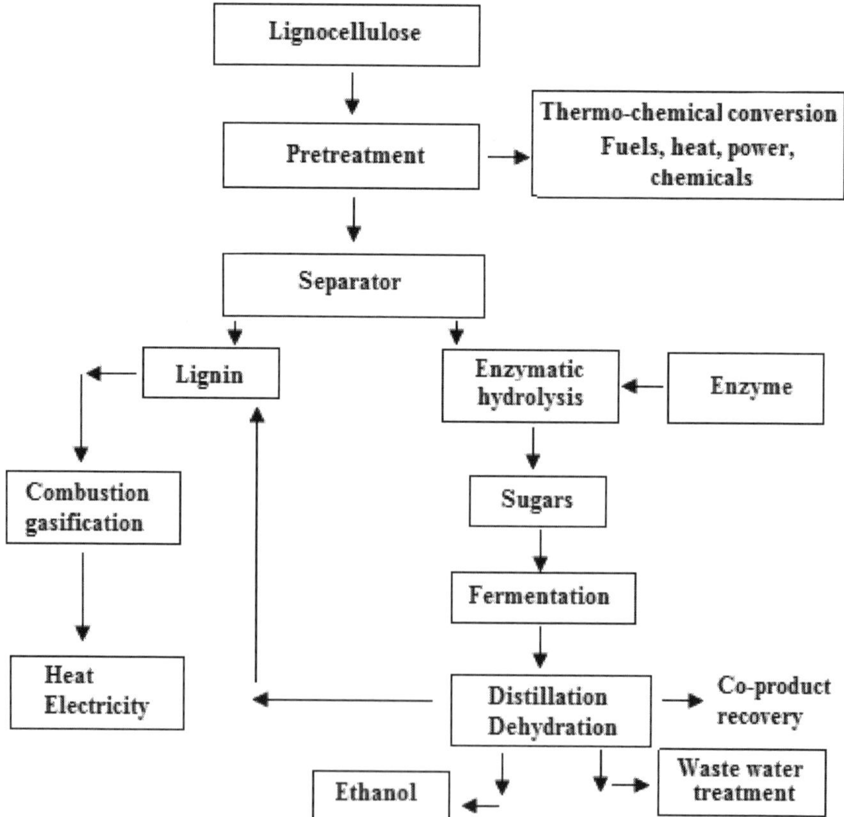

Figure 9.7. Flowchart for ethanol production from cellulosic biomass.

Table 9.8. Annual bagasse production for several countries

Country	Sugarcane Bagasse Production (Million tonnes)
Brazil	181
India	101.3
China	80
Thailand	20
Mexico	15
Colombia	7
Philippines	5.1
US	3.5

Thermo-chemical Method

We have seen in a previous chapter that several fuels can be produced from biomass through thermo-chemical process known as gasification. Clearly, the gasification process does not rely on chemical decomposition of the cellulose chain (cellulolysis) into sugar molecules. Instead, the carbon in the raw material is converted into synthesis gas using partial combustion. The carbon monoxide, carbon dioxide and hydrogen are then fed into a special kind of fermenter. Instead of sugar fermentation with yeast, this process uses *Clostridium ljungdahlii* bacteria which ingest carbon monoxide, carbon dioxide and hydrogen to produce ethanol and water. The process comprises the following three steps:

Gasification: Complex carbon based molecules are broken apart to access the carbon as carbon monoxide, carbon dioxide and hydrogen as per the following reaction.

$$\text{Biomass} \rightarrow CO + H_2 + CO_2 + CH_4 + H_2O + H_2S + NH_3 + C_xH_y + \text{Tar} + \text{Char}$$

Syngas conversion conditions (in terms of pressure, temperature, and catalyst) as well as its composition (in terms of H_2/CO and CO_2) determine the various biofuels produced. Therefore, syngas has to be produced under most suitable operative conditions to enhance biofuel production.

Fermentation: The syngas (a mixture of carbon monoxide and hydrogen) is conditioned to remove impurities such as tars and methane, and to adjust the hydrogen to carbon monoxide ratio to 2:l. The syngas is then reacted over a catalyst (vary for various end products) at elevated temperatures and pressures to form methanol and other fuels (Figure 9.8, Spath and Dayton, 2003). In one of the reactions syngas is converted to methanol by using predominantly copper-based on alumina support catalysts through an exothermic equilibrium limited synthesis process in the at pressure range of 50–100 bar and temperatures range of 200–300°C. The following reactions occur.

$$CO + 2H_2 \leftrightarrow CH_3OH$$

$$CO_2 + 3H_2 \leftrightarrow CH_3OH$$

$$CO + H_2O \leftrightarrow CO + H_2$$

Ethanol from syngas is directly obtained by employing catalysts such as Mo, Rh, K, Cu, Zn, and Fe. This process is facilitated in the pressure range of 1–100 bar and temperatures range of 230–300°C. The predominant reaction for ethanol production consists of CO hydrogenation, and to some extent CO_2 hydrogenation.

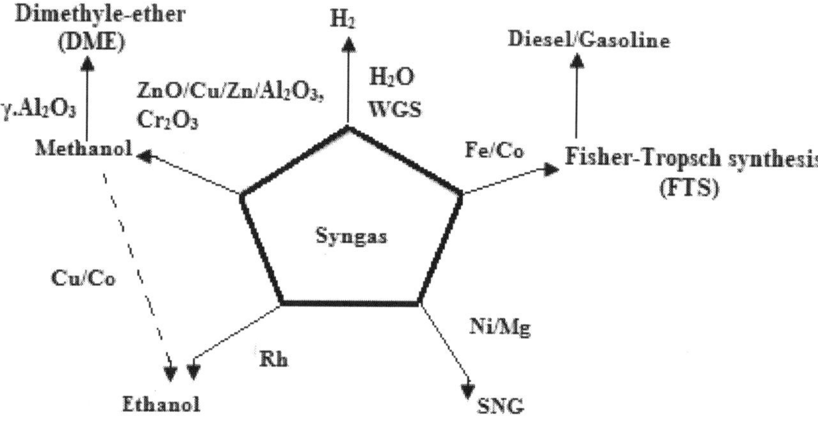

Figure 9.8 Syngas conversion routes to biofuels

$$2CO + 4H_2 \leftrightarrow C_2H_5OH + H_2O$$

$$2CO_2 + 6H_2 \leftrightarrow C_2H_5OH + 3H_2O$$

Ethanol can also be produced through methanol synthesis followed by methanol homologation, according to the following exothermic reactions catalyzed by Cu/Co catalysts.

$$CO + 2H_2 \leftrightarrow CH_3OH$$

$$CH_3OH + CO + 2H_2 \leftrightarrow C_2H_5OH + H_2$$

Synthesis of synthetic natural gas (SNG, methane) by syngas can be achieved by using the same catalysts as in the steam reforming reactions, mainly Ni on alumina. Other catalysts, such as Ru, Co, and Fe, can also be used. The reactions involved are shown below:

$$CO + 3H_2 \leftrightarrow CH_4 + H_2O$$

$$CO_2 + 4H_2 \leftrightarrow CH_4 + 2H_2O$$

Apparently, thermo-chemical gasification of biomass and wastes combined with simple robust conversion of CO and H_2 by autotrophic acetogenic bacteria provides a versatile and potentially economical process for the production of fuels and chemicals.

Pros and Cons of Cellulosic Ethanol

Pros

- GHG emissions are reduced by 85 to 94 per cent compared to petroleum-based fuels
- Cellulosic ethanol production uses biomass feedstock, which are not suitable for human consumption. Thus, feedstock costs are low or are almost free. In some cases one may even get some money in exchange for example in the case of MSW, as the municipalities would otherwise have to pay for landfill space
- Estimated positive net energy balance for the production of cellulosic ethanol is 2 to 12 or even 36 times the input energy
- Fast growing energy crops can be grown in degraded/poor soil
- Only few or no changes in the engines need to be made to adopt them for ethanol
- As ethanol is highly biodegradable, any spills are far less worrisome than petroleum spills

Cons

- Most cellulosic ethanol plants are in pilot scale and have encountered some difficulties in upgrading to commercial scale
- Most plants so far have substantial component of government grants providing the key capital. Failure to attract private funding may slow or eventually stop the construction of cellulosic ethanol plants
- On the basis of 2009 flex fuel vehicles, km covered per liter with E85 is lower than with petro-diesel. It has been observed that 1.35 to 1.40 liters of E85 covers the same mileage as 1.00 liter of gasoline
- The cost of enzymes is crucial to the production of cellulosic ethanol. In spite of great deal of research in this area, the cost of enzymes is still quite high
- Ethanol has a shorter shelf and tank life than gasoline because it can absorbs water as a result of which it is difficult to ship it through pipelines

CONCLUDING REMARKS

Work on biofuels began around 1930 but got a major boost only during the second World War, when many countries started producing alcohol using waste agricultural products for use in automobiles as a substitute for gasoline. It was used as a 10-15 per cent blend with the conventional fuels or sometimes as a complete replacement of the conventional fuel. In the last few decades biofuels have once again caught the attention of the world. More and more people are veering around the view that recent pace of developments in the field of biofuels will continue in times to come. Most people now realize that biofuels possess most of the desirable attributes of conventional sources of energy. Moreover, raw material is locally available throughout the world. Ministry of New and Renewable Energy, New Delhi released the National Policy on Biofuels during 2009 to promote biofuels in India. The Policy extends the scope of raw materials for ethanol production by allowing use of sugarcane juice, sugar containing materials like sugar beet, sweet sorghum, starch containing materials like corn, cassava, damaged food grains like wheat, broken rice, and even rotten potatoes unfit for human consumption. The Policy also encourages setting up of supply chain mechanisms for biodiesel production from non-edible oilseeds, used cooking oil, and short gestation period crops. The policy revised in 2018 advanced the target of 20 per cent blending of ethanol in petrol to 2025-26 from the earlier 2030. Besides, it added several other feed stocks for production of biofuels. Biofuels in India have special strategic importance as it fits well with the ongoing initiatives of the Government such as Make in India, *Swachh Bharat Abhiyan*, Skill Development and offers great opportunities to integrate with the ambitious targets of Doubling the Farmers Income, Import Reduction, Employment Generation, and Waste to Wealth Creation. Increasing use of biofuels in various sectors will be useful in the following ways.

Solution to energy crisis: Energy crisis is being felt around the globe because of the depletion of conventional non-renewable fuel sources such as coal, petroleum, and the like. Many energy-related organizations, businesses, and governing bodies have begun to recognize the importance of bio-based products and bio-energy. As a result, awareness and acceptance of biofuels as a viable potential substitute for petroleum has vastly improved. A more proactive approach and legislation in favour of these new environmentally friendly commodities in times to come can be foreseen.

Reduce import dependency: Dependence on import of crude oil is taking a huge toll on Indian economy. In fact, it is the most compelling reason to develop alternative energy sources. Biofuels besides meeting our future energy needs from local resources will also offer many economic and environmental benefits. Supply of 1500 million liters of ethanol can save over Rs. 40000 million of forex at 2017-18 prices. It is estimated that 10 million liter of E10 at current rates can save Rs. 280 million of forex.

Clean environment: A multi-option approach has been advocated to reduce CO_2 emissions. Switchover to clean carbon-neutral alternative fuels fits well in this approach. Biofuels can address the growing concerns on the harmful effects of greenhouse gases and global warming. It is estimated that 10 million liter of E10 can reduce around 20000 tonnes of CO_2 emissions. Conversion of agricultural residues/wastes to biofuels may reduce residue burning, which may further reduce green house gas emissions. It is now accepted that first generation bioethanol is able to reduce CO_2 emissions between 20 and 50 per cent in transportation sector alone compared to petroleum based fuels depending upon the energy efficiency of the bioethanol plants. Second generation bioethanol will be able to reduce CO_2 emissions by as much as 90 per cent with several technological advancements in the pipeline.

Health benefits: Prolonged reuse of cooking oil for preparing food, particularly for deep-frying is a potential health hazard and can lead to many diseases. Used cooking oil is a potential feedstock for biodiesel. Its use for making biodiesel will prevent diversion of used cooking oil in the food industry.

Municipal Solid Waste (MSW) management: It is estimated that 62 million tonnes of MSW is annually generated in India. Now we have a basket of technologies which can convert waste/plastic MSW to biofuels.

Infrastructural investment in rural areas: It is estimated that around Rs. 8000 million capital investments is needed to establish one 100 kilo-liter per day (klpd) bio-refinery. Thus, establishment of bio-refineries across the country will spur infrastructural investment in the rural areas.

Employment generation: One 100 klpd second generation (2G) bio-refinery can create 1200 jobs in plant operations alone. Besides, many more jobs may be created to manage the supply chain leading to village level entrepreneurship to emerge.

Additional income to farmers: Farmers are often at risk of not getting appropriate prices for their produce and residues during the surplus production phase. Agricultural residues/waste which otherwise are burnt by the farmers can be converted to ethanol by adopting 2G technologies of bioethanol production. Since ethanol production is a year round activity, it will help in price stabilization of agricultural products and the residues eventually leading to farmers fetching good price for their products.

Fuel environmentally friendly cars: Consumers today are increasingly making environmental friendly choices, including automobiles they use. In response to this new consciousness, engines that run on biodiesel fuel or a blend of petro diesel and biodiesel are being manufactured. Automobile industry is now offering hybrid vehicles that operate on a combination of fuels and electricity. Increasing investments on service stations that deal in alternative biofuels and recharging are needed to support these attempts.

QUESTIONS (THEORY)

1. Name and describe various kinds of feedstocks used to produce biofuels.
2. What is meant by biodiesel? What are its advantages and limitations?
3. Name various methods of biodiesel production. Briefly discuss each of them.
4. List and describe various steps in the transesterification of vegetable oils.
5. Draw a neat diagram of the flow process to manufacture biodiesel from *Jatropha curcus*. Briefly explain the various steps.
6. What do you understand by bioethanol? List its advantages and limitations?
7. Write a short note on characteristics of methanol and ethanol.
8. Name various steps to manufacture ethanol from starchy feed stock (grains). Draw a neat diagram showing the flow process of manufacturing ethanol from corn.
9. Draw a neat diagram showing the flow process of manufacturing first generation ethanol from sugarcane.
10. Draw a neat diagram showing the flow process of manufacturing second generation ethanol from cellulosic material. Briefly discuss the various steps of the process.
11. Briefly discuss the thermo-chemical method of methanol and ethanol production. List the pros and cons of cellulosic ethanol.

Chapter 10

Solar Energy Measurement and Applications

THE SUN AND THE EARTH

The Sun is an extremely powerful energy source. It is a large sphere of very hot gases having the diameter of 1.39×10^6 km. The energy in the sun is generated through the process of nuclear fusion taking place deep in the Sun. Protons of hydrogen atoms in this process violently collide in the Sun's core and fuse to create a helium atom. It is possible because of the extreme pressure estimated to be the equivalent of 250 billion atmospheres (25.33 trillion KPa), and temperature of about 15.7 million Kelvin that exists within the core. Since about 620 million metric tons of hydrogen is fused every second, it emits an enormous amount of energy in the form of light and heat. The process, also known as a proton-proton (PP) chain reaction, happens mostly in the inner 30 per cent radius of the Sun. The rest of the Sun is heated by the energy transferred from the core through the successive layers, eventually reaching the solar photosphere and escaping into space as sunlight or the kinetic energy of particles. The sun releases energy at a mass-energy conversion rate of 4.26 million metric tons per second, which produces the equivalent of 384.6 septillion watts (3.846×10^{26} W). At the surface of the sun, the temperature drops to about 5,700 K. It is why; the radiation coming from the Sun as viewed from the earth appears to be essentially equivalent to that coming from a black surface at 5762 K. The intensity of solar radiation, also known as electromagnetic radiation, at the surface of the Sun is about 6.33×10^7 W/m^2. Lastly, we have the visible surface of the sun *i.e.* the photosphere. It is from here that the sunlight and heat radiated and convected to the surface propagates out into space and reaches Earth's atmosphere and the other planets of the solar system. Temperatures in the layer range between 4,500 and 6,000 K.

The earth, having the diameter of 1.27×10^4 km, lies within sun's habitable zone with the mean distance between the two being 1.5×10^8 km. This means that it is neither too close nor too far to receive the Sun's abundant energy in the form of light and heat. Sun light, termed as solar energy, is by far the largest source of energy received by Earth. All life on the earth depends on this energy. In fact, the Earth would be barren without light and infrared radiation. Green plants make food using sun light through the process of photosynthesis. Humans and animal get their food from plants or by eating other animals that feed on plants. The energy received from the Sun is absorbed by the Earth's air and crust, heating the planet and providing organisms with a source of energy. While every location on Earth receives some sunlight over the year, the amount of solar radiation that reaches any one spot on the Earth's surface varies widely. Nonetheless, the amount of sunlight that strikes the earth's surface in an hour and a half is enough to take care of the entire world's energy consumption for a full year. The solar energy that falls on India in one minute is enough to supply the energy needs of our country for a day. Although this amount of energy is quite significant, it does not mean that solar energy can easily provide all of our primary energy needs. For this to be true, solar equipment/panels that capture this radiation and convert it into electricity or other usable forms of energy should be 100 per cent efficient. Besides, other problems associated with solar energy such as cloudy days and the lack of reasonable ways to store excess energy collected on sunny days may limit the exploitation of full potential of solar energy.

Solar Radiation at the Earth's Surface

The Earth at top of its atmosphere continuously receives 1.8×10^{17} W of incoming solar radiation, which by nature is low density energy source. About 98 per cent of the total emitted energy lies in the spectrum range of 250 nm to 3000 nm. About half of the radiation is in the visible short-wave part, other half in the near-infrared part and some in the ultraviolet part of the electromagnetic spectrum. The solar radiation while traversing through the atmosphere is subjected to scattering, reflection and absorption by air molecules, aerosols and clouds (Figure 10.1). Nearly all the ultraviolet radiation having wavelength less than 0.286 nm is absorbed by oxygen and ozone (O_3, ozone layer in stratosphere). Water vapours and carbon dioxide absorb some of the energy in the infrared range. Thus, only half of the solar radiation at the top of atmosphere reaches the Earths' surface (Figure 10.1). *Solar radiation that has not been absorbed or scattered and reaches the ground directly from the sun is called "direct radiation" or beam radiation.* Atmospheric conditions can reduce direct beam radiation by 10 per cent on clear, dry days and by 100 per cent during thick, cloudy days. Part of the solar radiation is scattered *i.e.* its direction is changed by droplets in clouds, atmosphere molecules, and dust particles. This diffused radiation comes to the earth from all parts of the sky (Not shown in Figure 10.1). *Diffuse radiation is the solar radiation received from the sun after its direction has been changed by reflection and scattering by the atmosphere.* The sum of the beam and diffuse radiation flux is referred to as total or global radiation. The radiation budget represents the balance between incoming energy from the Sun and outgoing thermal (long-wave) and reflected (short-wave) energy from the Earth. Globally, the budget is balanced although locally there may be some diurnal, seasonal and annual variations in the budget. Tropical areas get more than they release, while higher latitudes of the winter hemisphere release more than they receive.

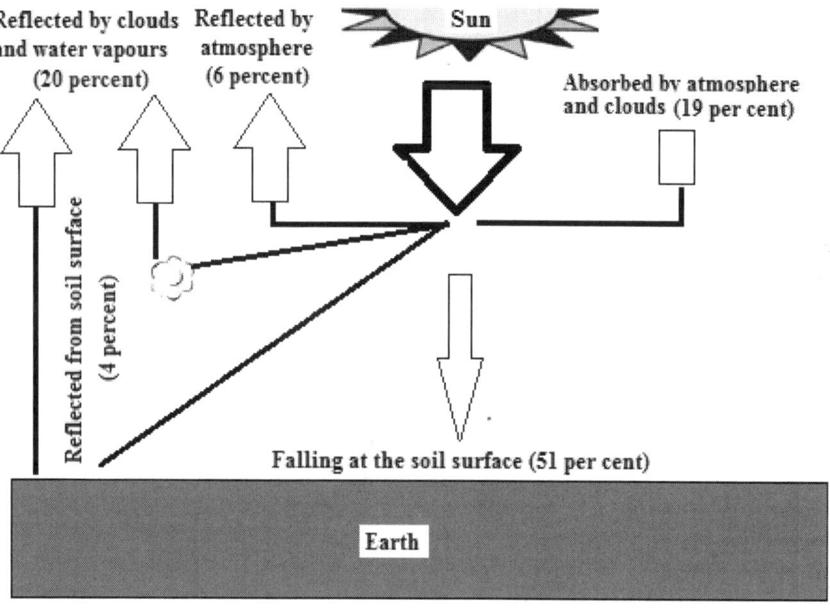

Figure 10.1. Solar radiations, atmosphere and Earth interactions.

Solar Constant

The solar constant (SC) is defined as the average radiation intensity falling on an imaginary surface perpendicular to the Sun's rays at the edge of the Earth's atmosphere. For easy remembrance, its value is taken as 2 cal/min-cm^2; being about 3 per cent greater than the average value during the summer and about 3 per cent smaller during the winter in the northern hemisphere because the earth's orbit around the sun is not exactly circular (Kumar *et al.*, 2021http://www.agrifair.in/). With appropriate conversions, it can be shown that the yearly average value of the solar constant equals 1389 W/m^2. However, the World Radiation Center (WRC) has adopted a value of 1367 W/m^2, with an uncertainty of the order of 1 per cent. As already stated, this value may vary because the output of the Sun changes by about ±0.25 per cent due to Sun spot cycles. Moreover, the value of this constant is being constantly upgraded as more data are being produced. Recent estimate of the value of the Sun's constant is 1353 W/m^2, which is equal to 116.5 Langley per hour (1 Langley being equal to 1 cal/cm^2 of solar radiation received in one day) or 429.2 Btu per ft^2 per hour. Apparently, the word 'constant' is also a little misleading. Because of the Earth's elliptical orbit the intensity of the solar radiation falling on the Earth changes by about 7 per cent between January 1st, when the Earth is nearest to the Sun, and July 3rd, when the Earth is farthest from the Sun. In terms of the amount received on the earth surface, it works out to be 345 W/m^2, as the earth surface is spherical (area = $4\pi r^2$). The solar constant, 1389 W/m^2 is divided by 4 to arrive at this value. The radiant power incident on the entire surface of the earth becomes (345 W/m^2) × (5.10×10^{14}) = 1.77×10^{17} or ≈ 5×10^{31} erg/year. The number 5×10^{31} erg/year provides a reasonable resource estimate for solar energy availability over the entire planetary surface. Note that radiation data for solar electric (photovoltaic) systems are often represented as kilowatt-hours per square meter (kWh/m^2). Radiation data for solar water heating and space heating systems are usually expressed in British thermal units per square foot per hour (Btu/ft^2 per hr).

Greenhouse Effect

A large fraction of the solar energy in the form of infrared, visible, and UV radiations reaching the Earth is reflected back into space in the form of infrared waves. As they pass through the atmosphere, they are intercepted by greenhouse gases, such as water vapour, carbon dioxide, methane, nitrous oxide besides others. Greenhouse gases trap the reflected heat as these gases act like the glass walls of a greenhouse. This greenhouse effect keeps the Earth warm enough to sustain life.

SOLAR RADIATION AND MEASURING INSTRUMENTS

The two common methods to characterize solar radiation are the solar radiance and solar insolation. *The solar radiance is the rate at which radiant energy is incident per unit area of the surface. It is an instantaneous power density in units of kW/m^2.* The solar radiance varies throughout the day. It is 0 kW/m^2 at night and a maximum of about 1 kW/m^2. The solar radiance is strongly dependant on the location on the Earth's surface and local weather. Solar radiance measurements consist of global and/or direct radiation measurements taken periodically throughout the day. The measurements are taken using either a pyranometer (measuring global radiation) and/or a pyrheliometer (measuring direct radiation).

Solar insolation, the other form of radiation data used in system designs, is defined as *the total amount of solar energy received at a particular location during a specified time period, often in units of kWh/m^2- day*. The total solar irradiation received in a day can vary from 0.5 kWh/m^2-day in the UK winter to 5 kWh/

m^2-day in the UK summer and can be as high as 7 kWh/m^2-day in desert regions of the world, such as regions of Nigeria and the Sahara in Algeria (Punter, 2011). Many tropical regions do not have large seasonal variations and receive an average 6 kWh/m^2-day throughout the year. Note that the units of solar insolation and solar irradiance are both a power density (the hours in the numerator for solar insolation are a time measurement as is the day in the denominator), yet solar insolation is quite different than the solar irradiance. The solar insolation is the instantaneous solar irradiance averaged over a given time period. Solar insolation data are commonly used for simple PV system designs while solar radiance is used in more complicated PV system performance, which calculates the system performance at each point of the day.

Thermopile Pyranometer

The thermopile pyranometer has been named after the sensor used in the instrument, which is based on thermopiles. A pyranometer measures either global or diffuse radiation over a hemispherical field of view. Basically, the pyranometer consists of two concentric hemispheres, 30 and 50 mm in diameter respectively (Figure 10.2). These are made of optical glass having excellent transmission characteristics. They protect the disc surface from the bad weather and also shield the thermopile sensor from convection. A pyranometer limits the spectral response from 300 to 2,800 nanometers cutting off the part above 2,800 nm but preserves the 180° field of view. The thermopile sensor used in

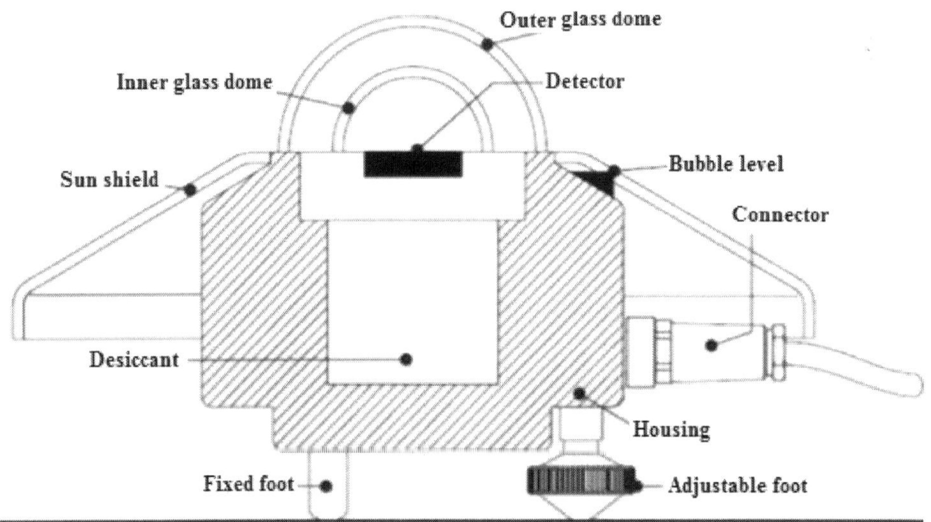

Figure 10.2. A line diagram of pyranometer.

the instrument has a black coating, which heats up when exposed to solar radiation. Its temperature increases until its rate of heat gain by solar radiation equals its rate of heat loss by convection, conduction and radiation. Active (hot) junctions of the thermopile located beneath the black coated surface are heated by the radiation absorbed by the black coating. The passive (cold) junction of the thermopile is fully protected from solar radiation and is in thermal contact with the pyranometer housing. It serves as a heat-sink. As a result an electromotive force (emf) is generated. The small voltage so generated is in proportion to the temperature difference between the black coating surface and the instrument housing, being of the order of 10 μV (micro-volts) per W/m^2. On sunny days the output may be around 10 mV (milli-volts). This can be read and recorded or integrated over a period of time, a measure of the global radiation. Accuracy of the instrument is about ±2 per cent.

Pyrheliometer

A pyrheliometer specifically measures the direct beam solar irradiance, often abbreviated as DNI: direct normal incidence, with a field of view limited to 5°. It is achieved by the shape of the collimation tube, with precision apertures and the detector design. It has a solar tracking system that keeps the instrument aimed at the sun throughout the day. The main components of a pyrheliometer are a quartz window, a black absorber, a thermopile, the pyrheliometer tube which defines the field of view and in some cases additional electronics (Figure 10.3A). Sights are included to enable the instrument to be pointed correctly. Pyrheliometer irradiance sensors also incorporate thermopiles, the sensors based on the Seebeck- or thermoelectric effect. The window (Figure 10.3A, 10.3B) on a pyrheliometer acts as a filter that transmits solar radiation with wavelengths from roughly 200 nm to about 4000 nm (this contains the near-infrared, visible, UV-A and part of the UV-B radiation), but blocks thermal radiation with wavelengths longer than 4 μm. A pyrheliometer is often used with a pyranometer (Figure 10.3A). The direct radiation E, together with diffuse radiation E_d, gives the total amount of solar energy available on the Earth's surface, the global irradiance $E_g\downarrow$.

$$E_g\downarrow = E\cos(\theta) + E_d$$

Here θ is the angle between the surface normal and the position of the sun in the sky. Pyrheliometer measurement specifications are subject to International Organization for Standardization (ISO) and World Meteorological Organization (WMO) standards. The field of view of the instrument is limited so that radiation coming only directly from the sun is measured. As per WMO convention, all modern pyrheliometer use the same field of view, characterized by an opening half-angle of 2.5°. It means the measurement of direct solar radiation includes some circumsolar radiation. Pyrheliometer is the most accurate instrument to measure sunshine duration. The WMO defines *sunshine hours as "the sum of the time intervals (in hours) during which the direct normal solar irradiance exceeds a threshold of 120 W/m².*

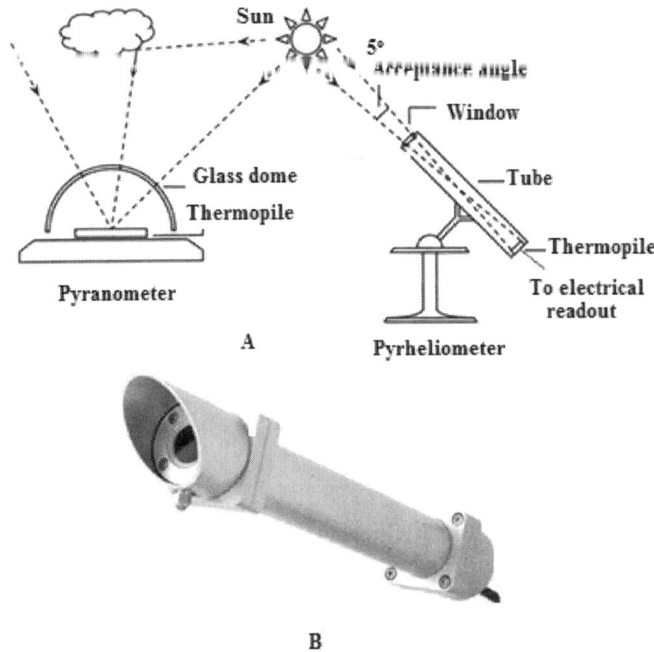

Figure 10.3. Schematic line diagram of a pyranometer and a pyrheliometer (A) and a photographic view of the pyrheliometer (B).

Sunshine Recorder

Sunshine duration is determined by a sunshine recorder. The recorder, also known as Campbell-Stokes recorder, measures the number of hours in the day during which the sunshine is above the threshold value. It is less accurate than a pyrheliometer, but is less expensive and is widely used for this purpose. A sunshine recorder consists of a glass sphere (Figure 10.4). It focuses the Sun's rays on a point on a card strip held in a groove in a spherical bowl mounted concentrically with the sphere (Figure 10.4). Whenever there is bright sunshine, the image formed is intense enough to burn a spot on the card strip. As the sun moves across the sky, the image also moves along the strip. Thus, the length of the burnt trace obtained on the strip is proportional to the duration of sunshine. Data collected in this way are used to determine the solar insolation by comparing the measured number of sunshine hours to those based on calculations after applying several correction factors. Older recorders required a human observer to interpret the results. Modern recorders use electronics and computers to precisely interpret the data to avoid any errors of human interpretation. Newer recorders can also measure the global and diffuse radiations.

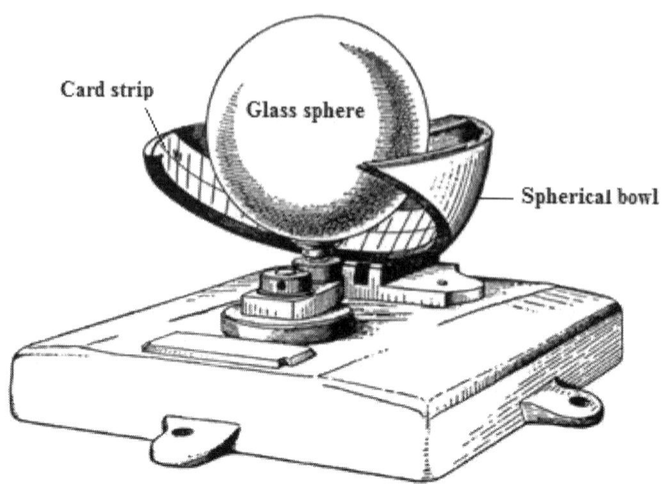

Figure 10.4. A view of a sunshine recorder.

Availability of Solar Energy in India

India is known as a solar belt country. On an average it receives 3.0 to 7.5 kWh/m^2 energy in a day in various parts of India. Solar radiations of the order of 3.0–6.5 kWh/m^2-day (10.8–23.4 MJ/m^2-day) are received over 90 per cent of India during the summer season. However, in northern parts of the country this value can reach a maximum of 7.5 kWh/m^2-day (27 MJ/m^2-day) in the month of May (Ramachandra *et al.*, 2011). Mean annual Global Solar Radiant Exposure over India is 15.40 MJ/m^2-day at Srinagar to 20.99 MJ/m^2-day at Bhavnagar. Mean annual Diffuse Solar Radiant Exposure over India is 5.40 MJ/m^2 at Srinagar to 8.76 MJ/m^2 at Chennai. This solar radiation potential can be utilized in desalination, solar-thermal collectors, heating and day-lighting of building and Photovoltaic (PV) Cells *etc.* The estimated solar energy potential of the country is 200 MW/km^2, which can be utilized both for thermal and electrical applications. Based upon availability of land and solar radiation, National Institute of Solar Energy assessed the potential solar power in the country to be around 750 GWp (gigawatts peak). State-wise estimates of solar energy potential in the country are given in Table 10.1.

Applications

Technological processes to utilize solar energy can be categorized under the three heads.

Heliochemical: The energy in this process is absorbed by the chemical reaction hence the process is called Heliochemical. The plants absorb solar energy to produce various products through the process of photosynthesis.

Heliothermal: This process is used for heating of building, water *etc.*

Table 10.1. State-wise estimated solar energy potential in the country

State/UT	Solar Power Potential (GWp)
Andhra Pradesh	38.44
Arunachal Pradesh	8.65
Assam	13.76
Bihar	11.20
Chhattisgarh	18.27
Delhi	2.05
Goa	0.88
Gujarat	35.77
Haryana	4.56
Himachal Pradesh	33.84
Jammu and Kashmir	111.05
Jharkhand	18.18
Karnataka	24.70
Kerala	6.11
Madhya Pradesh	61.66
Maharashtra	64.32
Manipur	10.63
Meghalaya	5.86
Mizoram	9.09
Nagaland	7.29
Odisha	25.78
Punjab	2.81
Rajasthan	142.31
Sikkim	4.94
Tamil Nadu	17.67
Telangana	20.41
Tripura	2.08
Uttar Pradesh	22.83
Uttarakhand	16.80
West Bengal	6.26
UTs	0.79
Total	**748.98**

Helioelectrical: The solar energy serves as an input to photovoltaic converters and output power in the form of electrical energy is used for various purposes.

Solar energy in heliothermal and helioelectrical processes is commonly harvested through passive and active systems. Passive solar technologies do not use any external device. Instead, they take advantage of the local climate to heat structures during the winter, and reflect heat during the summer. In such systems, orientation of the structures, their design and/or the materials used in construction optimize the use of heat or light directly from the sun. For example, we often talk of orientation while constructing the greenhouses. Similarly, materials for the greenhouses are selected to take full advantage of the sun energy. Active solar technologies on the other hand use electrical or mechanical devices to actively convert solar energy into another form of energy, most often heat or electricity. Photovoltaic power stations have been built all over the world including India. These power stations generate hundreds of megawatts of electricity to meet the energy requirement of homes, businesses, schools, and hospitals. A major advantage of photovoltaic technology is its scale. Solar cells can be small enough to power smaller devices such as calculators and parking meters. They can work as standalone units to supply electricity for an individual structure say irrigation pumps or building. They can be placed along roads to light highways. On larger scale, grid connected solar power systems can supply electricity to industrial hubs. While a concise list of some applications of solar energy is included in this section, major discussion on some of the important applications will follow in subsequent chapters.

- Solar water heating
- Solar cookers
- Solar distillation
- Solar pumping
- Solar heating of buildings
- Solar drying of agricultural and animal products
- Solar furnaces
- Solar electric power generation
- Solar thermal power production
- Solar refrigeration
- Solar green houses
- Electric fences
- Water treatment
- Power applications in satellites

Heating of buildings: Solar energy is used for heating the buildings by directly admitting the solar radiation into them through appropriate designs and location of windows. The other way is to use separate solar collectors which may heat either water or air or storage devices which accumulate the collected solar energy for use at night and during inclement days. The heat from these collectors or storage devices can be transferred by conventional equipment such as fan, ducts, air outlets, radiators and hot air registers *etc.* to warm up the living spaces of a building.

Solar furnaces: Extremely high temperatures can be produced in solar furnaces compared to the range achievable in laboratories with flames and electric currents. It is achieved by concentrating the solar radiations onto a specimen using a number of heliostats (turn-able mirrors) arranged on a sloping surface. The solar furnace is especially useful for metallurgical and chemical operations. An important future application of solar furnaces is the production of nitric acid and fertilizers from air.

Solar cooking: A variety of fuel like coal, kerosene, cooking gas, firewood, dung cakes and agricultural wastes are used for cooking purposes. Supply of these fuels is either decreasing/deteriorating (wood, coal, kerosene, cooking gas) or are too precious to be wasted for cooking purposes (cow dung can be better used as manure for improving soil fertility). This necessitates the development of solar cookers for using solar energy for cooking purposes.

Solar thermal power plants: Solar thermal power production means the conversion of solar energy into electricity through thermal energy. In this process, solar energy is first utilized to heat up a working fluid, gas, water or any other volatile liquid. This heat energy is then converted into mechanical energy in a turbine. Finally a conventional generator is coupled to the turbine to convert the mechanical energy into electrical energy.

Solar ponds: A solar pond is a natural or artificial body of water utilized for collecting and absorbing solar radiation and storing it as heat. It is a shallow (5-10 cm deep) pond having a radiation absorbing (black plastic) bottom. It has a curved fiber glass cover that permits the solar radiation to enter but reduces losses by radiation and convection (air movement). A bed of insulating material under the pond is used to minimize loss of heat to the ground. Solar energy stored in solar ponds is used for many applications such as space heating, industrial process heating and to generate electricity by driving a turbine powered by evaporating an organic fluid with a low boiling point.

Solar refrigeration: Space cooling is one of the promising applications of solar energy to provide comfortable living conditions (air-conditioning). It can also be used to run a refrigerator for keeping food products at low temperature to increase their shelf life. Solar energy can be utilized for refrigeration purposes from one of the three methods namely: solar electric method, solar mechanical method and solar thermal method. In solar electric method, the solar energy is directly converted to DC current by an array of solar cells known as photovoltaic (PV) panel. In solar mechanical method, the mechanical power required to drive the compressor is generated by solar driven heat power cycle. Rankine cycle is commonly used for the heat power for this process. A typical solar thermal refrigeration system consists of four basic components - a solar collector array, a thermal storage tank, a thermal refrigeration unit and a heat exchange system to transfer energy between components and the refrigerated space.

Greenhouses: Agricultural greenhouse is a structure covered with transparent material (glass or plastic) that acts as a solar collector and utilizes solar radiant energy to improve plant productivity. Solar radiations can pass through the greenhouse glazing but the thermal radiations emitted by the objects within the greenhouse cannot escape through the glazed surface. As a result, the radiations are trapped within the greenhouse resulting in an increase in temperature. Heating, cooling and ventilating devices can be provided to control the temperature inside the green house. Moreover, the greenhouse structure being a closed boundary structure, the air inside the greenhouse is enriched with CO_2 because there is no mixing of the greenhouse air with the ambient air. All these features help to sustain plant growth at a higher level than in the open.

HEAT TRANSFER PROCESSES

Heat transfer is the study of transmission of thermal energy from a high temperature region/body to a low temperature region/body on account of temperature difference. Transfer of heat between two regions/bodies maintained at different temperatures can occur in three different modes namely: Conduction, Convection, and Radiation. In the conduction and convection modes, heat flows from high temperature to low temperature region/body whereas in radiation mode, transfer of heat takes place from both the bodies towards each other. Even in this case, net transfer of heat is always from high temperature body to low temperature body.

Conduction

Conduction is a process of heat transfer from a high temperature region to a low temperature region within a body or between different bodies which are in direct physical contact. The process is more common in solids in which atoms are tightly bound. When heated, they do not leave their sites; but vibrate about their respective equilibrium positions with increased kinetic energy (KE). Being in contact with their nearest neighbouring atoms, they pass on some of their kinetic energy to them. These atoms further transfer some KE to their neighbours and so on. This process continues and kinetic energy is transferred from one atom to another. As average kinetic energy is proportional to temperature, the heat is transferred from one place to another and the process is known as conduction. The basic law of heat transfer by conduction was proposed by the French Scientist J. B. J. Fourier in 1822. One dimensional conduction rate equation described by the Fourier Law is written as:

$$Q_x = - k\, A\, dT/dx \tag{10.1}$$

Where, Q_x is the heat flow, W, k is the thermal conductivity of the material, W/(m-K), A is the cross-sectional area in the direction of heat flow, m^2, and dT/dx is the temperature gradient, K/m. Minus sign appears because the heat flows from higher to lower temperature. It can be shown that quantity of heat transferred from one face to another per unit time depends upon the temperature difference, ΔT ($T_h - T_c$) *i.e.* the temperature difference at the hot and cold ends respectively, the cross-sectional area, A and the inverse of the length, L as per the following relation.

$$Q_x = KA\,(T_h - T_c)/L \qquad (10.2)$$

Convection

Heat transfer by convection occurs when a fluid (Liquid or gas) comes in contact with a solid through direct contact and a temperature difference exists between them. *Heat transfer by convection occurs under the combined action of heat conduction and mixing motion.* In convection, molecules of fluids receive thermal energy and move up bodily. Thus, in convection heat is transferred by mass motion of a fluid such as air or water when the heated fluid is caused to move away from the source of heat carrying energy with it. Convective heat transfer rate is governed by Newton's law of cooling and is expressed as:

$$Q = h\,A\,(T_2 - T_1)^a \qquad (10.3)$$

Where, h is convective heat transfer coefficient in W/m^2-K, A is heat transferring area, m^2, T_2 and T_1 are temperatures of surface and the fluid respectively, a = 1 but sometimes it is taken as 5/4.

Radiation

Heat exchanged between two bodies or mediums, which are separated and are not in contact with each other, is called radiation heat transfer. Radiation heat transfer does not require presence of an intervening medium between the two bodies as in the case of conduction and convection. It takes place most effectively in a vacuum. Most of the energy received on the earth from the Sun is through radiation heat transfer. Energy emitted per unit area as thermal radiation is called emissive power of a body. The maximum energy emitted as radiation by a body at a particular temperature is governed by Stefan-Boltzmann law which is expressed as

$$E_b = \sigma\,AT^4 \qquad (10.4)$$

Where, E_b is the energy emitted per unit time, W, A is the surface area, m^2, T is the absolute temperature of the body, K and σ is Stefan-Boltzmann constant which is equal to 5.67×10^{-8} W/(m^2 - K^4). At a given temperature, maximum radiations are emitted by an ideal emitter called black body. The energy emitted by non-black bodies is less as compared to the ideal body when both are maintained at the same temperature. All bodies radiate energy and receive energy emitted by other bodies simultaneously. Consider heat exchange between two black bodies maintained at temperatures T_1 and T_2 respectively such that body 1 is completely enclosed by body 2. The net heat transfer by radiation from body 1 to body 2 is given as

$$Q_{1\text{-}2} = \sigma A_1(T_1^4 - T_2^4) \tag{10.5}$$

QUESTIONS (THEORY)

1. Write a descriptive note on solar radiation at the earth's surface
2. Define and briefly illustrate the term solar constant.
3. Write short notes on: thermopile pyranometer, pyrheliometer, sunshine recorder and greenhouse effect.
4. List and briefly describe the various applications of solar energy.
5. Discuss various heat transfer processes highlighting the process of heat transfer by radiation.

Chapter 11

Solar Collectors and their Applications

SOLAR COLLECTORS

Conversion of solar energy to thermal energy is accomplished using a special kind of heat exchanger known as solar collector. Thus, a solar collector is a device that collects and/or concentrates solar radiations from the Sun. The device is primarily used for active solar heating and is an engine of any solar heating system. It absorbs the incoming solar radiation, converts it into heat, and transfers this heat to a fluid (usually air, water, or oil) flowing through the collector. The solar energy so collected is either used directly to heat the water or carried by the circulating fluid to space conditioning equipment, or to a thermal energy storage tank from where it can be drawn for use at night and/or on cloudy days. The purpose of heating may be for domestic use as in solar water heaters or for industrial use as in solar thermal power production. Accordingly, there can be wide variations in the design, complexity and cost of solar collectors.

Types of Collectors

Solar collectors are classified according to their solar power collecting characteristics, the way they are mounted, and the type of transfer fluid used in them. On the basis of collecting characteristics these are classified as non-concentrating and concentrating types (Figure 11.1). *In a non-concentrating collector the absorbing surface for solar radiations (the area that intercepts the solar radiation) is the same as the absorber area (the area that absorbs the radiation).* Flat plate collectors (FPC) and evacuated tube collectors (ETC) are essentially non-concentrating type collectors. These collectors are mainly designed for hot water and industrial process heat applications, which require energy delivery in the temperature range of 60-250°C. These collectors use both diffuse and beam solar radiation and do not have any system to track the sun. These collectors are simple in construction than concentrating collectors and are easy to maintain. *A concentrating or focusing collector usually contains reflectors or employs other optical means to concentrate the energy falling on the aperture onto a heat exchanger of surface area smaller than the aperture. It increases the radiation flux.* Different types of concentrating type collectors include compound parabolic concentrator (CPC), parabolic dish concentrator, parabolic trough concentrator and solar tower concentrator (Figure 11.1).

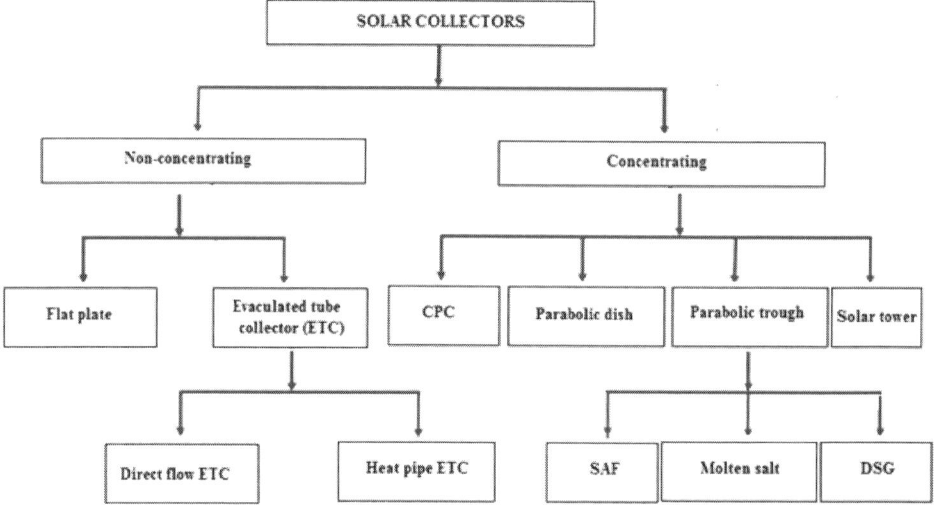

Figure 11.1. Categorization of solar collectors.

On the basis of mounting, collectors are categorized in three categories namely stationary, adjustable with tilt angle (measured from the horizontal) adjustment with changing solar declination, and the collector that tracks the sun. Tracking of the sun employs either an alt-azimuth mounting or equatorial mounting both increasing the absorption of the solar irradiation. An alt-azimuth mount is a simple two-axis mount for supporting and rotating the collector about two perpendicular axis – one vertical and the other horizontal. Its primary axis (the azimuth axis) is vertical to the ground. The secondary axis, often called elevation axis, is then typically normal to the primary axis. Rotation about the vertical axis varies the azimuth (compass bearing) of the pointing direction of the instrument. Rotation about the horizontal axis varies the altitude angle (angle of elevation) of the pointing direction. An equatorial mount has a declination axis allowing the collector to be moved north and south, and a right ascension or polar axis allowing the equipment to be moved east and west. It is similar to the alt-azimuth mount. The key difference is that the polar axis of an equatorially mounted collector must point at or be polar-aligned to the north celestial pole, the point in the sky around which all the other stars appear to rotate.

A collector uses either a liquid or a gas as the transfer fluids. The most common liquids are water or a water-ethylene glycol solution. The most common gas is air. Parabolic trough collectors depending upon the heat transfer fluid (HTF) are further sub-grouped as synthetic aromatic fluid (SAF), molten salt and direct stream generation (DSG) (Figure 11.1).

Flat Plate Solar Collector

The most common type of solar thermal collector is the flat plate collector, first developed in the 1950s. These collectors are primarily used to heat water for domestic purposes. A flat-plate collector essentially consists of (Figure 11.2 top):

- The absorber plate normally 20 gauge copper, steel, aluminium or GI corrugated sheet plate in thermal contact with the tubing. Both the plates and tubes are painted black or with special coatings to absorb and retain heat better than the traditional black paint
- 12 mm galvanized iron (GI) pipe tubes, fins, passages or channels all integral to the collector absorber plate or connected to it. They carry water, air or other fluid
- A transparent cover which may be one or more sheets of low iron solar safe glass or radiation transmitting plastic film or sheet
- Rock or mineral wool 2.5 to 8.0 cm thick provided on the back and sides for insulation to reduce the thermal losses
- The wooden or metal container that encloses the components to protect them from the adverse effects of weather and for safety

Solar radiation passes through the transparent cover and hits the absorber plate. It converts the absorbed radiation into thermal energy. Heated plate transfers the heat to either water (or air) held in the tubing being in thermal contact with the plate. Heated water is stored in an insulated hot water storage cylinder. The efficiency of a flat-plate solar water collector depends upon the thermal conductivity of the absorbing plate as it governs how best the heat from the absorber sheet is transferred to the absorber fluids in tubes. The efficiency of a standard liquid flat-plate collector can be improved by reducing the optical and thermal losses. This can be achieved by: i) multiple glazing with anti-reflective glass, ii) filling a hermetically sealed flat plate collector with a noble gas or iii) evacuating a hermetically sealed flat plate collector. Reflectors can be attached to the edges of the flat collector to further improve the efficiency (Figure 11.2 bottom). In such cases, some of the irradiation falls directly on the collector while some fall on the side mirrors and reflected to the collector. This arrangement has not proved much effective.

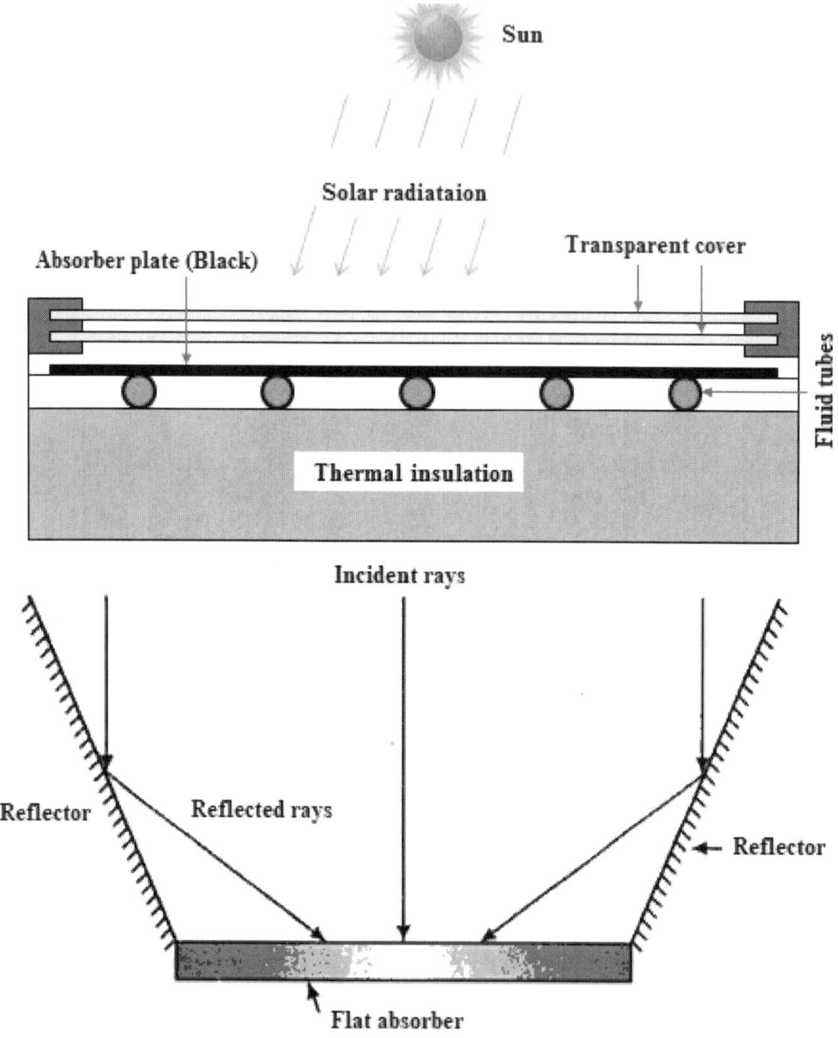

Figure 11.2. Schematic diagram of a flat plate solar collector.

Two basic types of flat plate collectors are the liquid type and air type. In the former, fluid is water or mixture of water and an antifreeze oil *etc.* Heat transfer medium in the air type collectors is air. These are mostly used for agricultural drying and space heating applications. The basic advantages of air type are their low sensitivity to leakages and no need for an additional heat exchanger for drying and/or space heating applications. These collectors require higher flow rates and large heat transfer area mainly because of the low heat capacity of the air and the low convection heat transfer coefficient between the absorber and the air.

Applications of Flat Plate Collectors

The flat plate solar collectors with improved designs are gaining popularity and are widely used for the following applications.

- Used for water heating in homes, hotels, hospitals and industries *etc.*
- Used for space and air heating
- Used for making distilled water
- Used for drying agricultural produce
- Used for wood seasoning

Losses in Flat Plate Collector

Cosine loss factor: For maximum power collection, the surface of collector should receive the sun rays perpendicularly. If the angle between the perpendicular to the collector surface and the direction of sun rays is θ, then the area of solar beam intercepted by the collector surface is proportional to cosθ (https://www.vssut.ac.in/lecture_notes/lecture1428910296.pdf).

Reflective loss factor: The collector glass and the reflector surfaces collect dust, dirt, moisture *etc.* The reflector surface gets rusted, deformed and loses its shine. Accordingly, the efficiency of the collector is significantly reduced with passage of time.

Evacuated Tube Collectors

An evacuated tube collector consists of a number of rows of parallel transparent glass tubes. The cylindrical shape of the glass tubes ensures that the angle of the sunlight is always perpendicular to the heat absorbing tubes. It enables the collector to perform well even when sunlight is low *i.e.* early in the morning or late in the afternoon. The evacuated tube consists of two concentric tubes, an inner and an outer tube. The outer solar tube is made of borosilicate glass, which is very low in iron and allows 98 per cent of light energy to pass through it. The inner tube is coated with a highly absorptive low emissivity coating material. It helps in absorbing more Sun's energy. Besides, it can withstand temperatures in excess of 300°C without any deterioration. The air between the two tubes is evacuated before the tubes are sealed together at the end. The vacuum so created improves the thermal performance of the evacuated tubes, as it eliminates the convection losses while improving thermal insulation.

Three common types of evacuated tube solar collectors having similar basic features, as discussed in preceding paragraph, are in use (Mishra and Saikhedkar, 2014).

- Water-in glass evacuated tube solar collector
- U-type evacuated tube solar collector
- Evacuated tube heat pipe solar collector

Water-in Glass Evacuated Tube Solar Collector

Water-in glass evacuated tube solar collector, also known as open vacuum tube system, consists of a vacuum glass tube collector as described previously, an insulated storage tank and connecting accessories. The water is heated directly with in the vacuum tube (Figure 11.3). Thermo-siphon process caused by density difference of cold and hot water makes the hot water to rise and be replaced by the cold water. As the cycle repeats, water heats up through thermo-siphon action based on natural convection. The insulated storage tank reduces heat losses so that water stays hot for use later in the day.

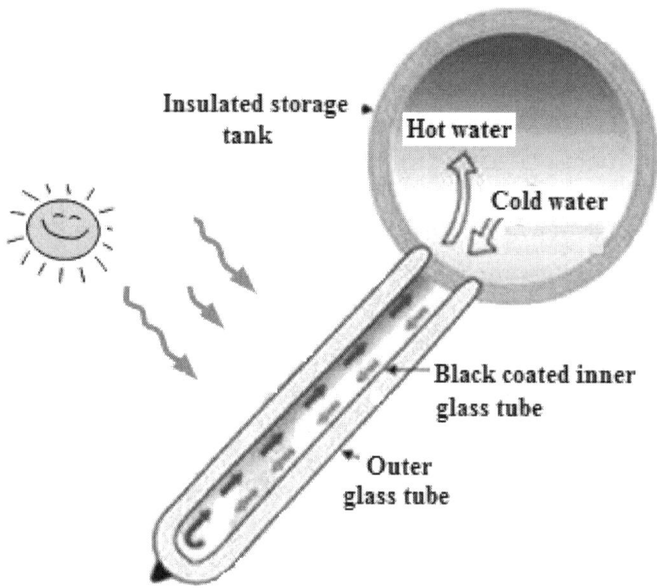

Figure 11.3. Water-in glass evacuated tube solar collector.

U-type Evacuated Tube Solar Collector

The construction of evacuated U-tube collector is almost similar to that of water-in-glass collector. Besides what has been stated previously, it has a circular fin to store and conduct the collected heat and a U-tube of copper within the evacuated tubes (Figure 11.4). As the sun rays pass through outer glass tubes, the radiations absorbed by the selective coating are converted to thermal energy. The transfer of heat to the water takes place by means of convection. As the water temperature in the tubes rises, it naturally moves up to the top of collector and goes to storage through the evacuated collector tube. Simultaneously, the cold denser water moves to bottom through the U-tube. The cycle continues. A key feature of this system is that the evacuated tube solar collector is given a little tilt because the straight tube would not let the water flow within the tube.

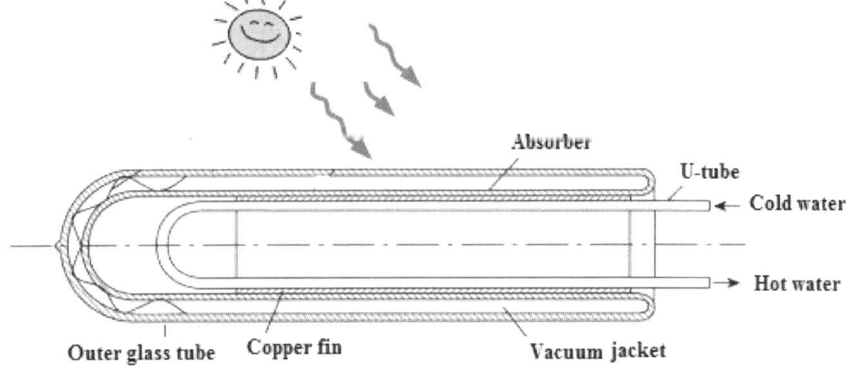

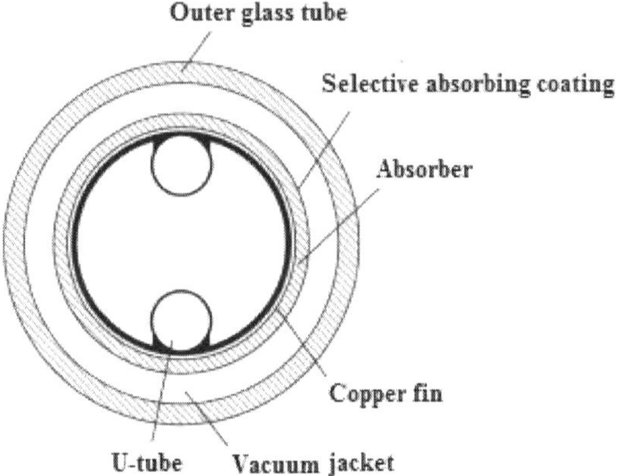

Figure 11.4. Line diagrams showing two sections of a U-type evacuated tube solar collector.

Evacuated Heat Pipe Technology

In this system, a sealed hollow heat pipe, usually made of copper, is attached to a heat absorbing reflector plate (fin) within the vacuum sealed tube. Copper is preferred material to increase the collector efficiency even under cold climatic conditions. Air from the hollow heat pipe is evacuated and a small quantity of liquid such as alcohol or purified water with some special additives is put inside the tube (Figure 11.5). The vacuum enables this liquid to boil at a lower temperature than at normal atmospheric pressure. When solar radiation hits the surface of the absorber plate inside the tube, the liquid in the heat pipe quickly turns into a hot vapour type gas. Gas vapours rise up to the top portion of the pipe and heat it to a very high temperature. The top portion of the heat pipe and the evacuated tube is connected to a copper heat exchanger called the manifold (Figure 11.5). When the hot vapours still inside the sealed heat tube enters the manifold, the heat energy of the vapours is transferred to the water or glycol fluid flowing through the manifold. Hot vapours are condensed back to liquid as they lose energy. The liquid flows back to the heat pipe and is reheated. The heat pipe and the evacuated tube collectors are mounted to have a minimum tilt angle of about 30° so that internal liquid of the heat pipe easily returns to the hot absorber plate at the bottom of the tube. This process continues inside the sealed heat pipe as long as the sun shines.

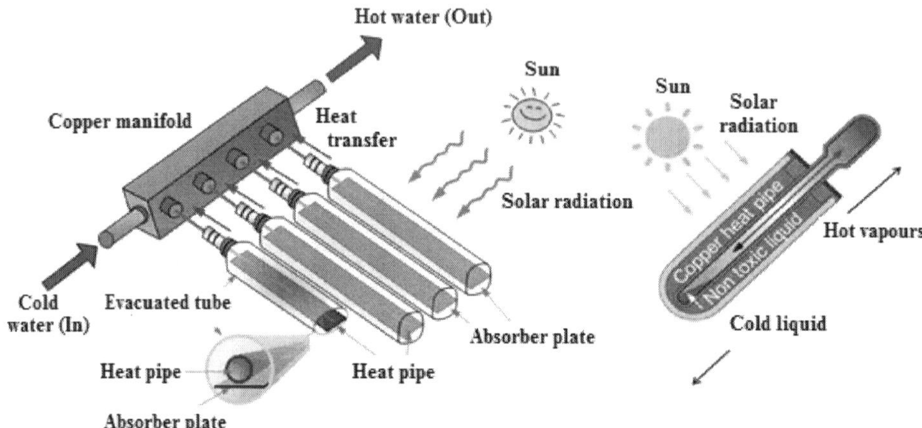

Figure 11.5. Evacuated heat pipe water collector (left) and the view of a single pipe (right).

Advantages of Evacuated Heat Pipe Technology

- ✰ It can be used under extremely hot to extremely cold weather. The vacuum ensures its operation under cold weather conditions
- ✰ Its installation is much easier than direct flow collectors because of the dry connection between the absorber plate and the manifold
- ✰ Evacuated tube solar collectors require less space
- ✰ Heat loss during day time is negligible
- ✰ Because of its tubular design it is able to absorb solar radiation from multiple angles
- ✰ The loss in vacuum remains limited to the individual tube in the event a tube is cracked or broken. The tube can be exchanged without emptying or dismantling the entire system. This flexibility makes it ideal for closed loop solar designs as the modular assembly can be easily installed. The system capacity can be easily expanded by adding as many tubes as one wants

Limitations of Evacuated Heat Pipe Technology

- ✰ The initial cost is high as compared to other collectors

- ☆ Construction of evacuated U-tube collector is little bit complicated requiring proper design and handling
- ☆ Improper handling and maintenance might result in high temperatures in the tube resulting in cracking of the tube due to thermal shocks

CONCENTRATING COLLECTORS

A reflecting (or refracting) surface (concentrator) is introduced between the solar radiation and the absorber of a concentrating or focusing collector. This surface collects the solar radiations and transmits high intensity solar energy on the energy-absorbing surface. Focusing collectors are capable of increasing the intensity of energy flux from a low value of 1.5 to 2 to high values of the order of 10,000 so that the temperature in the range of 450 to 500°C are achieved. Therefore, concentrating solar collectors are generally used for high-temperature applications, including steam production for generating electricity and thermal detoxification. *Concentration ratio of a concentrator is given by the ratio of the concentrator aperture to energy absorbing area of the receiver. It is used to determine the effectiveness of the concentrator.* Concentrating collectors have been categorized as line focus collectors such as parabolic trough collector and point focus collectors such as power tower collector discussed in the following sections.

Advantages of Concentrated Solar Collectors

- ☆ Zero fuel cost
- ☆ Can produce both electricity and heat
- ☆ Innovative designs ensure round-the-clock availability of electricity
- ☆ No carbon emission
- ☆ Employment generation
- ☆ Can be up-scaled to take advantage of economy of scale

Disadvantages of Concentrated Solar Collectors

- ☆ High cost
- ☆ May aggravate water shortage
- ☆ Environmental issue specifically related to wildlife
- ☆ Poor performance on cloudy days
- ☆ Applications limited to areas with high direct insolation such as arid and desert regions

General Construction and Types

Focusing collector comprises a receiver (absorber) and a concentrator. Concentrator can be cylindrical or parabolic and can be continuous or segmented. Receiver can be convex, flat, cylindrical or concave and can be covered with a glazing or may be left uncovered. Radiation flux on the receivers can be increased using wide variety of ways and means such as:

- Use of lenses or reflectors
- Varying the mounting and/or orienting the systems
- The concentration of direct along with diffused radiations
- By varying the materials of construction

Some collectors resemble flat plate collectors and can be mounted directly on to building surfaces. But in most cases concentrating collectors are mounted on free-standing structures to allow sun tracking. Five types of concentrating solar collectors are in common use.

- Parabolic trough collector
- Parabolic dish collector
- Power tower receiver
- Fresnel lens collector
- Compound parabolic collector

Parabolic Trough Collector

Augustin Mouchot was the first to develop a parabolic trough solar collector in France in 1866. Parabolic trough collector (PTC), also known as line focus collector, uses highly reflective materials to collect and concentrate the heat energy of the solar radiation. The main components of a PTC are: reflector, receiver through which heat transfer fluid flows (hot water pipe in Figure 11.6), support structure, piping and tracking system (Figure 11.6). The main working component is the parabolic shaped reflective sections shaped like a long trough. A pipe that carries water is positioned along the focal line of the reflector. The sunlight collected by the reflective material is focused onto the pipe that heats its contents. These troughs can be extremely effective particularly those that can pivot, and track the Sun to ensure maximum collection of sunlight. The sophisticated versions keep tracking the sun during the day along a single axis. These are high-powered collectors and are mostly used for generating steam for solar thermal power plants. The concentration ratio is usually about 70-80. A working fluid (*e.g.*, molten salt) flowing through the receiver is heated between 150 and 350°C, which is then used as a source of heat for generating electricity.

Parabolic Dish Collector

A parabolic dish collector or dish stirling has a stand-alone parabolic reflector. It consists of mirrors (or other reflective surfaces) in the shape of a parabolic dish (Figure 11.7). It concentrates solar radiations on a receiver placed at its focal point. The working fluid in the receiver is heated in the temperature range of 250°C to 700°C for use in a Stirling engine to produce electricity. Parabolic dish collectors are suitable for small-scale power generation (up to

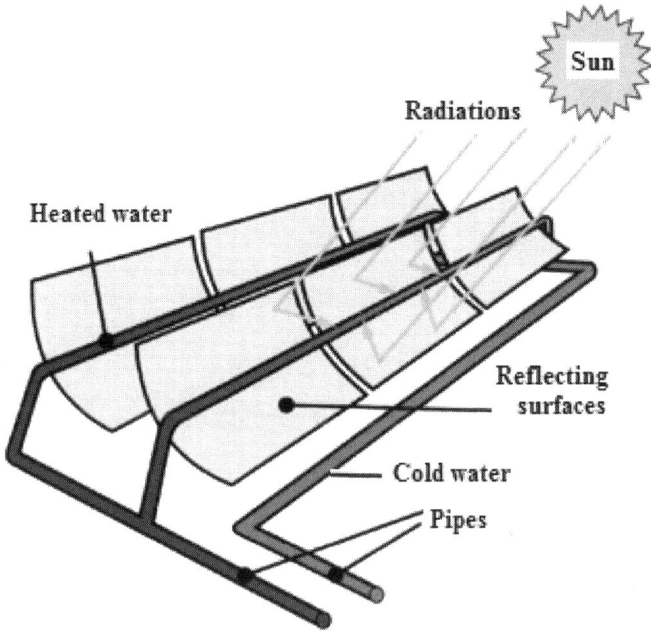

Figure 11.6. Schematic diagram of a line focus parabolic trough solar collector.

a few kW) because of the limitations of size and the small quantity of fluid. The reflector tracks the Sun along dual axis. Parabolic-dish systems have high solar-to electrical efficiency of about 32 per cent. The system being modular in nature can be easily scaled up.

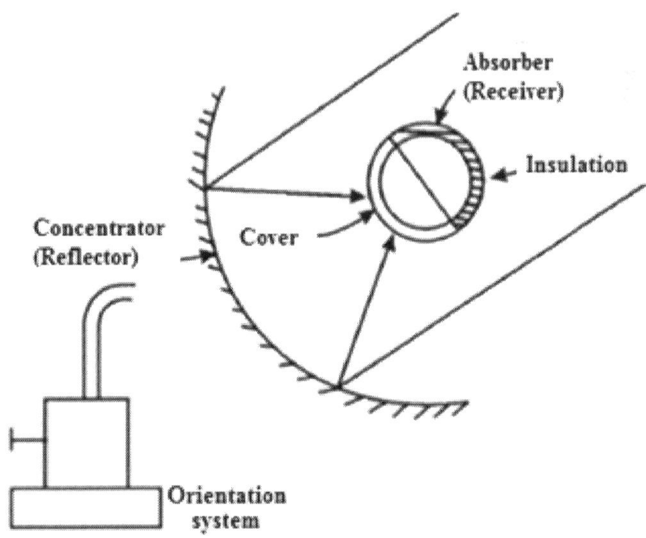

Figure 11.7. A line diagram of a parabolic dish collector.

Power Tower Collector

Power tower collector (PTC) comprises large number of flat, sun-tracking mirrors, known as heliostats. These heliostats focus sunlight on a receiver at the top of a tall tower (Figure 11.8). The heliostats are generally located in the horizontal plane, but under favourable situation can follow the existing terrain. In the typical central receiver, a heat-transfer fluid is used to heat a working fluid, which in turn is used in a conventional turbine generator to produce electricity. Some power towers use water/steam as the heat-transfer fluid. Other advanced designs use high temperature molten salts or sand-like particles to maximize the power cycle temperature.

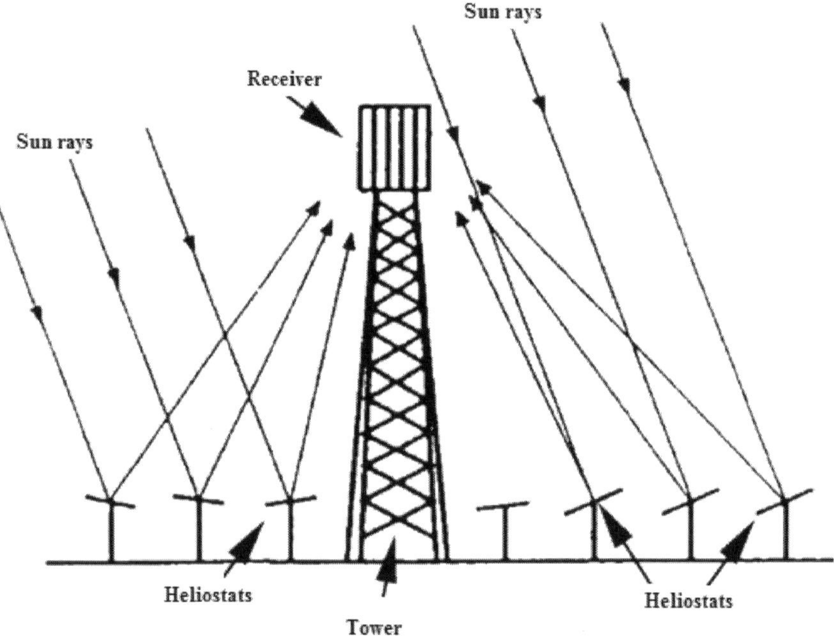

Figure 11.8. A schematic view of a power tower concentrator.

Linear Fresnel Collector

Linear Fresnel collector (LFC) is quite similar to PTC, but the parabola is divided into many small, nearly flat, and long mirrors (Figure 11.9). Each mirror moves independently, but all the mirrors move simultaneously to concentrate sunlight on the linear receiver (absorber) located at the optical focus. The concentration ratio is usually in the range of 50-60. Typically, the temperature can reach as high as 400°C. LFCs only capture direct irradiation (diffuse radiation is not utilized). Efficiency of Fresnel collectors is low when compared with other methods. Recently some new models with ray tracing capacity have been tested and have higher output than the standard version.

Nonetheless, LFCs have certain advantages over PTCs. Flat mirrors in a LFC accommodate more reflective surfaces in the same amount of space than a parabolic reflector. They capture more sunlight besides being much cheaper than parabolic reflectors. Wind effect is minimal because mirrors are small and close to the ground. The main disadvantages over PTCs result from significant shading in the early morning and late afternoon. Moreover, their efficiency is low as stated previously.

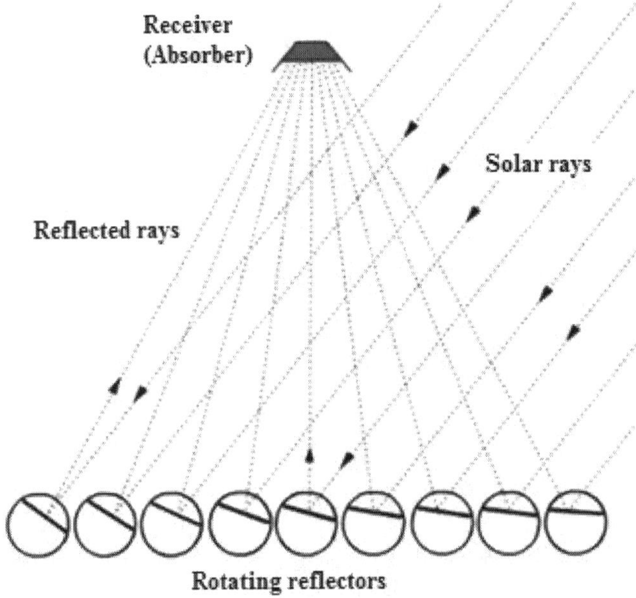

Figure 11.9. A schematics of linear Fresnel collector.

Another version of this type of collector is circular fresnel lens concentrator, generally used where high flux is desired, such as with silicon solar cells or with gallium arsenide solar cells as receiver. It is divided into a number of thin circular zones. The tilt of each zone is so adjusted that optically the lens approximates a thin spherical lens. The concentration ratio may be as high as 2000, but is less than that obtained from a paraboloidal reflector. In solar cell applications, tracking is required to keep the small solar image centered on the receiver.

Compound Parabolic Collector

The name compound parabolic concentrator has emerged from the fact that the two parabolic reflectors which funnel the radiations from aperture to absorber belong to different parabolas. It consists of mirrors shaped in the form of segments of parabolas that are attached to the receiver (collector). Entering sun rays either hit the receiver directly, or hit the receiver after being reflected by the parabolic segments (Figure 11.10). Compound parabolic collector (CPC or Winston collector) is capable of concentrating solar radiation by a factor of

10 even without diurnal tracking. Properly designed CPCs can absorb both direct and diffuse irradiations.

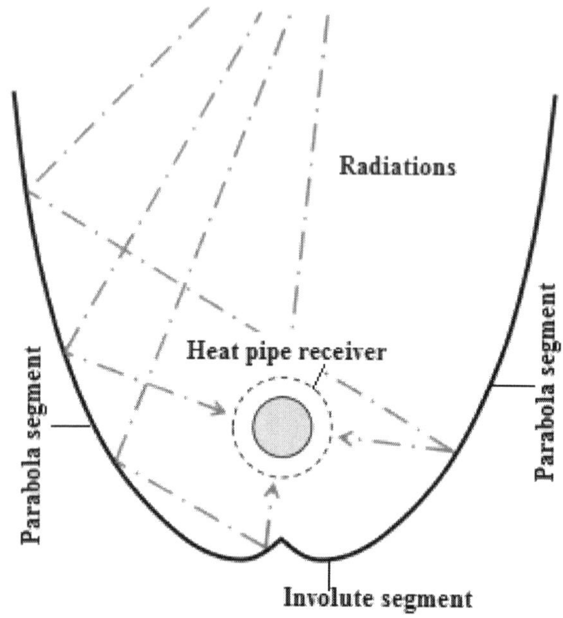

Figure 11.10. Line diagram of a compound parabolic collector.

It may not be out of place to mention that point focus collectors and similar devices can also be utilized to concentrate solar energy for use with concentrated photovoltaic. In such cases, the Sun's energy is converted directly into electricity with high efficiency photovoltaic cells designed specifically to harness concentrated solar energy.

SOLAR SPACE HEATING

Significant amount of energy in cold climatic regions is used in space heating. In the Indian context, space heating becomes important during winters especially in the north and north eastern regions. Space heating using solar energy under such situations can save lot of energy besides cost saving to the consumers. Space heating can be achieved by any of the following two methods.

- ☆ Passive heating
- ☆ Active heating

Passive Heating of Buildings

Passive heating employs the use of natural sources of heating, cooling and ventilation to create comfortable conditions inside the buildings. It harnesses the environmental conditions such as solar radiation, cool air and air pressure differences to control the internal environment. Obviously, there is no need to use any mechanical or electrical systems. Passve heating is categorized as direct gain and indirect gain.

Direct Gain

South-facing glass transmits insolation directly into the living space in a direct gain passive solar building, where it is absorbed. Part of the absorbed heat is released immediately to the indoor air by convection from the heated surfaces of the space. The rest moves into the thermal mass by conduction for release back to the indoors late in the day, or at night (Figure 11.11). An advantage of a direct gain passive system is its relatively low construction cost and rapid warm-up in the morning. Disadvantages include overheating during midday, excessive light glare in the living space, and accelerated solar degradation of materials like wood and fabrics near the solar glazing. Additionally, carpets and furniture tend to prevent solar gain from reaching the thermal mass, exacerbating overheating and reducing effectiveness of the thermal mass.

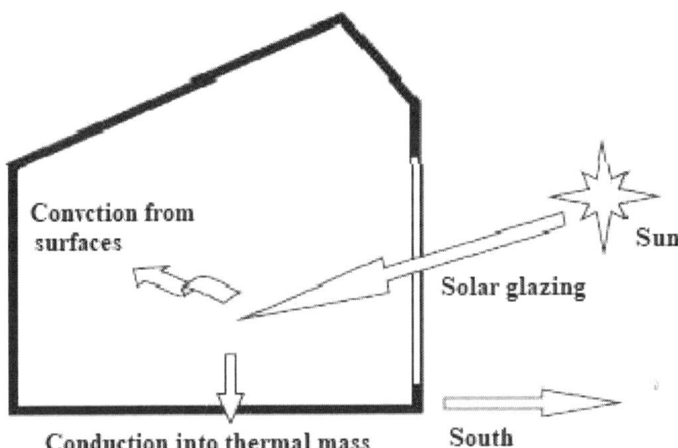

Figure 11.11. Direct gain solar building for winter heating in northern hemisphere.

Indirect Gain, Trombe Wall

Solar chimney (or thermal chimney) is a form of passive heating that regulates the temperature of the building and provides appropriate ventilation. Essentially, solar chimneys are hollow structures that connect the inside part of the building to the outside part of the building (Figure 11.12). It is usually a tall and wide structure constructed to face the sun. It is generally placed on the south-facing wall in the northern hemisphere. It is painted black to minimize sunlight reflection off the chimney to ensure that more heat is transferred to the air inside the building. Since this system was designed by Prof. Trombe, the whole unit consisting of the storage wall with vents and the glazing is also referred as a Trombe wall. In all, it has four vents but only 2 to 3 are used at a time during heating/cooling. The column of air inside the chimney gets heated as and when solar radiation hits the chimney wall. The top exterior vent A of the chimney remains closed (Figure 11.12 left). The heated air is forced back into the living space through vent B. It provides a type of convective air heating. As the air cools in the room it is pulled back into the solar chimney through vent C. The air column gets heated once again and this cycle is repeated. The vent D remains close during the whole cycle. Cooling the space using Trombe wall is slightly different than heating. The vent A at the top of the chimney is opened and vent B is closed (Figure 11.12 right). The vent D at the opposite end of the chimney also remains open. This provides an opening between air inside the building and the air outside to facilitate ventilation. Vent D is normally located in such a way that the air pulled in through comes from a shaded and cool area. Presence of overhang on the roof of the house and greenery on this side can prove to be quite useful. When solar radiation hits the side of the chimney, the column of air inside the chimney is heated. Since the vent at the top of the chimney is open, heated air is pulled up and allowed to move out of the chimney. It helps to pull outside air through the vent D creating a sort of draft that provides cool, fresh air into the building. As the air inside heats up, it goes into the chimney through vent C. This cycle continues during cooling operation. This system can also be used to improve passive ventilation in buildings where cross ventilation is quite restricted.

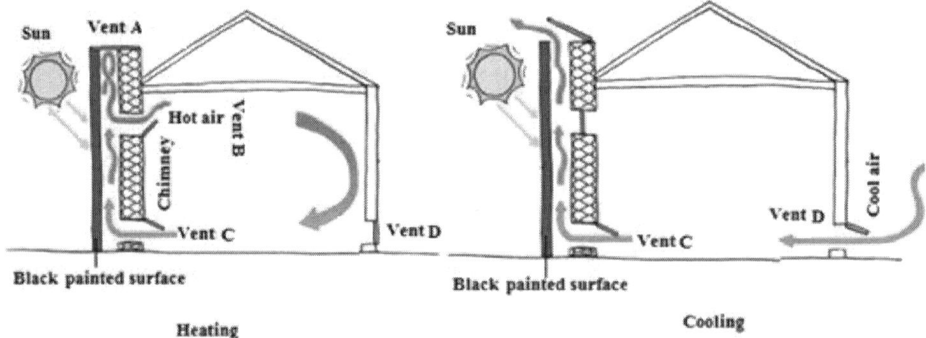

Figure 11.12. A line diagram of a solar chimney showing the process of heating (left) and cooling (right) of buildings.

Active Heating

An array of collectors is arranged on the roof/side of the building in active space heating. Air gets heated up in the flat plate collectors. Hot air is circulated as shown in Figure 11.13 (left). Heating system can also be used with liquid as the heating medium. In this case, solar collector has an insulted storage tank, auxiliary heater, pump/blower and heat exchanger (Figure 11.13 right). The liquid from storage tank is transmitted by pump/blower to heat exchanger. The heat exchanger blows out hot air and heats up the surrounding living space of the building. Again the cool water passes to the storage tank, which moves to flat plate collectors for heating. An auxiliary heater can also be provided for space heating to take care of periods when the solar energy is not available.

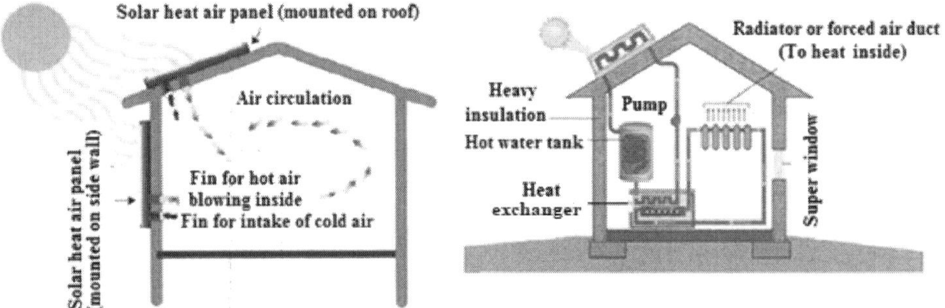

Figure 11.13. Active space heating using solar air panel (left) and a solar heat collector with liquid (right).

SOLAR WATER HEATING

The main components of a roof top domestic water heater are listed below (Figure 11.14).

1. Solar collector
2. Insulated water tank
3. Water inlet
4. Water outlet
5. Tank support
6. Frame
7. Liquid pipe
8. Water line

Normally a non-concentrating flat-plate collector, discussed in a previous section of this chapter (Figure 11.2), is used in domestic water heaters The collector is kept on an angle iron stand such that the flat plate is at 11° slope facing south (Figure 11.14). The hot water storage tank is typically a double walled tank, the inner tank made of copper or stainless steel to ensure long life. The outer tank may be made of stainless steel sheet, painted steel sheet or aluminium. The space between the inner and the outer tanks is filled with some insulation material to minimize heat losses. The capacity of the tank is in proportion to the collector area. A common thumb rule is to use 50 L of storage for every m^2 of collector area. Too large or too small tanks reduce the heating efficiency. The storage tank is kept just above the collector with ends of water pipes connected to it. The system is generally installed on the roof or open ground. The collector faces the sun and is connected to a continuous water supply. The black absorbing surface (absorber) inside the flat plate collector absorbs solar radiation and transfers the energy to water flowing through the tubes. The heated water is stored in the tank for further use. Circulation of water from the tank through the collectors and back to the tank continues automatically due to density difference between hot and cold water (thermo-siphon effect). The water stored in the tank remains hot overnight as the heat losses from the insulated storage tank are in the range of 4 to 8°C.

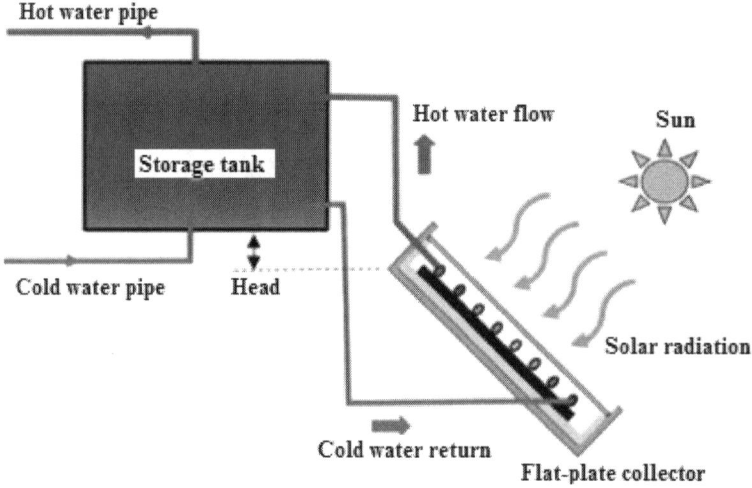

Figure 11.14. A flat plate domestic solar water heater.

General Guidelines

- The system should be mounted on a rigid structure firmly fixed to the roof to prevent damage during high winds
- The temperature of water in the system is governed by the combination of collector area and the tank capacity. Typically it would be 60-80°C, which is much hotter than the bathing water temperature (around 40°C)
- Normally a 100 L system is considered optimum for a family of 4 adult members. A typical rooftop, solar water heater of this size costs approximately Rs. 13,000-15,000. Solar water heaters (SWHs) of 100-300 L capacity are used for domestic applications
- It requires little/no maintenance
- Electrical heating elements controlled by thermostats can be provided as an option in the tank to take care of those days when sun is not there or demand of water is high

- ☆ A 100 L capacity SWH for residential use can annually save about 1500 units of electricity compared to an electric geyser
- ☆ SWHs have a life span of 15-20 years with a payback period of about 3-7 years depending upon the system of heating being replaced. It is 3-4 years when electricity is replaced, 4-5 years when furnace oil is replaced and 6-7 years when coal is replaced

Solar Distillery

Poor quality water is a major issue in arid, semi-arid and coastal regions as good quality potable water is inadequate. Distillation process is too expensive and the fuels required for this purpose are scarce. Under this situation, solar distillation units can be used to convert saline/polluted water into potable water using abundant sunlight available in these areas. Solar distillation is the process of removing undesirable chemicals, biological contaminants, suspended solids and gases from the contaminated water. It is by far the most reliable, least costly method providing 99.9 per cent true purification of most types of contaminated water. Distilled water from a solar distillery can be used for drinking, in lead acid batteries, laboratories, hospitals and to produce commercial products such as rose water. The basic principle of solar water distillation is simple as it replicates the way nature makes rain. The Sun's energy heats water to the point of evaporation. As the water evaporates, water vapours rise and condense on the glass surface. The water so condensed is then collected in a storage system.

Single basin stills are commonly used being cheaper. The system consists of a simple basin enclosed in a thermally insulated wooden box with a cover (Figure 11.15). The top cover is transparent, resistant to water absorption and tilted at an angle so as to catch maximum sunlight. The cover material should also be smooth and easy to clean. A glass cover is mostly used as it meets most of these requirements. The material of the still basin should be able to absorb heat. The material should resist heat transfer to the surroundings, as the efficiency of the system depends upon the heat trapped inside. It should also be able to support the cover and prevent the water from spilling. Some such materials are leather sheet, silicon, reinforced plastic or steel plate. System is provided with an inlet for impure water and an outlet for overflow. Solar rays enter the underlying still basin through the cover. The untreated water stored in the basin heats up and then evaporates. The evaporated water is trapped by the cover. The condensed water flows down the sloping roof and is collected in the trough placed at the bottom and from there into a water storage tank. The per liter cost of distilled water with solar distillation system is much cheaper than distilled water obtained by other electrical energy-based processes. A major limitation is its low efficiency typically about 25 per cent. Daily output is a function of solar irradiation and is greatest in the early evening when the feed water is still hot but when outside temperatures are falling. The minimum size of the still is about six square meters to produce enough water for a single person for a day.

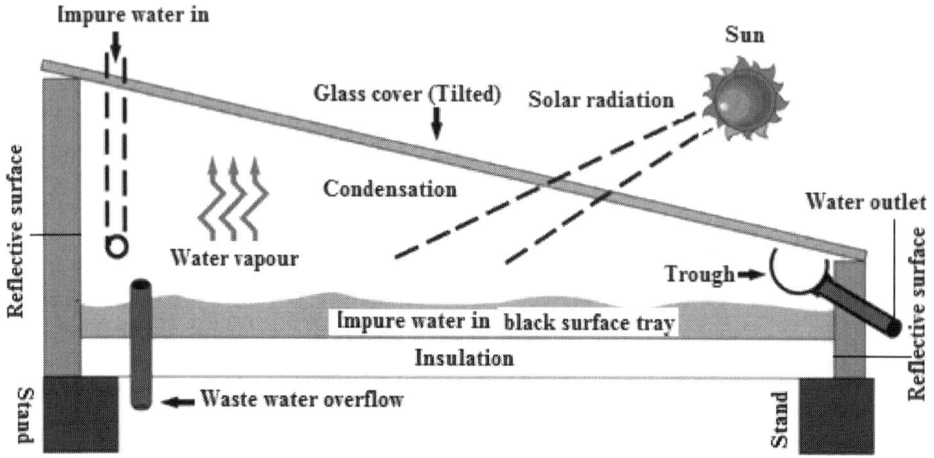

Figure 11.15. Construction details of a single basin still.

Enhancing Efficiency of Solar Distillation

Some passive and active methods are employed to improve the efficiency of solar distillation. Passive methods to improve efficiency are:

- Use of dye, charcoal, black gravel, rubber and floating perforated black plate to increase the solar absorptivity of water
- Make the side walls reflective
- Provide good insulation, keep the water depth in the basin low, and ensure vapour tightness

Active methods include the following.

- Use of solar collectors or waste heat to heat the basin water
- Use of internal and external condensers
- Creating vacuum inside the solar still to enhance the evaporation/ condensation processes
- Cool the glass cover to increase the temperature difference between the glass and the water in the basin. It helps to increase the evaporation rate

New Designs and Improvements

Several kinds of new solar distillation units having high efficiencies have been designed. One of them is a multiple solar distillation system having two or more compartments. The condensing surface of the lower compartment is the floor of the upper compartment. The heat given off by the condensing vapours provides energy to vaporize the feed water in the above compartment.

Therefore, efficiency of such units is higher than a single-basin still typically being 35 per cent or more but the cost and complexity are correspondingly higher. Second kind is a wick still. Here the feed water flows slowly through a porous, radiation-absorbing pad (the wick). The wick can also be tilted so that the feed water presents a better angle to the sun (reducing reflection and presenting a large effective area). The basin wick-type solar stills with jute and charcoal as wick materials are quite simple to construct. The wick material can be blackened for more absorption of solar radiation. The wick material in the basin sucks water and the upper surface of the wick material is always wet during peak sunny hours as a result of capillary action. The water is evaporated and the water vapours condensed on the condensing surface are collected as in the normal solar distillation still. Feed water in the wicks being much less, it gets heated quickly as well as attains a higher temperature. For the same output, some designs of simple wick stills are claimed to cost even less than a basin still.

Advantages

Solar stills have major advantages over other conventional distillation/ water purification/de- mineralization systems.

- Produces pure water
- No prime movers required
- No conventional energy required
- No skilled operator required
- Local manufacturing/repairing possible
- Low investment low-maintenance system
- Can purify highly saline water (even sea water)

Disadvantages

- The main disadvantage is its low productivity or high capital cost per unit output of distillate. Rate of distillation is usually slow around 6 L of water per sunny day
- It is not suitable for large consumptive needs
- The materials required for the still may not be locally available
- The waste stream of the distillation having high concentrations of salts and pollutants can be a potential source of environmental pollution. Therefore, proper care has to be exercised for its disposal
- Solar energy is only available during the day

Anticipated Problems

- Poor fitting and joints can result in cold air from outside to leak into the still

- Cracking, breakage or scratches on glass may reduce solar energy transmission to the basin or even allow the outside air to enter into the still
- Growth of algae, deposition of dust, and bird droppings on the cover, *etc.* may reduce the transmission of energy to the basin. Therefore, stills need regular cleaning every few days
- Damage over time may occur to the blackened absorbing surface
- Salt accumulation will occur at the bottom. It needs to be periodically cleaned

Solar Cooker

The working principle of a solar cooker is based on the fact that whenever electromagnetic radiations emitted by the Sun strike any matter, the energy causes the molecules of the matter to vibrate. The excited molecules jump to higher energy levels. This activity generates heat. To put the principle into practice, solar cookers have some means to concentrate (reflection, or reflectance), absorb (ability to attract or hold heat) and retain (means capacity to retain) heat. A mirror surface with high spectral reflection is used to concentrate and channelize light from the sun into a small cooking space. Metallic blackened pots are used to absorb and retain heat. For household cooking applications, solar cookers are designed to raise the temperatures in the range of 65°C to 400°C.

Box-type Solar Cooker

A box-type cooker is the most commonly used solar cooker. The primary elements of this cooker are an insulated box, metallic cooking pots, double glass cover on the box, and a reflecting mirror fitted on the underside of the lid of the box (Figure 11.16).

Black box: It comprises of an outer and an inner box. The outer box is an insulated metal or wooden box which is painted black from the inside so that it is able to absorb maximum heat. Inner box is made from sheet metal and painted with black colour, which works as absorber plate to receive heat energy

Glass cover: A cover made of two sheets of toughened glass held together in an aluminium frame is used. The glass cover besides enhancing light transfer from the top to the pans also decreases the convection energy losses. It improves heat holding capacity of the cooker. The thick glass sheet does not allow the entrapped heat to escape. As a result, it helps in raising the temperature in the box to sufficiently high degree to cook the food, vegetables, meat and rice.

Plane mirror reflector: The plane mirror reflector is fixed on the underside of the lid of the box with the help of hinges. The mirror is positioned so as to allow the reflected sunlight to fall on the glass cover of the box. To achieve this, mirror reflector has the provision for positioning it at any desired angle to the box.

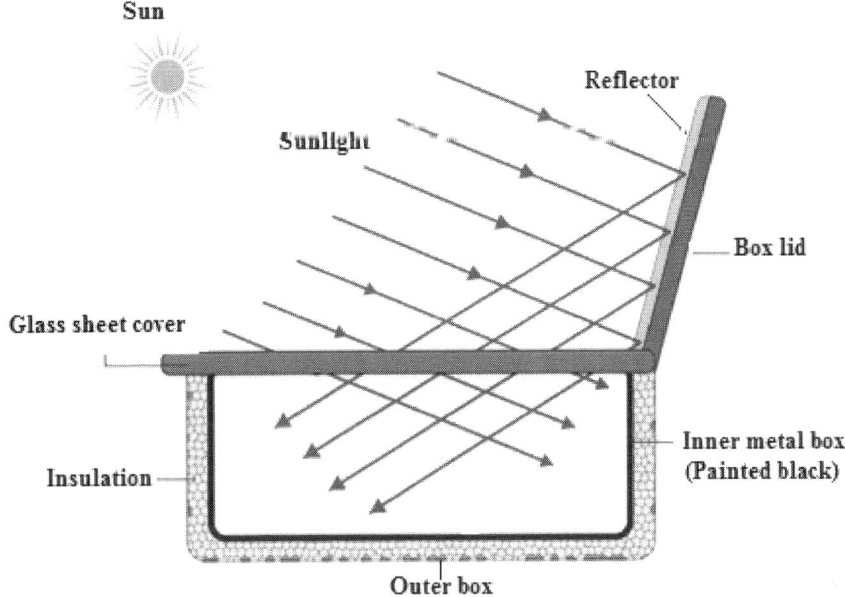

Figure 11.16. Construction details of a box-type solar cooker.

Cooking containers: A set of aluminium pots and pans matte black in colour are used for cooking. It helps to maximize absorption of solar energy. Moreover, pots and pans are made of materials that conduct and retain heat to maximize conversion of sun energy to heat.

Working

When the solar cooker is placed in the sunlight, the angle of the plane mirror reflector is adjusted in such a way that the strong beam of sunlight falls on the glass cover. The incoming solar radiation passes through the glass cover and strikes the blackened cooking pots and the cooking tray. The heat is absorbed by the blackened surface and is transferred to the food inside the pots to facilitate cooking. Up to four black painted vessels can be placed inside the box. The box type solar cooker takes about 1.5 to 2 hr to cook vegetables, rice, *dals*, cake *etc.*

Besides the box-type cookers, other kinds of solar cookers namely panel cookers and parabolic solar cookers are also used. The panel cooker is quite similar in principle and operation as the solar box-type cooker. But a panel cooker has a large reflective panel instead of an insulated box. Parabolic cooker (sometimes called curved concentrator cooker) can reach much higher temperatures and can cook quickly, but require frequent adjustment and supervision for their safe operation.

Advantages of Solar Cooker

- ☆ Solar cookers do not require any fuel resulting in cost saving in cooking the food
- ☆ Unlike other fuels, no environment polluting fumes or gases are produced resulting in reduced carbon footprint
- ☆ Maintains quality of cooked food
- ☆ Retains the nutritive value of food
- ☆ No soot formation
- ☆ No health problem expected as there is no smoke

Disadvantages of Solar Cooker

- ☆ Solar cookers cannot be used on days with cloudy weather
- ☆ Solar cookers take longer to cook food than a conventional stove or an oven
- ☆ Some solar cookers are affected by strong winds, which can slow down the cooking process
- ☆ It is difficult to cook some thick food materials such as large roasts and loaves of bread

SOLAR PONDS

Solar ponds can be effectively used to generate thermal energy in many industrial applications such as process heating, refrigeration, water desalination, production of magnesium chloride, bromine recovery from bittern, and enhancing salt yield in salt farms. The concept of harnessing energy from the Sun through solar ponds requires an isolated body of water. This body collects and stores the solar energy. In general practice, solar energy warms a body of water exposed to the Sun, but the water loses this heat through the processes of convection and evaporation. Warm water below the surface expands and begins to rise as hot water is less dense than cold water. The cold water being heavier moves down to replace the warm water. It creates a natural convective circulation that dissipates the heat to the air through the processes of convection, or evaporation. A solar pond is designed in such a way that it reduces either convection or evaporation such that heat collected in the body of water is retained within the body. The heat is then harnessed for desired applications. In general, solar ponds are divided into convective solar ponds and non-convective solar ponds. Convective solar ponds have further been grouped into shallow and deep solar ponds.

Convective Shallow Solar Pond

A shallow convective solar pond consists of pure water enclosed in a large narrow plastic bag measuring 5 cm × 60 cm having the water depth in the range of 5-10 cm. The bag is blackened at the bottom from inside that helps to absorb sunlight. A layer of insulation beneath the plastic bag minimizes heat loss to the ground. A glazing sheet made of PVC or clear acrylic panel covers the pond to minimize radiative losses when there is no sun (Figure 11.17). The Sun heats the water in the bag during the day. At night, the hot water is pumped into a large heat storage tank to minimize heat loss for use during the next day. The basic principle of the convective solar pond is to traps the heat by preventing evaporation although no attempt is made to prevent convection. One of the major limitations is due to excessive heat losses during pumping the hot water to the storage tank.

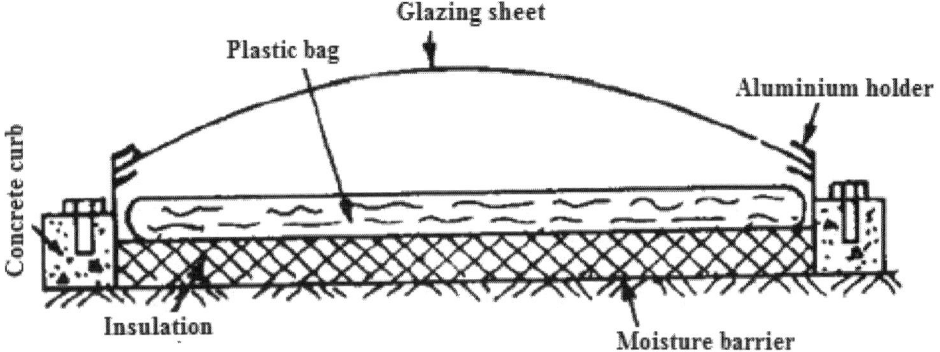

Figure 11.17. Schematic diagram of a shallow solar pond.

Convective Deep Solar Pond

A deep convective pond has a black coloured impermeable liner at the bottom that helps in absorbing sunlight as well as preventing seepage of water. It is filled with fresh water. Transparent double-glazing plastic film made of polyvinyl chloride (PVC) and/or clear acrylic panels are used to cover the pond to avoid water losses through evaporation (Figure 11.18). Some insulating material may also be placed on top of the glazing material to help reduce heat losses during night or when there is no sun. The incident energy absorbed by the blackened bottom transfers the heat by convection. Depending upon the capacity of the pond, large quantities of water can be heated in such kinds of solar ponds. Therefore, water from a deep convective pond need not be pumped in and out of storage as is the case with shallow solar ponds. The water from the pond is withdrawn using batch methods such as closed cycle continuous flow or by open cycle continuous flow.

Non-Convective Solar Pond

Non-convective solar ponds are artificially constructed ponds where significant temperature rise is allowed to take place in the lower regions of the ponds by preventing convection currents. These ponds are also termed as salt-gradient solar ponds as convection currents are controlled by dissolving salt in the bottom layer of the pond making it too heavy to rise.

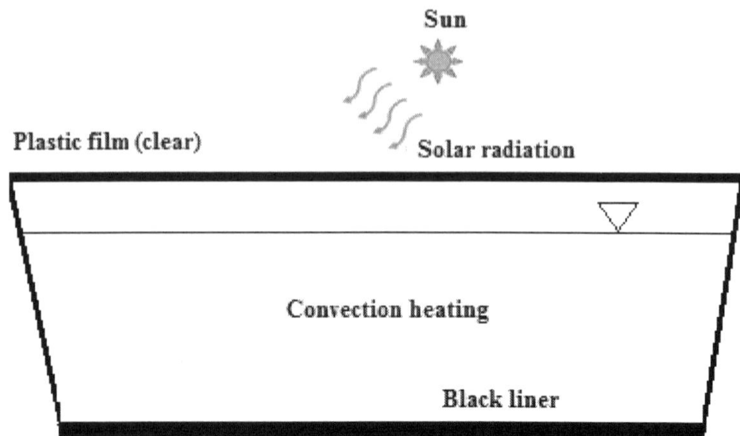

Figure 11.18. Schematic diagram of deep convective solar pond.

Suitable Sites

Although, non-convective solar ponds can be constructed anywhere, it is much more economical to construct them at places where there is good supply of sea water and/or availability of low cost salts and bittern, ample supply of water for filling and flushing, high solar radiation, and availability of land at low cost (Murthy and Pandey, 2000). Coastal areas in Tamil Nadu, Gujarat, Andhra Pradesh, and Odisha are ideally suited for such solar ponds. In general, the site for the construction of solar pond should have the following attributes.

- Be close to the point where thermal energy from the pond is to be utilized
- Be close to a source of water for flushing the surface mixed layer of the pond
- The thermal conductivity of the soil is not too high
- The water table is not too close to the surface

Construction and Working

Practically speaking, a solar non-convective pond is a mass of shallow water about 1 to 2 m deep with a large collection area. At the bottom of the pond, a thick plastic liner is laid, which acts as an insulator to allow high temperature to develop (Figure 11.19 top). Materials used for liner include butyl rubber, black polyethylene and hypalon reinforced with nylon mesh. Salts like magnesium chloride, sodium chloride or sodium nitrate are dissolved in the water, the concentration varying from 20 to 30 per cent at the bottom to almost zero at the top (http://www.teriin.org/tech_solarponds.php/November 2022). As the salt content increases with increasing depth, a salinity or density gradient is created. Thus, a solar pond essentially consists of three zones. The top zone is the upper convective zone (0.3-0.5 m), where the water is at normal atmospheric temperature and has little salt content (< 5 per cent). The storage zone or lower convective zone lies at the bottom say 1.5-2.0 m for a 2 m deep pond (Goutham and Siva Krishna, 2019). The salt concentration in this zone is more than 20 per cent. The deeper this zone, the more is the heat stored. Both the salt concentration and temperature are constant in this zone. In between these zones is a separating non-convective zone (0.5-1.5 m). The non-convective zone is much thicker and occupies about half or more than half the depth of the pond. Both the salt

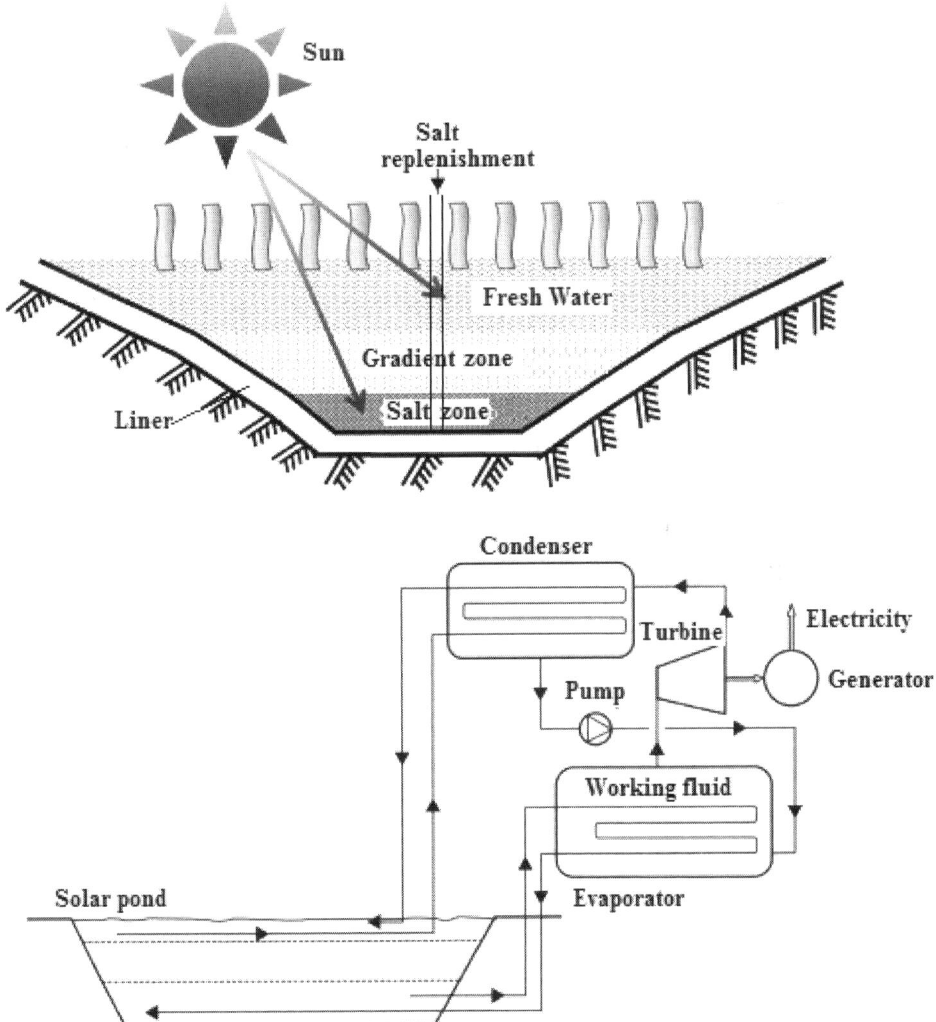

Figure 11.19. A line diagram of a non-convective pond showing 3 salt zones (top) and application of stored thermal energy to produce electricity (bottom).

concentration and the temperature increase with the depth in this zone. Some part of this zone also stores some thermal energy. To understand the working principle, let us consider the layer in the non-convective zone. Water below the layer cannot rise, as the water in this layer has less salt content and is lighter than the water below. Similarly, the water from the above layer cannot move down as the water in this layer is heavier due to its higher salt content. Thus, the gradient zone acts as a transparent insulator permitting sunlight to reach the bottom zone and at the same time entrap it there. The side walls of the pond can be insulated with glass wool *etc.* When solar radiation strikes the pond, most of it is absorbed by the surface and is transmitted to the bottom of the pond increasing the temperature of the dense salt layer. But the density differences keep the layers of the solar pond intact by not allowing it to rise. The energy in this system can be stored below the boiling point of brine, the temperature of the layer rising as much as to 95°C. The trapped solar energy is then withdrawn from the pond in the form of hot brine from the storage zone without disturbing the non-convective salt gradient zone. It can be accomplished either by pumping the brine through an external heat exchanger or an evaporator is used to remove the heat from the bottom layer. Another method to extract heat is with the help of a heat transfer fluid as it is pumped through a heat exchanger placed at the bottom of the pond. The water outlet is provided at about the same height as the water inlet. Hot brine can be withdrawn and cold brine returned in a laminar flow pattern because of presence of density gradient. Provision to add salt to the bottom layer as and when required is kept in the pond (Figure 11.19 top). It can also be used to produce electricity. The schematic diagram of this set-up shown in Figure 11.19 (bottom) illustrates the system but detailed discussion is avoided being beyond the scope of this book (https://courseware.cutm.ac.in/courses/renewable-energy-green-technology).

Advantages

- Up-scaling for large scale energy generation can be possible through ponds covering large surfaces
- Diffuse radiation (cloudy days) is fully used
- Thermal storage is achieved at low cost
- Low investment costs per installed collection area
- Reflectors can be put on the sloping sides to achieve high temperatures

Applications

Heating and cooling of buildings: It is ideal solution for heating at high latitude stations and for several cloudy days, because of its large heat storage capability in the lower convection zone of the pond

Power production: A solar pond can be used to generate electricity by driving a thermo-electric device or an organic Rankine cycle engine - a turbine powered by evaporating an organic fluid (Figure 11.19 bottom). It has great promise in areas where soil and terrain conditions allow for construction of large solar ponds necessary to generate meaningful quantities of electrical energy.

Industrial process heat: Industrial process heat is the thermal energy used directly in the preparation and treatment of materials and goods manufactured by industry. Several studies have shown that the heat from solar pond is economically competitive with oils and natural gas.

Desalination: The low cost thermal energy can be used to desalt or otherwise purify water for drinking or irrigation. The multi-flash desalination plant is suitable for places where potable water is in short supply but brackish water is in plenty. It has been estimated that about 4700 m^3/day distilled water can be obtained from a pond of 0.3 km^2 area.

Heating animal housing and drying crops on farms: Low grade heat can be used in many ways on the farms having enough land for solar ponds. Several small ponds in Ohio, Iowa and Illinois are being used to heat green houses.

Heat for biomass conversion: Site built solar ponds can provide heat to convert biomass to alcohol or methane. Although no solar ponds have been used for this purpose so far, yet it can prove to be an ideal combination of two renewable-energy technologies.

Improving the Efficiency of Non-convective Solar Ponds

Dust, Debris and Algae Control

The thermal efficiency of a solar pond is strongly dependent upon the clarity of the water in the pond, which can be mainly reduced by the presence of algae or dust. Bits of debris, dust or leaves lighter than water floating on the surface should be skimmed off to improve the penetration of sun rays. Dust and debris that sink to the bottom may not adversely affect the absorption of solar radiation at the bottom of the pond. Alum is commonly used to make the dust floating in the gradient zone to settle at the bottom. Bleaching powder or copper sulphate is used to control algal growth. Alkaline water in pond may pose some difficulties in the dissolution of copper sulphate.

Rainfall

Rainfall can be beneficial or detrimental to the operation of a solar pond. If the rainfall is not too heavy, it may help to maintain the density of surface layer at low value. In such cases, flushing the surface layer to maintain the density at low levels may not be required. Heavy rainfall on the other hand can penetrate to the gradient zone and dilute the water. The studies have revealed that raindrops can penetrate to about 50 cm from the surface. It is therefore desirable to have thicker surface zone in areas with heavy rainfall particularly during the monsoon season in India.

SOLAR POND AT BHUJ

The Bhuj Solar Pond is a research, development, and demonstration unit. The pond is 100 m long and 60 m wide having a depth of 3.5 m (http://www.teriin.org/tech_solarponds.php/November 2022). Constructed for Kutch Dairy, Bhuj in 1987 is a collaborative effort between Gujarat Energy Development Agency, Gujarat Dairy Development Corporation Limited, and TERI under the National Solar Pond program of the Ministry of Non-Conventional Energy Sources. To prevent seepage of saline water, a specially developed liner comprising locally available material is used. The pond after construction was filled with water and 4000 tonnes of common salt was dissolved in the bottom layer to make dense brine. A salinity gradient was established and wave suppression nets, a sampling platform, diffusers for suction and discharge of hot brine, *etc.* were installed. TERI carried out execution, operation, and maintenance of the Bhuj Solar Pond in the initial years. This pond has been successfully supplying enough process energy to the dairy since September 1993. Presently, it is the largest operating solar pond in the world.

APPLICATIONS IN SPACE COOLING AND REFRIGERATION

Space cooling is one of the promising applications of solar energy to provide comfortable living conditions (air-conditioning) or for increasing the shelf life of food products through refrigeration. Since the energy of the sun is received as heat, absorption refrigeration system is the obvious choice, which requires most of its energy input as heat. A diagram of a simple solar operated absorption refrigeration system is shown in Figure 11.20. The heated water from a flat

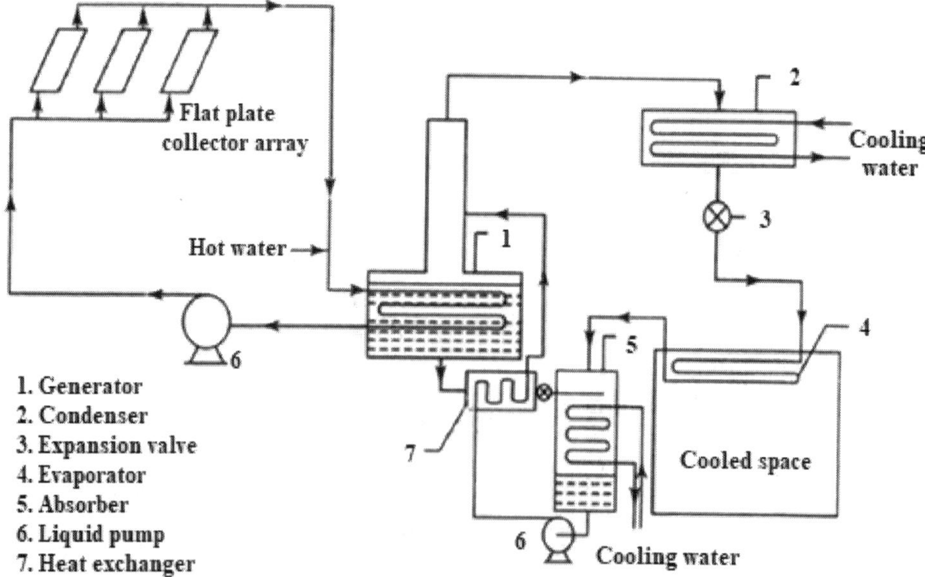

Figure 11.20. A line diagram of solar absorption refrigeration system.

plate collector array is passed through a heat exchanger called the generator where transfer of heat takes place to a solution rich in refrigerant (absorbent + refrigerant). The common refrigerant absorbent liquids are ammonia-water, and water-lithium bromide. The later is mostly used in air conditioning. Refrigerant vapours boiled off at a high pressure go to the condenser where these are condensed into a high pressure liquid. The high pressure liquid is throttled to a low pressure and temperature in an expansion valve that passes through an evaporator coil. The refrigerant vapours absorb heat and cooling is obtained in the space surrounding the coil. The refrigerant vapours are now absorbed back into a solution mixture withdrawn from the generator. The refrigerant concentration is weak in this solution and it is pumped back into the generator, thereby completing the cycle. Detailed information on this system is being withheld being beyond the scope of the book.

QUESTIONS (THEORY)

1. What is meant by a solar collector? Name the various kinds of solar collectors.
2. Briefly explain the flat plate solar collector, evacuated tube solar collector, water-in glass evacuated tube solar collector and U-type evacuated tube solar collector.
3. Explain the evacuated heat pipe technology. List its advantages and limitations.
4. Briefly discuss the general construction of concentrating collectors. Name various types of these collectors. List the advantages and disadvantages of concentrating collectors.
5. Explain the parabolic trough collector, parabolic dish collector, power tower collector, linear Fresnel collector, and compound parabolic collector.
6. Differentiate between active and passive heating of air space. Draw a neat line diagram to explain the passive heating of buildings using Trombe wall.
7. Explain the construction and working of a flat plate domestic solar water heater.
8. Explain the construction and working of a solar distillery unit.
9. Explain the construction and working of a domestic solar cooker.
10. What is a solar pond? Differentiate between convective and non-convective solar ponds.

11. Explain the working of a non-convective solar pond. How its efficiency can be improved?

12. Briefly discuss the application of solar power in refrigeration.

Chapter 12

Solar Dryers

Solar energy has been used for drying agricultural and animal products since pre-historic times, but in a most primitive manner. Sun energy, being the cheapest source of energy for drying, has been widely used to preserve and save food products from damage so that they can be used when these are not easily available. Conventional method of drying is to spread the material in a thin layer on the ground, mats, concrete floors *etc.* or into trays. In rack type (shade type) drying, moist food material is kept in racks. The major disadvantages of traditional drying methods are as follows.

- Lot of space is required for spreading the produce
- Large labour force is required for handling the material
- Chances of product degradation due to rain, air humidity and dew
- Accumulation of dust and atmospheric pollutants on the products
- Losses due to over-drying or in terms of inferior quality product fetching lower price
- Drying of the product may not be uniform or it may not attain the desired moisture content
- Since it takes long to dry the products, they are prone to attack by fungi and bacteria especially the fruits and vegetables
- Very often, the products are eaten and spoiled by birds

SOLAR DRYERS

The difficulties encountered in conventional drying can be overcome by using solar dryers. *A solar dryer is a food dehydrator system that uses solar energy to remove the moisture from the produce.* The dryers can be used to dry cereal crops, fruits, vegetables (potato-chips, peas), fodder (*berseem*), spices (ginger, pepper, dry fruits (cashew-nuts), medicinal plants, tobacco curing, timber, veneer drying and other innumerable products. Spray drying of milk and fish drying are examples of solar dried animal products. A dryer functions on the principle of the greenhouse effect, meaning the heat generated within the system evaporates the moisture in the produce. The main advantages of drying with dryers are listed as under.

Safe storage of produce: The produce can be dried to optimum moisture content. Thus, it can be stored for long without any deterioration in its quality. The possibility of quality degradation or germination of the cereal is almost eliminated.

Less molds and/or aflatoxins: The products at high moisture content are prone to damage by molds or aflatoxins. Such products pose risks to human and animal health. The low moisture content of the products after drying eliminates the risk of the onset of harmful agents such as molds or aflatoxins.

Less waste: Dried grains can be safely stored in bins resulting in significant reduction in the losses associated with movement of the product, minimum exposure to theft, and no losses caused by birds and rodents. According to an assessment made by ICAR on quantitative harvest and post-harvest losses of major crops and commodities in India, post-harvest losses vary from 1-16 per cent (Table 12.1). Most post-harvest processing in India can be classified as primary processing having lower value-addition compared to secondary processing. Even the primary processing of drying alone can reduce post harvest losses by 2-3 per cent. Secondary processing is icing on the cake as it has numerous additional advantages. It can lift millions out of poverty and malnutrition.

Table 12.1. Cumulative harvest and post-harvest losses of various agricultural commodities

Crops	*Cumulative Wastage (Per cent)*
Cereals	4.65 – 5.99
Pulses	6.36 – 8.41
Oil seeds	3.08 – 9.96
Fruits and vegetables	4.58 – 15.88
Milk	0.92
Fisheries (Inland)	5.23
Fisheries (Marine)	10.52
Meat	2.71
Poultry	6.74

Higher productivity and better quality: Scientifically designed grain dryers speed-up the drying process, allow greater flexibility and help to attain greater productivity. Whereas natural drying may take several weeks, large tonnage of product can be dried in few hours with the help of dryers. Even the quality of the product is maintained. Solar drying especially of fruits improves their quality as the sugar content increases upon drying. Uniform drying and better quality product improves marketability of the products.

More value and profits: The time flexibility gained by the farmer (can wait for most favourable periods for marketing the produce) for selling his produce combined with better quality products helps fetching good prices and maximize profits.

Categorization

Drying is the removal of moisture from food products as a means of food preservation, improve shelf life and prevent bacterial growth. Two types of solar dryers namely natural convection solar dryer and forced convection solar dryer are used for drying various products (Figure 12.1). Natural convection solar drying process is much slower than the forced convection types. These together have been further categorized into six types namely cabinet or direct type dryer, greenhouse dryer, tunnel type solar dryer, indirect type solar dryer, mixed solar dryer and hybrid solar dryer *etc.* (Figure 12.1). Indirect solar dryers are also categorized in various groups but are not covered in this chapter being beyond the scope of this book.

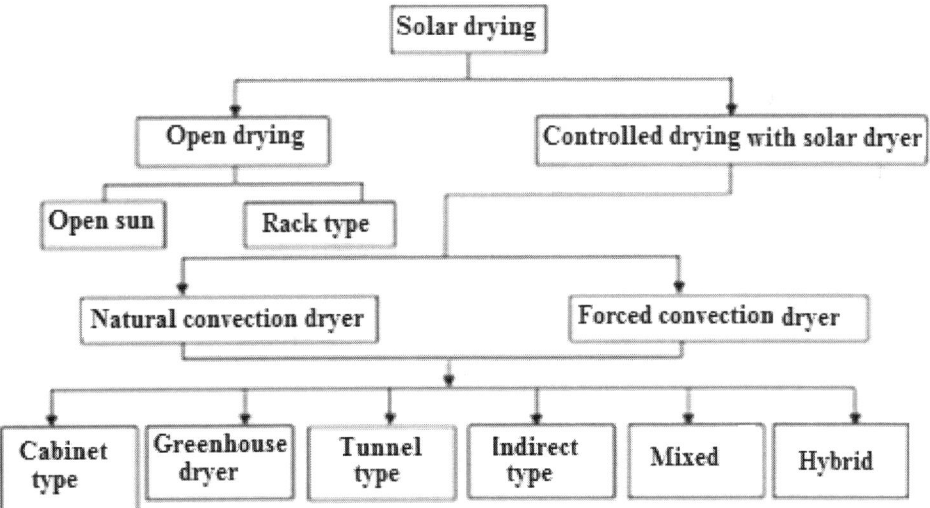

Figure 12.1. Categorization of solar dryers.

DIRECT TYPE OR CABINET SOLAR DRYER

A cabinet type solar dryer, also known as box type dryer, is simplest and cheapest of all the dryers. It is mostly used for small scale drying and can either be fixed type or portable type. Generally, it has an area of about 3 × 5 m^2 with a glass sheet fixed at the top at an angle of about 0 to 30° (Figure 12.2). The cabinet is painted black on the inner side. Suitable openings (holes) at the bottom and top ensure natural air circulation (Figure 12.2). Products are placed on a wire meshed black trays. The perforated trays or racks are designed to provide controlled exposure to solar radiations. Transparent glass transmits solar radiation directly into the enclosure, which is absorbed by the product and the surrounding internal surfaces of the enclosure. As a result, air inside the enclosure is heated in the temperature range of 50-80°C. The heated air helps to remove the moisture from the product. The drying time varies from product to product depending upon its initial and desired moisture contents and several other parameters. Products like dates, apricots, grapes, chilies, turmeric *etc.* can be dried in theses type of dryers. Design and construction is made to minimize convective heat losses to attain higher temperature in the drying chamber. Other direct solar dryers based on similar principle are glass roof solar dryer and greenhouse dryer.

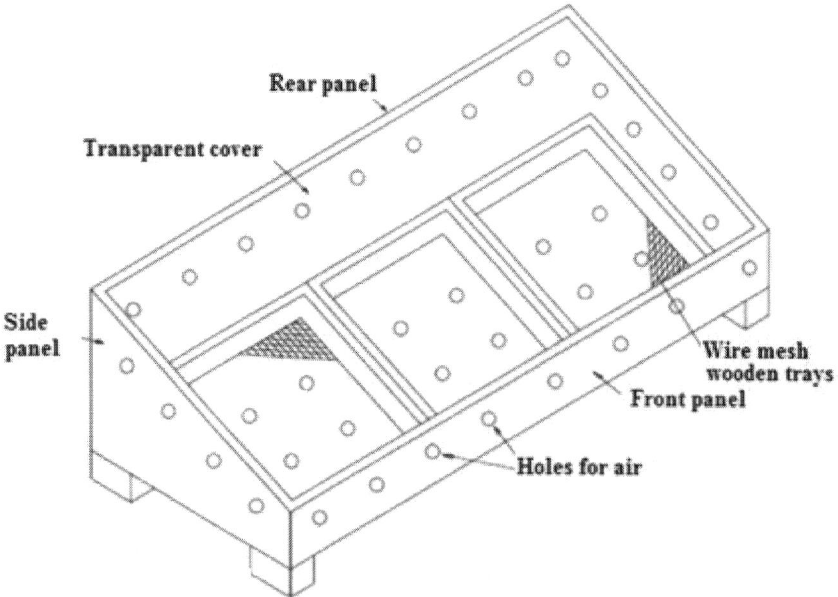

Figure 12.2. A line diagram of wooden cabinet dryer.

Advantages

- ☆ Minimum contamination of the product due to enclosure with transparent cover

- Better quality product than the one from open sun drying
- Protection from rain, dew, *etc.*
- Product handling is minimized as drying and storage occur in the same bin
- For the same capacity, it is less costly to fabricate than indirect dryers

Disadvantages

- Time required for drying is almost similar to sun or natural drying
- Quality and colour of the product deteriorate when the product is directly exposed to sunlight
- Improper removal and accumulation of vapour inside affects the drying rates
- Useful only for small scale applications
- Transmissivity of the glass cover reduces after some time as moisture condenses inside the glass cover
- Initial moisture content is normally limited to 22 to 24 per cent

NATURAL CONVECTION DRYERS

Natural convection solar dryers are also known as passive solar dryers. In these dryers, the air flow needed for drying is either from natural factors or from buoyancy *i.e.* due to density difference of hot air and ambient air (thermo-siphon effect). Heated air dries the product and the moisture laden air escapes out through the chimney placed at the top of the drying chamber. This group of dryers has been categorized in the following three types.

Conventional indirect type solar dryer (ITSD): It consists of a solar air collector (SAC) to heat the inlet air and a chamber to keep the food for drying. Natural circulation of air occurs as a result of thermo-siphon effect.

ITSD with chimney: The main components of this type are: SAC, drying cabinet and a chimney to create the draft force.

ITSD with chimney and heat storage system: These dryers are also provided with a chimney as in the ITSD with chimney along with some material that stores solar thermal energy. The stored thermal energy is utilized during night or when there is no sunlight.

Working Principle of ITSD

The ITSD shown in Figure 12.3 consists of two main components; a solar collector and the drying cabinet/chamber. The drying chamber has wire mesh trays to keep the moist products to be dried. Solar collector set-up consists of an absorber plate and a glass cover which is coupled to the drying chamber at a particular inclination, depending on the latitude of the site. The absorber is selectively coated with black colour for maximum absorption of solar radiation. When solar radiation is incident on glass cover, some amount of energy is absorbed by the absorber plate and a certain quantity is transferred by convection to heat the air which passes into the chamber. Hot air is allowed to flow over the food material. The food product in the drying cabinet absorbs the heat and loses some moisture with each pass.

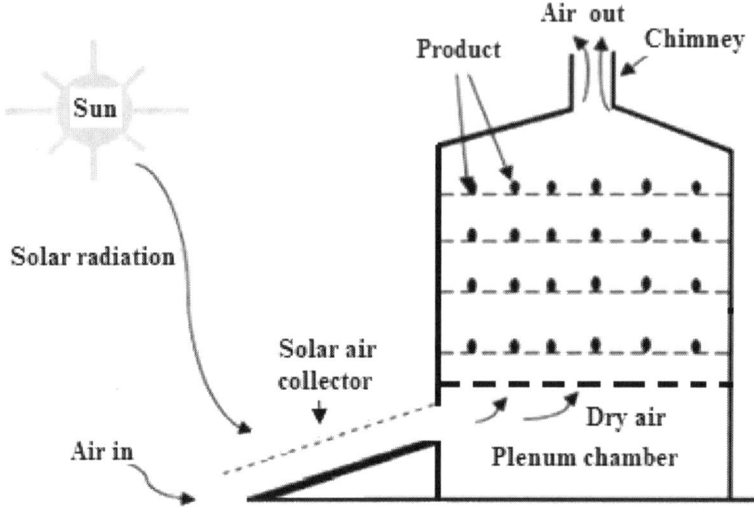

Figure 12.3. Line diagram of an ITSD solar dryer with chimney.

In ITSD with chimney dryer, the atmospheric air is heated in flat plate collector or concentrated type solar collector. The heating process is either passive or active. In the passive system, hot air flows through the plenum chamber into the cabin where products are stored on perforated trays. The drying air before it enters the grain bulk is delivered into the plenum chamber. The purpose of the plenum chamber is to let the air calm down before it enters the air distribution system in order to ensure an equal distribution of pressure and temperature of the drying air before it enters the drying section. The moisture in the grains is lost by convection and diffusion. The moist air is led out of the chamber through the chimney. ITSD with chimney and heat storage system is included in a forthcoming section of this chapter.

Advantages

- No contamination of the final product
- It is more efficient than direct type of solar drying
- Maintains the quality of product by avoiding direct exposure to solar radiations
- Time required for drying the product is less than direct solar dryers

- ✰ Final product is not affected by natural factors
- ✰ Moisture pick-up per kg of dry air is high
- ✰ Natural colour of the product is retained as the product is not exposed to ultraviolet radiation
- ✰ There is no heat damage to products
- ✰ Most suited for drying photo-sensitive materials such as lemon, cucumber, papaya, *etc.*

Disadvantages

- ✰ Initial cost is high
- ✰ Need regular maintenance
- ✰ Since the intensity of solar radiation varies with time, some thermal storage or auxiliary energy source is needed during cloudy days and in the evening
- ✰ Large surface area is required for the collector as solar radiation has low energy density

FORCED CONVECTION SOLAR DRYER (HOT AIR SYSTEM)

Active mode drying or forced circulation drying has an electric fan or blower to force the air into or out of the dryer. Therefore, drying rate can be controlled in this kind of dryer. It has three main components *viz.*, flat plate collector, blower and drying bin (Figure 12.4). The area of the collector may vary depending upon the capacity of the drying bin. The absorber is made out of 20 gauge corrugated galvanized iron (GI) sheet and is painted black. Another plain GI sheet placed 5 cm below the absorber plate creates air space for heating. This sheet is insulated at the bottom with glass wool and is supported by another plain GI sheet. The absorber is covered at the top with two layers of 3 mm thick plain glass. The unit is supported on all sides with wooden scantling and is placed at 11° to the horizontal facing south. Baffle plates are provided in the air space. The air space is open at the bottom to suck atmospheric air and at the top it is connected to a duct leading to suction side of the blower, which normally has 80 m^3/min capacity run by 3 HP electric motor. The delivery side of the blower is connected to the plenum chamber of a circular grain holding bin. Hot air from the collector is forced into the drying chamber. An auxiliary heating system can also be provided to supplement heat. The auxiliary systems of heating or some thermal storage devices for collecting extra energy during daytime can take care of the night operations. These dryers, either batch type or continuous flow type are widely used for drying high moisture paddy. The drying of freshly harvested paddy takes about 7-8 hours during which moisture content reduces from 30 per cent to 16 per cent on dry weight basis.

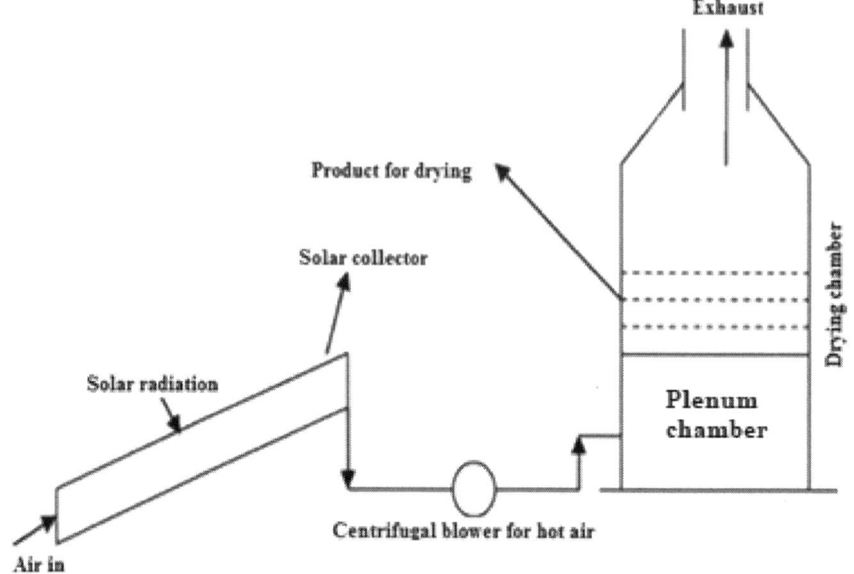

Figure 12.4. Forced convection solar dryer type.

Mixed Mode Solar Dryer

As the name indicates, energy required for the drying process is supplied through a combination of the solar radiation incident directly on the material to be dried and the circulation of pre-heated air from the solar air collector/heater. Some slight modifications are made in design of the drying chamber of a mixed mode solar dryer. The outlet vent is shifted to the side wall and the roof is made transparent. Thus, solar rays can directly enter the chamber to raise its temperature resulting in higher efficiencies (Figure 12.5). Thermal performance, cost, serving life of the system, durability, and ease of maintenance and installation are few factors that form part of the design procedure. The main advantage of a mixed mode dryer is its rapid rate of drying resulting in quick drying compared to other drying methods. Major disadvantages are its high capital and maintenance costs.

SOLAR TUNNEL DRYER

Solar tunnel dryer is a kind of polyhouse dryer where most food crops can be dried. The working principle of this dryer is based on green house effect. The system comprises a semi-cylindrical tunnel structure with hoops and laterals (Figure 12.6). Semi-circular portion of the dryer is covered with UV stabilized polyethylene film of 200 μ. Windows are provided on the sides to allow the ambient air to enter the dryer. Exhaust vents with adjustable butterfly valves are provided at roof top, which are equidistant from each other. An exhaust fan is also provided to evacuate the moist air that accumulates inside the dryer. A

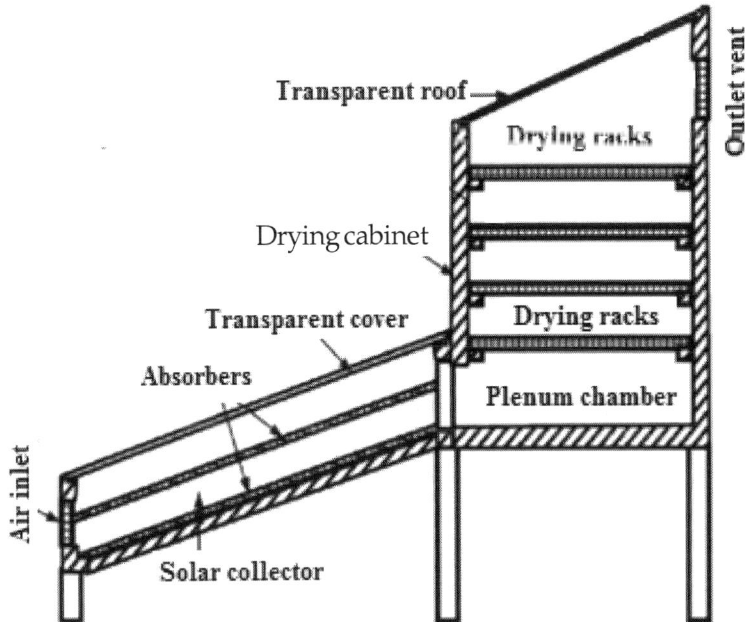

Figure 12.5. A schematic view of a mixed type dryer (Note the outlet vent on the right).

sliding door is provided to facilitate easy handling of the produce. Designs are made in such a way that no post is required inside the greenhouse. It allows better use of space inside the dryer. Cement floor inside is coated with black paint to improve the performance of the dryer. Metallic racks, trolleys and

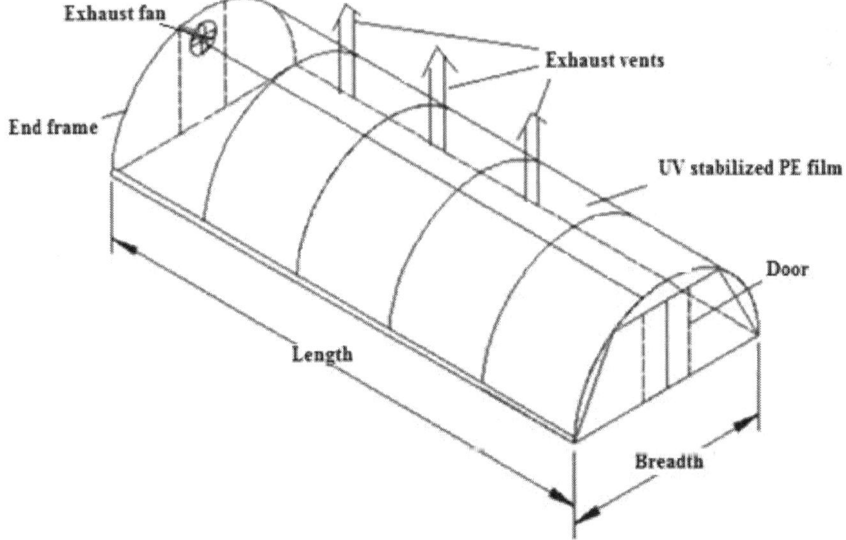

Figure 12.6. A line diagram of a tunnel type dryer.

trays are used to keep the products in layers for drying. Temperature in the tunnel is nearly 20-25°C higher than the ambient air. The dryer can be used to dry chilies, grapes, sapota, figs, mango, *amla*, fish *etc*. Studies have shown that a solar tunnel dryer of 18.0 × 3.75 × 2.0 m may handle 0.5 to 1.5 tonnes of product. Studies have also shown that drying time is reduced up to 50 per cent, while labour requirement is reduced to 40 per cent compared to conventional drying. An improved sophisticated tunnel dryer is shown in Figure 12.7. In this case, hot air from the solar collector is circulated in the tunnel while exhaust fan is operated with the help of photovoltaic solar system.

The main advantages of a tunnel dryer are:

- ✰ It is possible to instrument the dryer to control various drying parameters for various agro-products or during different seasons
- ✰ It can also be integrated with biomass hot air generation system or any other auxiliary heating system. Thus, drying can be performed both during the day and night

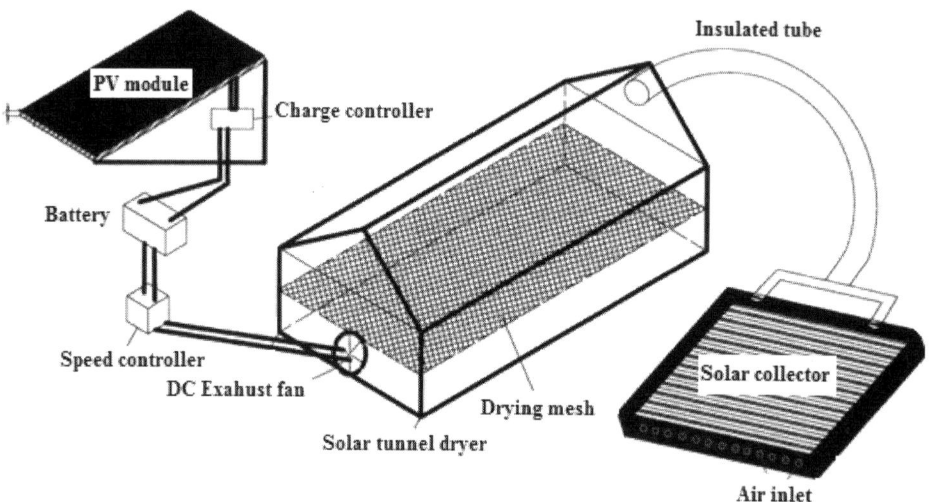

Figure 12.7. An improved solar tunnel dryer with solar collector and PV module.

THERMAL ENERGY STORAGE SYSTEMS

Solar heat can be stored in the form of specific and latent heat in the thermal energy storage (TES) materials/devices. It can be stored in the solar air collectors below the absorber plate (Figure 12.8a) or by placing the TES material in the heat exchanger (Figure 12.8b) or in the storage tank itself (Figure 12.8c). In the former, TES material is directly in contact with the absorber plate, and absorbs heat energy from the absorber plate as long as sun shines. During nights or when there is no sunshine, heat energy flows from TES to absorber plate (Lingayat

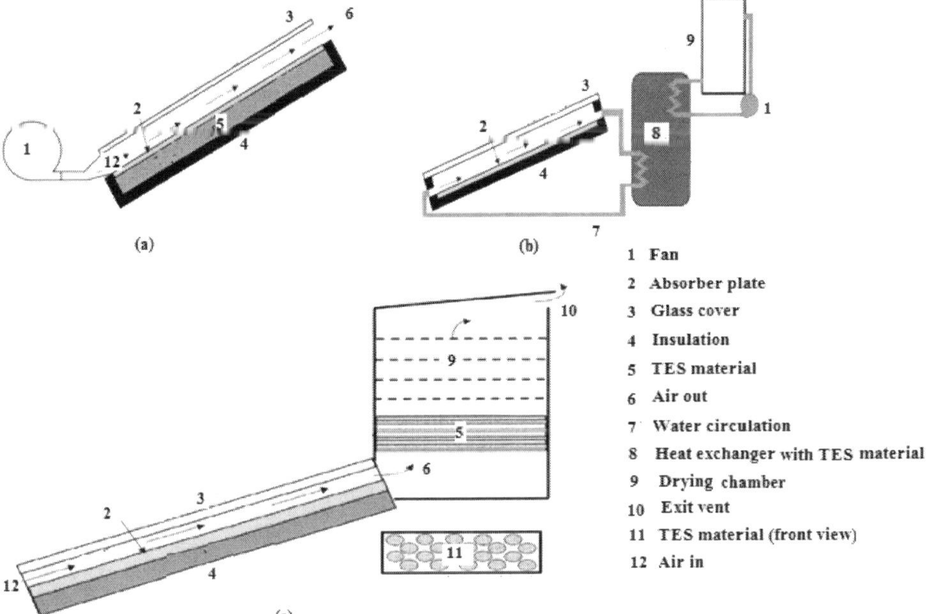

Figure 12.8. Dryers with thermal heat storage materials/devices: TES below the absorber plate (a), TES in the heat exchanger (b) and TES in drying chamber (c).

et al., 2020). When TES material is placed in the heat exchanger, solar flat plate water heater is used to heat the water. This heat is transferred from water to TES material which acts as a heat exchanger. The heat stored in the heat exchanger is utilized to heat the air for drying food products. When TES material is kept at the bottom of the drying chamber, hot air from the collector flows through the storage system. Some of the heat is absorbed by TES material. This heat is utilized for drying during off-sunshine hours. Following equations can be used to calculate the storage material required for storing known amounts of solar thermal energy.

The heat required Q in kcal/hr

$$Q = V \times \rho \times C_p \times \Delta T \tag{12.1}$$

Here V is air flow rate, m^3/hr, ρ is density of air, kg/m^3, C_p is the specific heat of air, and ΔT is the temperature rise. Volume of air to be handled, V per tonne of paddy can be determined from the difference in initial and final moisture contents, m by the following equation.

$$m \times \text{latent heat} = V \times \rho \times C_p \times \Delta T \times \text{efficiency} \tag{12.2}$$

Such that

$$V = m \times \text{latent heat}/(\rho \times C_p \times \Delta T \times \text{efficiency}) \tag{12.3}$$

The volume V′ of rock pile required for thermal storage of heat energy, Q is given as:

$$V' = Q/(\rho' \times C_p' \times \Delta T' \times e) \quad (12.4)$$

Here ρ' is the density of rock, C_p' is the specific heat of rock and $\Delta T'$ is the temperature increase in rock and e is the efficiency.

QUESTIONS (THEORY)

1. What are the major limitations of traditional drying processes? List the advantages of solar drying using solar dryers.
2. Explain the construction and working of solar cabinet dryer. List its advantages.
3. Draw a neat diagram to explain the working of indirect type solar dryer with chimney.
4. Explain the construction and working of tunnel type dryer.
5. Write short notes on forced convection solar dryer and mixed mode solar dryer.
6. Differentiate between natural convection solar dryers and conventional indirect type solar dryer, indirect type solar dryer with chimney and indirect type solar dryer with heat storage.

Chapter 13

Photovoltaic Cells and Practical Applications

PHOTOVOLTAICS

Photovoltaics (PV) comprises two terms namely photo means light and voltaic means electricity. *A photovoltaic (PV) cell, also known as solar cell, is a semiconductor device that directly generates electricity using the sun light.* Historically speaking, French physicist Edmond Becquerel discovered the process of using sunlight to produce an electric current in a solid material in 1839. The concept then was not fully comprehensible until the development of quantum theory of light and solid state physics in early to middle 1900s. It was only in 1883, New York electrician Charles Edgar Fritts constructed a selenium solar cell and filed a patent. In 1953, Bell Lab team comprising Chaplin, Fuller and Pearson produced the doped silicon solar cells having the conversion efficiency of 6 per cent. Since then, there is no looking back on the potential of this technology. Its first commercial use was made in powering orbital satellites of the US space programs in the 1950s.

Photoelectric Effect

The physical process by which a PV cell converts sunlight into electricity is known as the photoelectric effect. The electromagnetic radiations from the sun made of small energy particles called photons incident on a semiconductor material are either reflected, or absorbed, or passed through. The absorbed photons excite the electrons in the atoms of the PV cell causing ejection of electrons. *The energized electrons are known as photoelectrons and the phenomenon of emission of electrons is known as the photoelectric effect*. The electrons escaping from their original positions in the atoms of the semi-conductor material becomes part of the electrical flow or current in an electric circuit.

Construction of a PV Cell

A PV cell is made of a semiconductor material. The semi means that it can conduct electricity better than an insulator but not as good as a metal conductor. Semiconductor materials like Gallium Arsenide (GaAs), Cadmium Telluride (CdTe), Copper Indium Diselenide (CuInSe2, or CIS), Amorphous Silicon (a-Si), Selenium and Gallium have been used in the past for making PV cells. Currently, Silicon and Selenium are mostly used for this purpose. In fact, over 95 per cent of the solar cells produced worldwide are composed of Amorphous Silicon (a-Si). Solar cells made of this material provide a combination of high efficiency, low cost, and long life. These modules last for more than 25 years, and still produce more than 80 per cent of their original power. Other semiconductor materials are being tried to manufacture solar cells, some of them having the potential to surpass silicon cells in performance at lesser cost. These cells might become viable competitors to silicon cells in near future.

All silicon crystalline solar cells are made of two types of semiconductors, called n-type and p-type silicon (Figure 13.1). These are prepared by doping the silicon with specific chemical materials. Note that Silicon has a total of 14 electrons with 4 of these electrons being in the outermost shell. The n-type silicon is made by including atoms that have one additional electron in their outer shell than silicon. Phosphorus and arsenic for example have five electrons in their outer energy level compared to four in the silicon. Phosphorous is commonly used for doping to prepare n-type silicon. It bonds with silicon atoms, but one electron remains free to move inside the silicon structure as it is not involved in bonding. The p-type silicon is produced by adding atoms–such as boron or gallium–that have one electron less in their outer energy level than silicon. Boron commonly used for this purpose forms bonds with the surrounding silicon atoms, but an electron vacancy or hole is created. Thus, p-type silicon is a semiconductor rich in holes. These two kinds of semiconductors are placed in contact with each other to form a solar cell. At the region of contact, the extra electrons in the n-type semiconductor combine with the extra holes in the p-type semiconductor (Figure 13.1a). When this combination of electrons and holes happens, the loss of negatively charged electrons leaves positive ions in the n-type material, and the depletion of holes leaves negative ions in the p-type material. This creates an area around the junction, called the depletion zone, in which electrons fill the holes (Figure 13.1a). This build-up of charges near the interface creates an electric field between the n-type and p-type materials that stops any more combination of electrons and holes (Figure 13.1a). If an electron from the n-type material tries to penetrate through this barrier, it would be repelled by the negative charges of the p-type material and the holes would be repelled by the positive charges of the n-type material. This barrier acts as a wall that keeps the electrons to the n-side and the holes to the p-side. This barrier is also called p–n junction.

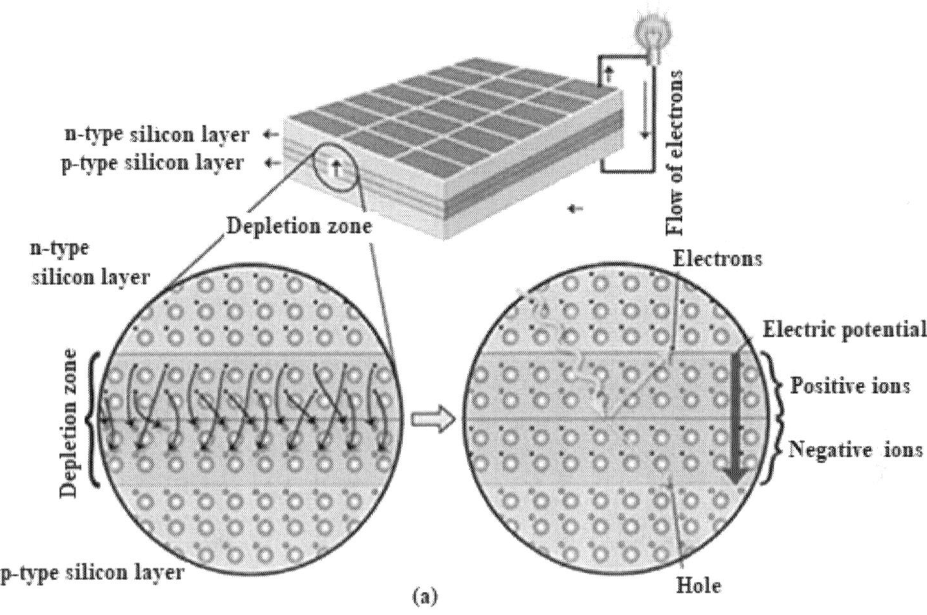

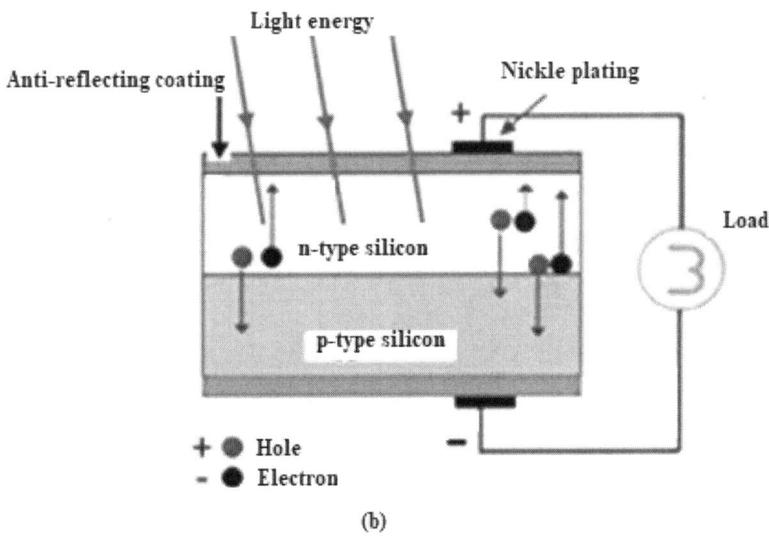

Figure 13.1. Construction of a silicon solar cell depicting the depletion zone (a) and a complete cell with anti-reflecting coating and output terminals (b).

Depending on the type of cell architecture, either n-type or p-type doped silicon wafer is used as the base or 'substrate' of the cell. A grid of very thin conductive contact strips are deposited on top of the wafer. The contact grid must maximize the surface area of the silicon wafer that is exposed to the sunlight in order to collect as much light energy as possible. The conductive grid across the top of the cell is necessary so that the electrons have to travel shorter distance through the silicon as and when an external load is connected. An antireflective coating is placed on top of the contact grid in the n-type semiconductor (Figure 13.1b). It allows the solar cell to absorb much of the sun's energy because the light energy reflected away from the surface of the cell is significantly reduced. It boosts cell efficiency. Finally, a glass or transparent plastic layer is attached to the top of the cell with transparent adhesive. It protects the cell from inclement weather. The metal rings placed around n-type and p-type material act as positive and negative output terminals respectively (Figure 13.1b). Solar cells are typically 100 cm^2 to 225 cm^2 in size and their total thickness may actually be about the thickness of an eggshell.

Working of PV Cell

As stated in the previous section, a solar cell consists of a layer of p-type silicon placed next to a layer of n-type silicon. When sunlight strikes the p–n junction with energy greater than the band gap energy of silicon, electrons are promoted from the valence band to the conduction band, generating an electron–hole pair with different potential energy. Since an electric field has been created at the junction, the electron will move towards the n-side and the hole towards the p-side. Thus, electrons accumulate in the n region, creating a negative charge, and holes accumulate in the p region, creating a positive charge. As the concentration of electrons becomes higher on one side, *i.e.* n-type side of the junction and concentration of holes becomes more on another side, *i.e.* the p-type side of the junction, the p-n junction will behave like a small battery cell. A voltage is developed between the contacts of n and p regions. When a load is connected to the solar cell, free electrons flow out through the negative contact, through the load and back into the positive contact on the bottom surface, and into the p region where they recombine with holes. The sunlight energy continues to create new electron-hole pairs and the process goes on. A voltage, known as photo voltage, is set-up. Combining the current produced by many such cells makes useful amount of electrical current and voltage. Since the current produced by a solar cell or a solar panel is direct current (DC), solar arrays are connected to an inverter to convert the DC into alternating current (AC). These panels are used to supply power to a wide variety of systems, from calculators, streetlights, parking meters and household appliances to space stations. They are also used in conjunction with other sources of energy, such as wind, natural gas and nuclear energies.

Voltage and Current

The output voltage and current from a single unit of the cell is quite less. A single photovoltaic solar cell, no matter how large it is, can produce an open circuit voltage of about 0.5 to 0.6 volts at 25°C (typically around 0.58 V). Unlike voltage, the DC current (I) vary in direct relationship to the amount or intensity of the sunlight (photon energy) falling onto the face of the PV cell. Besides, the output current is directly proportional to the surface area of the cells as larger the cell, more the light energy enters the cell. The output voltage and current of a solar cell is also temperature dependent. For the constant light intensity, the output voltage decreases as the temperature increases but the current is affected only by a small amount. The optimum operating voltage of a PV cell under load is about 0.46 volts at the normal operating temperature, generating a current in full sunlight of about 3 A. Thus, power output of a photovoltaic solar cell is equal to the product of voltage times the current (V × I) is about 0.46 × 3 = 1.38 W DC. Since the wattage of 1 cell is very less, different combinations of cells are used to increase the output efficiency. The PV cells can be combined in three possible ways. If more than two cells are connected in series, then the output current of the cell remains the same, but voltage is doubled. In the parallel combination, the voltage remains the same, but the magnitude of current is doubled. Magnitudes of both the voltage and the current increase in the series-parallel combination although their extent depends upon the actual architecture used. It is why; the solar panels are made by using the cells in a series-parallel combination.

PV Module and Panel

In order to increase power output, many single PV cells are connected together to form modules. A module holds about 40 cells. The combination of the solar modules together is known as the solar panel/array (Figure 13.2). About 10 modules are mounted in a PV array. The modular nature enables

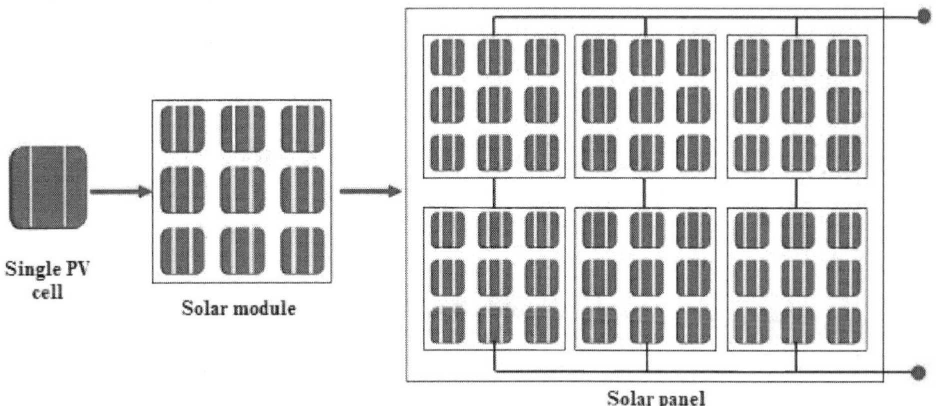

Figure 13.2. A PV cell, a module and a solar panel.

designers to build PV systems having power output that suits different types of applications. About 10 to 20 PV arrays can provide enough power for a household. Hundreds of arrays can be interconnected to form a single large system to serve large utilities or for industrial applications. Typical solar panels have an average efficiency of 12 per cent, while the best commercially available panels can have efficiency of around 20 per cent. The power of the sun at high noon on a cloudless day at the equator is about 1 kW/m^2 on the Earth's surface on a plane perpendicular to the sun's rays.

PHOTOVOLTAIC SYSTEM

A complete PV system besides PV modules/panels consists of support structures, wiring, storage, conversion devices, *etc.* It means everything else except the PV modules. Two major types of PV systems currently available in the market are: flat plate type and concentrator type. In a flat plate system the PV modules are fixed on a rigid and flat surface to capture sunlight. The flat plate systems are typically less complex but to have similar output need to employ larger number of cells. A concentrator system uses some kinds of lenses or concentrators to concentrate sunlight on the PV cells to increase their power output. These systems require smaller cell areas but require more sophisticated and expensive tracking systems. Moreover, such systems fail to work under cloudy conditions because these cannot focus diffuse sunlight.

A complete flat photovoltaic system is shown in Figure 13.3. It consists of the following major components.

Solar panel: It is a combination of several PV cells.

Maximum power point tracker: The efficiency of the PV panel depends on the intensity of sunlight falling on it. To absorb maximum sunlight, the panel needs to be moved along with the sun. Therefore, the maximum power point tracker (MPPT), a special kind of digital tracker, is used to tracks the movement of the sun.

Charge controller: The charge controller regulates the voltage drawn from the panel. It also protects the battery from overcharging or overvoltage.

Inverter: The inverter converts the direct current into the alternating current and vice versa. This kind of conversion is necessary because some utilities work on DC while others on AC (Subramanian and Sampathrajan, 2020).

Solar battery: A solar battery is an energy storage device that is paired with off-grid and hybrid solar systems. The batteries provide power back-ups during cloudy days or at night when the solar system is unable to generate electricity. Four main types of solar batteries in use are: lead-acid batteries, lithium ion batteries, nickel-cadmium batteries and flow batteries. Each of these batteries has their own set of unique characteristics. Many factors are taken into account while selecting the best one for a given application. It is beyond the scope of this book to discuss the construction, working and pros and cons of each kind of battery. It may suffice to mention that batteries are available in various sizes and capacities. Accordingly, a battery that may serve the purpose for which solar system has been designed and provides the required back-up is selected.

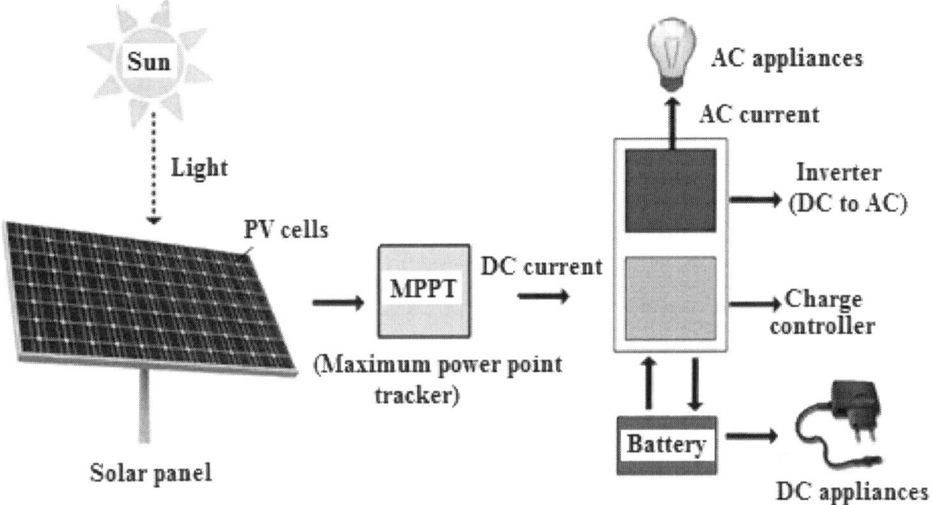

Figure 13.3. Block diagram of a complete PV system.

Classification of PV Systems

Stand-alone PV system: Stand-alone solar systems operate independent of the electric utility grid. These are generally designed and sized to supply certain DC and/or AC current to sustain connected loads besides charging the batteries to ensure required back-up power. The charge controller regulates the current output and prevents the voltage level from exceeding the maximum value needed to charge the batteries.

Grid interactive PV system: A grid-connected PV system is connected to the utility grid. Thus, it has a grid interface system, inverters and a power conditioning unit. They range from small residential and commercial roof top systems to large utility-scale solar power stations. A grid connected system may also include an integrated battery system of very small capacity. The battery in this system only serves as a backup for essential monitoring and control functions of the system.

Hybrid solar PV system: The certainty of meeting load demands at all times is greatly enhanced by using the hybrid systems. These systems refer to the combination of more than one power source. Most hybrid systems use diesel generator or wind turbine power systems as second source along with PV system, which remains the primary power source.

Advantages of Solar Cell System

- It is a source of green energy from a renewable source
- Maintenance cost is less
- Simple to operate
- It does not generate noise and harmful emissions
- It does not need water or fuel to generate electricity
- The lifespan of these cells is around 30 years
- No pollution concerns are associated with it
- Electricity bills can be reduced

Disadvantages of Solar Cell System

- High cost of installation
- It has low efficiency
- Weather dependent, the energy cannot be produced on a cloudy day as well as during night
- Occupies more space
- Storage of solar energy is expensive

CALCULATION OF ENERGY GENERATION THROUGH PV PANEL

The amount of energy that a solar panel produces depends on several factors such as the size and types of the panel, orientation of solar panel, its location and the weather conditions. On a daily basis, it is governed by the hours of sunlight received each day at the location and the Watts each panel can produce. We use the following simple formula to calculate the solar panel output.

Panel output (Daily watt hours) = STC rating (Watts peak) × Peak sun hours in a day × 0.75

On an average rating under standard test condition (STC) varies from 100 to about 400 Watts depending upon the manufacturing industry and is stated by them in their literature. Standard test conditions (STC) are: radiation = 1000 W/m^2, cell temperature = 25°C, Wind speed = 1 m/s, sea level air mass (AM) = 1.5. Here 75 per cent is used to adjust for the factors that affect the output of the solar panel. Peak sun hours can be determined by the longitude and latitude of the site or using sun hour maps. For short durations such as a month, multiply this output by the number of days in the month to get monthly Watt hours.

To estimate the electricity generated by a photovoltaic system on a long-term basis say a year, following formula is used.

$$E = A \times R \times H \times PR \tag{13.1}$$

Here E is energy, kWh, A is total area of the panel, m^2, R is solar panel yield or efficiency, H is annual average solar radiation on tilted panels (shadings not included), kWh/m^2-year and PR is the performance ratio, constant for losses (range between 0.5 and 0.9, default value = 0.75). The yield of the solar panel, R is given by the ratio:

$$R = Wp/(\text{Area of the panel} \times \text{standard radiation}) \quad (13.2)$$

For example, the solar panel yield of a PV module of 250 Wp (Watt peak) with an area of 1.6 m^2 at standard radiation of 1000 W/m^2 is 0.156 or 15.6 per cent. It is possible to calculate H by knowing the global annual irradiation incident on PV panels with the specific inclination (slope, tilt) and orientation (azimuth). PR estimates the quality of a photovoltaic installation as it gives the performance of the installation independently of the orientation and inclination of the panel. It includes all losses which depend on the size of the system, technology used and the site. The DC current generated undergoes a series of losses before it can finally become AC current. Example of losses that governs the PR value are: shadow losses, temperature losses, DC cables losses, AC cables losses, inverter losses and losses due to dust *etc.* Some normal values of the losses that affects PR value (depends on the site, the technology, and sizing of the system) are:

- ☆ Inverter losses (4 to 10 per cent)
- ☆ Temperature losses (5 to 20 per cent)
- ☆ DC cables losses (1 to 3 per cent)
- ☆ AC cables losses (1 to 3 per cent)
- ☆ Shadings (specific to each site)
- ☆ Losses for weak radiation (3 to 7 per cent)
- ☆ Losses due to dust, snow (2 per cent)
- ☆ Other Losses (to be estimated for each installation)

Number of Solar Panels Needed

Following step-wise procedure is used to calculate the number of panels required to support the desired load.

1. Find out the total power and energy consumption of all loads that need to be supplied by the solar PV system. Add the Watt-hours needed for all appliances together to get the total Watt-hours per day which must be delivered to the appliances.

   Energy demand Watt-hour/day = Σ (Power rating in Watt × Duration of operation in hours)

2. Multiply the total appliances Watt-hours per day times 1.3 (the energy lost in the system) to get the total Watt-hours per day which must be provided by the panels.

Total energy demand Watt-hour/day = Energy demand Watt-hour/day × 1.3

3. Divide the total Watt-hours per day needed from the PV modules (from item 1 and.2) by PGF to get the total Watt-peak rating needed for the PV panels needed to operate the appliances. Different size of PV modules will produce different amount of power. To find out the sizing of PV module, one needs to know the total peak watt produced, which depends on size of the PV module and climate of site location. For this purpose one needs to consider panel generation factor (PGF), which is different for each site. For example, PGF for Mataram, Indonesia is 3.19, Kota, India is 5.56, Morocco is about 4, Nigeria's capital is 3.625, USA = 3.22, EU = 293, and Thailand is 3.43.

   Total W_{Peak} of PV panel capacity = (Total energy demand Watt-hour/day)/PGF

4. Divide the answer obtained in the previous step by the rated output Watt-peak of the PV modules given by the manufacturer. Increase any fractional part of result to the next highest full number. It is the minimum number of PV panels for DC. The panels may be increased considering conversion and other losses. Note that systems with more PV modules will perform better besides increasing the battery life. With fewer PV modules, the system may not work at all during cloudy periods and battery life may be shortened.

   Number of PV panels needed= Total W_{Peak} of PV panel capacity/W_{peak} of one panel

SOLAR LANTERN

The name solar lantern has been derived from the common man's lantern that uses a wick and kerosene oil for lighting. Solar lantern is probably the simplest application of solar photovoltaic technology (Figure 13.4). It is most suited for remote areas where power supply is irregular and scarce. It is a preferred solution during power cuts even in the urban areas because of its simple mechanism.

The casing: Casing of the lantern is made of metal, plastic or fiberglass. Battery, and the charging and control circuit are housed inside the casing making use of the inside space in the casing. A lamp holder is mounted on the top of the casing. One CFL/now LED bulb of appropriate rating is attached with this holder. The lamp is covered from all sides by transparent glass or fiberglass so that it can emit light in all directions. The top cover is non-transparent and is generally made of the same material as the casing. The hanger is fitted at the top cover of the lantern. A plug point is provided for charging the battery.

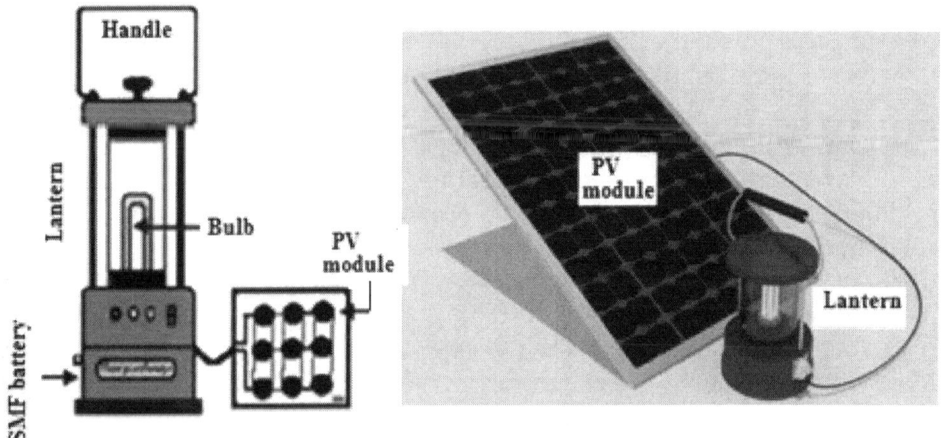

Figure 13.4. A line diagram (left) and a photographic view of a solar lantern (right).

Solar module or PV module: Solar module used for charging a single solar lantern has rating of 8, 10 or 12 Watts. The module is kept on the roof top at a tilt angle so that the module receives maximum intensity of sunlight for maximum duration. The lanterns are connected to module through a wire socket kept in the room or at other convenient place. In general, users do not install the solar module on roof top. They keep the module in sunlight on daily basis. The module is connected with lantern's battery terminals for charging purpose (Figure 13.4 right). Few specifications of the PV module are as follows.

Model	*Lamp (CFL)*	*Battery*	*PV Module*
I-A	5 W	12 V, 7 Ah at 20°C	8 to 99 Watts (Peak)
I-B	5 W	12 V, 7 Ah at 20°C	8 to 99 Watts (Peak)
II-A	7 W	12 V, 7 Ah at 20°C	8 to 99 Watts (Peak)
II-B	7 W	12 V, 7 Ah at 20°C	8 to 99 Watts (Peak)

Protection and indication system: The electronic circuit of the lantern has a cut-off provision that prevents excess charging or over discharging of the battery. An electrical fuse protects it from any fault resulting from short circuiting. Two light emitting diodes (LED) are provided on the casing to indicate charging and deep discharging states of the solar lantern.

Battery: Maintenance free sealed 12 V- 7 Ah capacity dry type lead acid battery is used in solar lantern. A 12 V- 2.3 Ah sealed maintenance free dry type lead acid battery is also used for LED solar lanterns.

Inverter: In lanterns with CFL lamps, an inverter is used to invert the DC of the battery to AC. Now there is no need to have an inverter since DC LED bulbs are fast replacing the CFL bulbs. A major benefit of LED bulbs is that they consume less power than CFL bulbs. Five mm white LED giving more than 30 lumens per Watt is most suited for LED based solar lanterns.

SOLAR STREET LIGHTING

Solar street lighting system is ideal for lighting in remote villages where grid power is yet to reach or is highly uncertain. It is an extended version of the solar lantern requiring a bigger PV module and battery. Battery storage back-up is sufficient to operate the light for 10-11 hr on a daily basis. The system is provided with an automatic ON/OFF time switch for dusk to dawn operation. It also has overcharge/deep discharge prevention cut-off with LED indicators. Major components of a street lighting system are (Figure 13.5):

- Solar panel(s)
- Solar rechargeable battery
- Smart solar controller with in-built wireless communication
- Remote management software
- LED light
- Pole
- Interconnecting cables

The performance of any functional off-grid system depends on the quality of each of these components, the technology used in the system and the management of the solar street light over time.

PV module: Selection of the PV module for a street lighting system is based on number of solar panel(s), panel type *i.e.* mono-crystalline or poly-crystalline, power of solar panels *i.e.* total Wattage and mounting location *i.e.* on top of the pole or side of the pole. For a standalone street light post 74 Watt solar PV module should be reasonably good.

Solar battery: The main function of the solar battery is to store electricity from the solar panel during the day to provide energy for the LED lamp during the night. The capacity and lifetime of the battery is important as it provides back-up power even during low energy periods – days without sun. A 12 V, 75 Ah tubular battery with battery box, and the charge controller should suffice. It is desirable to bury the battery underground as solar batteries do not perform well if exposed to extreme temperatures. The soil provides optimal temperatures both during winters and summers.

Controller: The solar controller is the heart of the solar street lighting system. This device controls battery charging and switches on and switches off the light. Smart controllers have built in functionalities to control lighting, dimming and battery charging. During the day, it continuously charges the battery and prevents its overcharging or undercharging. During the night, it supplies the stored power to run the LED street lights. Smart controllers can support single load or multiple loads.

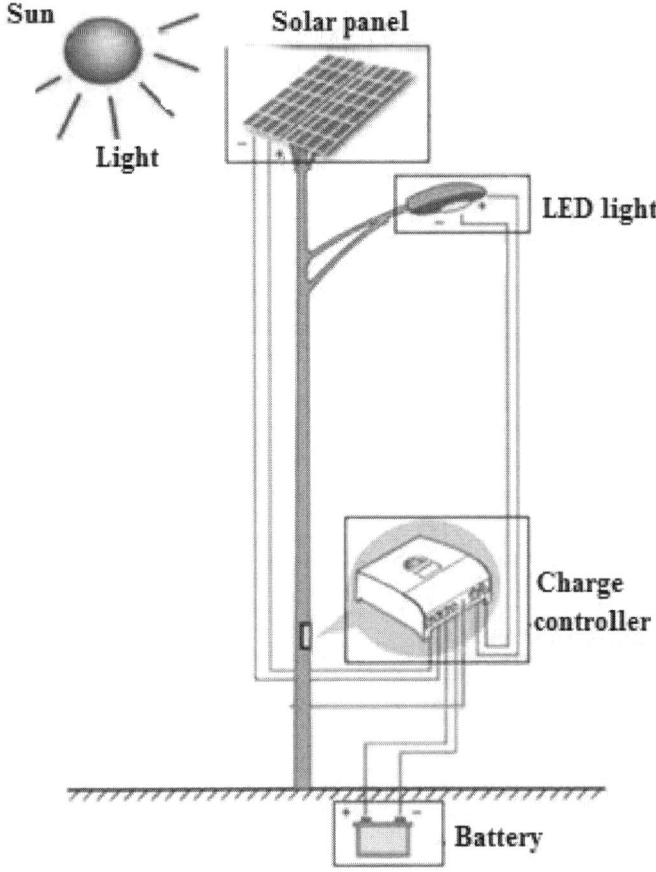

Figure 13.5. Block diagram of a street lighting system.

LED light: Low wattage LED lights are more efficient than traditional style fixtures. These have lower power consumption and at the same time provide higher luminosity with zero UV emissions. LED fixtures having an average life of 50,000 working hours are less prone to failures.

Pole: It is a mild steel lamp post above ground level with weather proof paint and mounting hardware.

The approximate cost for the most common specification is around Rs 24,000. It varies from one model to another and on the components quality and sizes.

Advantages

- ☆ No grid electricity is required

- ✰ Easy to install
- ✰ Simple to operate with low maintenance cost
- ✰ Eco-friendly

SOLAR WATER PUMPING SYSTEM

India is blessed with abundant solar energy with clear sunny weather for more than 300 days in a year. The annual range of solar radiation is from 4 to 6 KWh/m^2-day (ICID, 2019). These together with high utilization of ground water for agriculture in the country make solar-powered irrigation systems an attractive alternative to run ground water pumps in India. Out of total of nearly 30 million pumps operating in India, 21 million are grid operated, 8.8 million are diesel powered and only 0.13 million are solar-powered (https://www.ceew.in/sites/default/files/CEEW-Solar-for-Irrigation-Deployment-Report-17Jan18_0.pdf). It is heartening to note that more than 181,000 solar power pumps have been installed in the country over the past three years under the *Pradhan-Mantri Kisan Urja Suraksha evam Utthaan Mahabhiyaan* (PM-KUSUM) scheme. Out of 96,376 pumps sanctioned during 2017-18, state-wise details of the pumps installed are shown in Table 13.1 (https://mnre.gov.in/img/documents/uploads/file_f-1608040317211.pdf).

Table 13.1. State-wise details of solar PV pumps

Sl.No.	*State*	*Numbers**
1.	Andhra Pradesh	15000
2.	Bihar	225
3.	Chhattisgarh	15000
4.	Gujarat	3537
5.	Jharkhand	392
6.	Madhya Pradesh	14000
7.	Maharashtra	1000
8.	Odisha	530
9.	Punjab	2000
10.	Rajasthan	7134
11.	Tamil Nadu	525
12.	Uttar Pradesh	9588
	Total	**68931**

*The figures include pumps installed during 2017-18 and 2018-19.

A solar power pumping system is essentially a normal pump except that it derives renewable power from the sun through a solar system set-up. The system comprises solar panels that absorb sunlight, inverter to convert direct current into alternating current, mounting structure to hold the panels in place, batteries to store the extra power generated, grid box and other parts such as cabling, wiring, and nuts *etc.* The size of the system is designated in kW such as 1 kW, 3 kW, 5 kW, 10 kW *etc.* Besides, some systems are equipped with solar tracking system and integrated battery solution. The main features of this system are:

- Modular and hence easily upgradable
- Noise and pollution free operation
- Does not require any fuel as it uses the abundantly available sunlight
- Simple to install, operate and maintain
- Designed to give optimum output even during low sunshine periods

The pump: It is the heart of the solar water pumping system. Solar pumps are rated according to the volts that need to be supplied for its operation. A 12 volts pump is a small one, 24 volts is more a norm, while 48 volts and upwards require more power but are capable of pumping more water. Smaller wire sizes are used even in higher voltage systems without sacrificing power output from panels to pump. It is because wires are quite costly. Some pumps to function optimally may also require additional accessories such as filters, float valves, switches, *etc.* The manufacturer or distributor usually specifies as to what is required.

Solar Panel: A pump requires a certain amount of power to generate certain amount of pressure and flow. Thus, one should always size the PV array by the amount of power needed. It may be noted that most solar pumps actually require about 20 per cent more wattage than specified when panel is directly wired to the pump. Photovoltaic (PV) panels (Figure 13.6) are rated in watts of power it can produce. A larger panel allows the pump to turn on earlier and work until later in the day because it can work under relatively low light conditions. Since panels have now become much more affordable than before, one may go for higher wattage (ICID, 2019).

Inverter: The actual power supplied to the pump from the solar panel depends upon the power of the inverter and its efficiency. Most inverter comes with an efficiency of about 80 per cent. Therefore, a 3 kW inverter system will give up to 2.4 kW of peak power at a given point of time.

Linear current booster: A small device installed between the panels and the pump allows the pump to switch on during low light conditions. Linear current boosters are sized in accordance with the pump voltage and panel output.

Wiring accessories: These include minor components that will connect the panels to the pump. A combiner box is used to make wiring safe and simple when more than one panel is used. Circuit breakers are installed in the box to enable safe and quick shut-off of the panels as and when the system needs servicing. The circuit breakers at the discretion of the user can also serve as a switch for turning the pump on and off.

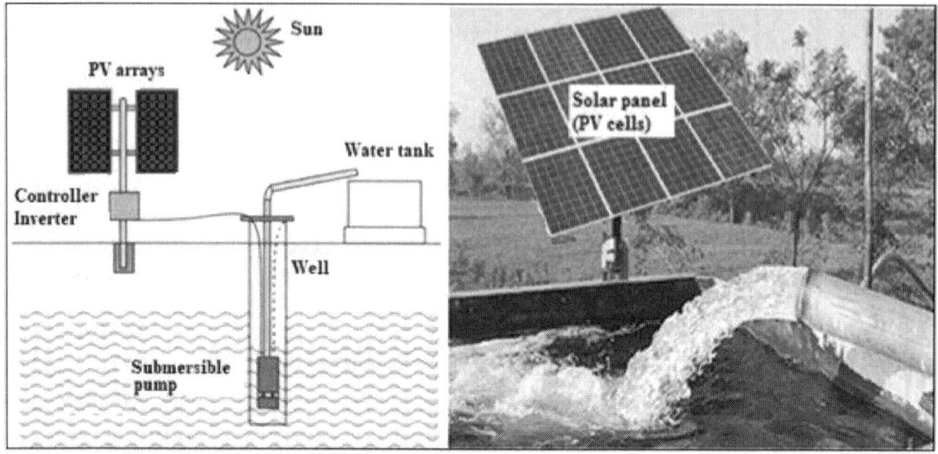

Figure 13.6. A schematic of solar pumping system (left) and a solar panel on a ground water well (right).

Battery: Many types of batteries are in use, but the lead-acid is the most commonly used battery in stand-alone PV systems. It differs from deep discharge batteries because they use heavier, thicker plates and strong inter-cell connections to withstand the mechanical stresses from deep discharges. Battery storage systems often have power ratings (the capability of the battery to supply power) in kilowatts (kW), typically in the range of 1-7 kW. The battery storage capacity is measured in ampere-hours (Ah) or kilowatt-hours (kWh). A 12-volt lead-acid battery rated at 480 Ah may only be capable of supplying 2.25 kWh of energy because of the limitations of depth of discharge. It is why; lithium-ion batteries are now becoming more popular because allowed depth of discharge is much higher than lead-acid batteries.

Solar tracking system: Trackers are used to track the sun movement. The trackers direct the solar panels or modules toward the sun by changing their orientation throughout the day following the sun's path. It helps to maximize energy capture. Single-axis solar trackers as the name suggests rotate on one axis moving back and forth in a single direction. Some single-axis trackers include horizontal, vertical, tilted, and polar aligned. Dual-axis trackers on the other hand make the solar panel to continually face the sun. These include tip-tilt and azimuth-altitude types consisting of servo motor, stepper motor, rain drop sensor, temperature and humidity sensor and an LCD.

A set-up of a solar power pumping system on canal irrigation in Haryana is shown in Figure 13.7 (left). On the right, a rain gun is being operated in an adjoining field for irrigation (Figure 13.7). Some advantages and disadvantages of solar power pumping systems are as follows (Hartung and Pluschke, 2018).

Figure 13.7. A solar-powered grid connected system on canal irrigation in Haryana (left) and a rain gun in operation (right).

Salient Features of the System

Some salient features that are incorporated in the system are listed as under.

- ✰ Automatic start and shut-off as per solar intensity
- ✰ Dry run protection (automatic switch off)
- ✰ Single phase 50 Hz AC input operating voltage range of 100-270 Volts
- ✰ Visual indication of faults through flickering LED displays
- ✰ Will not restart automatically unless attended for faults like dry run, phase imbalance, output short circuit
- ✰ Controller is designed to accept DC voltages

Advantages of Solar-Powered Irrigation

- ✰ Rugged construction with long life
- ✰ A reliable source of energy for pumping of irrigation water in remote areas, particularly those that are not connected to the electricity grid or where regular grid supply, liquid fuels and maintenance services are not guaranteed
- ✰ Solar power irrigation system can help to stabilize, increase and diversify production through improved access to energy and water
- ✰ Reduced dependency on fossil fuel can play an important role in climate change mitigation by reducing green house gases emissions in irrigated agriculture.

- Energy subsidies for fossil fuels can be reduced while offering an alternative to farmers and rural communities whose livelihoods may otherwise be negatively affected
- Since cost of solar panels is dropping over the years, it may result in reduced cost of water pumping in the long run. Solar systems are now economically viable and competitive with other sources of energy
- Potential for job creation in the renewable energy sector (producers, suppliers, services)
- Potentially more efficient use of water can be achieved when combined with drip or other water- efficient irrigation technologies
- Lower hourly yields over more hours per day results in gentler abstraction of sensitive groundwater resource, reducing risk of borehole collapse
- Easy to operate and maintain

Disadvantages

- Relatively high initial investment need innovative financing models (or subsidies) especially for small and marginal farmers
- Since optimal operation and maintenance requires a certain degree of technical knowledge and skill, farmers need to be trained and/or adequate services (extension services or private service suppliers) for this purpose need to be arranged
- There is a risk of over-abstraction of ground water and unsustainable water use as low energy costs can lead to wasteful use of water using irrigation methods that have low field water application efficiencies
- Taxes on imported equipment may distort and artificially keep prices high
- Production of PV panels requires some toxins and rare minerals. Mining and production of these tends to produce environmentally harmful wastes
- Panels need to be disposed of appropriately to avoid environmental damage
- Illegal pumping might increase and may lead to depletion and degradation of ground water resources
- Lack of codes and standards that guarantee quality of solar power irrigation system may pose few problems
- Systems are vulnerable to theft. Therefore, insurance agencies do not insure the systems, a pre-requisite for financing by the banks

Relative merits and demerits of solar power irrigation system and diesel pumps have been listed in the following Table (Abu-Aligah, 2011).

Type	Advantages	Disadvantages
PV pump	Unattended operation	High investments
	Low maintenance cost	Water storage may be required as the pump operates at sub-optimal level during cloudy times and during night
	Long life	Skill required for maintenance
Diesel pump	Low investment cost	High operational cost
	Fast and easy installation	High maintenance costs
	Familiarity of the farmers to technology	Short life
		Environmental issues with noise and air pollution

Applications

- Minor irrigation
- Drinking water for non-electrified villages and remote locations
- Horticulture, poultry farming, silviculture and pisciculture
- Domestic supplies for farm houses
- Wild life sanctuary
- Tourist resorts

SOLAR FENCING SYSTEM

Fencing is an integral part of agriculture to save the crops and other farm structures from intruding animals. Moreover, it is also a reliable and cost effective way to control animal movement and manage pastures. The objectives of fencing include:

- Keeping domestic animals within the designated areas
- Keeping wild animals out
- Separating different groups of animals
- Rationing of crops and pastures
- Fencing off sensitive areas, trees, rivers and roads

Conventional fencing includes poles and barbed wire fencing, and live fencing comprising thorny and bushy trees *etc*. These conventional fences are passive fences and cannot resist the animals to go in or out if they force their way in/out. On the contrary, electric fencing system is an active fencing system that gives a shock to the animals yet does not prove fatal to hm. It gives a sharp, short but a safe shock that creates psychological fear. Being a psychological barrier, it doesn't need great physical strength. Nonetheless, it must be well designed and constructed to absorb some pressure from animals, snow and wind. Solar fencing has come into operation and becoming popular because it is often difficult to arrange electricity at remote places. A fencing system can also be provided with an alarm that gets activated as soon as some tempering is done to the fencing system.

Basic Components

The basic building blocks of a solar power fence are (Figure 13.8):

1. Solar panel
2. Solar charge controller
3. Battery
4. Energizer
5. Earthing (Grounding system) and
6. Fence system wiring

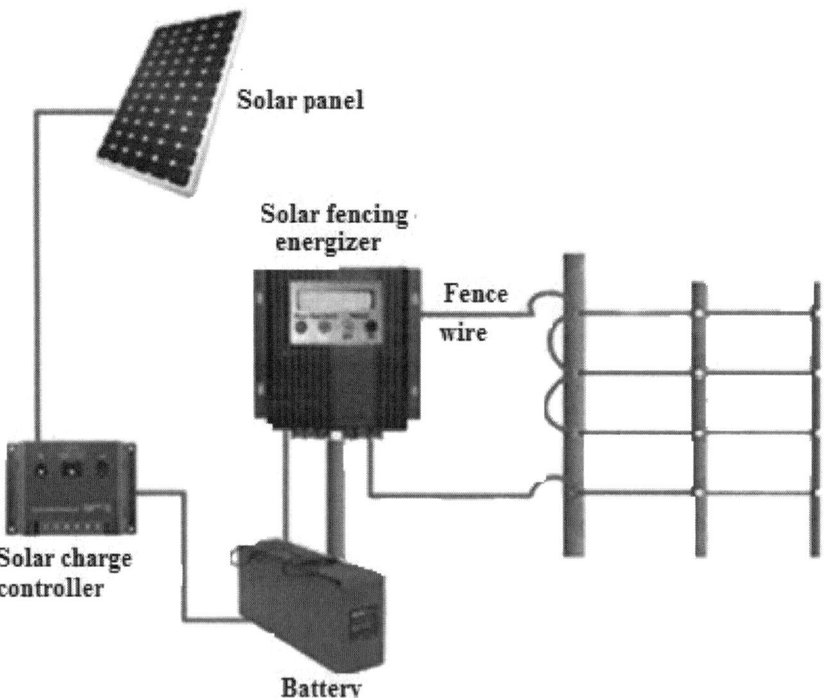

Figure 13.8. A schematic diagram of the solar fence system showing major components.

Solar Panel and Charge Controller

Solar-powered panel comprises multiple small photovoltaic (PV) cells connected in series. The PV cells in the panel absorb, collect and convert solar energy into electricity. The energy flowing through the charge controller gets stored in a battery. The solar panel keeps charging the battery throughout the day, which in turn keeps the energizer operational 24×7.

Energizer

Energizer, also known as controller or charger or fencer, is the heart of any power fence. Depending upon the source of power, energizers are categorized as mains energizers, battery energizers powered by a 12 V rechargeable battery, or a 9 volt disposable dry cell battery. Battery energizers are typically portable suited to remote areas far from power outlets. Solar energizers are used both in portable and permanent installations being an excellent choice for temporary fence applications. Multi-powered energizers allow the use of power from multiple sources. Main function of the energizer is to produce short and sharp pulses of about 8000 volts at regular intervals. Power stored in the capacitors of the energizer serves as the potential energy available for the single pulse per second to power the fence line. The higher the Joule rating of the capacitor, greater is the energizer's shock intensity as well as to push past through shorts caused by weeds and fence faults. The higher Joule rating also helps to take care of any future expansion of the fence line. A minimum of 15 J energizer is recommended. The selection of an energizer mainly depends on the kind of animals to control, length of the fence and number of strands. The energizer must have enough power for the length of fence and for the kind of animals being controlled. The energizer should be enclosed, be free from mechanical damage and should be kept away from inflammable materials and children.

Earthing System

The earth system acts as half the circuit of the fencing system. Electrons travel from the energizer, along the fence wires and back through the earth to the energizer to complete the circuit. The earth or ground system must be perfect to enable the pulse to complete its circuit and give the animal an effective shock. The number of earth stakes used for earthing may vary depending on the power of the energizer and the soil type. High powered energizers need more earth stakes than low powered energizers. Dry, sandy, rocky or frozen soils require more earth stakes than wet soils. Normally 3 stakes are used. Following equation is used to calculate the number of stakes where more than 3 stakes are used.

$$\text{No. of stakes} = \text{Stored Joule rating of the energizer}/5 \qquad (13.3)$$

A 4-3-2-1 rule is normally adopted for permanent live fence. It means the distance between stakes should be 4 m, 3 earth stakes minimum, 2 m minimum length of earth stakes and 1 wire connecting all earth stakes to energizer earth terminal (green terminal) (https://am.gallagher.com/-/media/Bynder/Animal-Management/Document/ALL-GGL-Brochures/Introduction-to-Electric-Fencing-Brochure-original.pdf).

Types of Fencing Systems

Depending upon the soil conditions, two kinds of fence wire systems are used.

- ✰ All live wire system
- ✰ Earth or ground wire return system

Both the solar-powered fencing systems work on a similar mechanism as other solar powered products. The solar module generates the DC energy and charges the battery. The output of the battery is connected to the energizer, which produces a short, high voltage pulse at regular rate of one pulse per second. The higher the Joule rating of the energizer, the bigger is the shock. Bigger the shock, more lasting is its impact on the memory of the animal to avoid intrusion in future.

All live wire system is best suited for wet and greener regions where earth has good conductivity. In this system, all fence wires are connected to the red (live) terminal on the energizer. The earth stakes are connected to the earth (green) terminal of the energizer. Whenever, an animal/intruder touches the live wire, it creates a path for the current through its body to the ground and back to the energizer via the earth system (Figure 13.9 left). It completes the circuit as a result of which intruder receives a shock. The degree of shock felt by the animal results from a combination of fence voltage and pulses time or energy. Note that the shock is delivered to the animal only when it touches the ground and the fence at the same time.

Earth return wire system is used where ground is unable to conduct enough power due to year-round dry, frozen or snow conditions. Since soils in such cases are not good conductors, the electrons spread out and travel over a wide area and are unable to conduct enough power. The working principle is almost same except that the fence wires comprise live and earth wires. Live wires of the system are connected to the read terminal of the energizer while the earth wire to the green terminal via the earth stakes. The shock to the animal is delivered as soon as the animal touches both the live and earth wires on the fence at the same time (Figure 13.9 right).

Features of Solar Fencing System

Solar fencing system is a cost effective solution to conventional fencing with the following additional features.

- ✰ It is easy to construct and needs minimal maintenance
- ✰ It is aesthetically pleasing then barbed wires or bushes

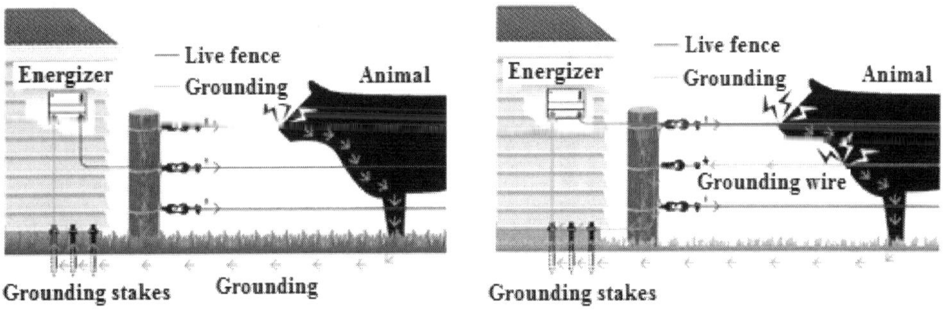

Figure 13.9. All live wire system (left) and earth or ground wire system (right).

- It is long lasting because no physical pressure is exerted on the system
- Intrusion by all domestic and wild animals can be economically controlled. Even the system can be designed to target specific species only

QUESTIONS (THEORY)

1. What do you understand by photovoltaic and photoelectric effect?
2. Explain in a detailed manner the construction and working of a photovoltaic (PV) cell.
3. What do you understand by PV cell, module and panel? How the PV cells can be combined in a module and how it affects the total voltage and current?
4. Explain the construction and working of a solar lantern.
5. Explain the construction and working of a solar street lighting system.
6. Discuss the solar water pumping system. What are its advantages and disadvantages?
7. Discuss the various components of a solar fencing system. Explain the working principle of the solar fence system.
8. Differentiate between all live wire system and earth or ground wire return system of solar fencing. Use a neat line diagram to illustrate your answer.

Chapter 14

Hydropower

HISTORICAL BRIEF

Power of moving water has been used by humans for more than 2,000 years. First references to watermills are available in Greek, Roman, and Chinese texts. The Egyptians used Archimedes water screws for irrigation during the third century BC. By the fourth century, watermills were found in Asia and northern Europe. In the early 11th century, William noted that thousands of watermills utilizing the power of streams and rivers are being used in England. Some water mills even utilized the power of tides. The evolution of the modern hydropower turbine began in the mid-1700s when a French hydraulic and military engineer, Bernard Forest de Bélidor, wrote the groundbreaking *Architecture Hydraulique.* In the late 1700s, an American named Oliver Evans designed a mill that combined gears, shafts, and conveyors. In 1826, a French engineer, Jean-Victor Poncelet designed an enclosed water wheel so that water flowed through the wheel instead of around it. This idea became the basis of the modern American water turbine. In the mid-1800s, James Francis, Chief Engineer of the Locks and Canal Company in Lowell, MA, improved the enclosed water turbine by reshaping the blades. It is known as the Francis turbine. Modern variations of this turbine are still in use in hydropower plants. In 1880, a dynamo driven by a water turbine was used to provide arc lighting- a technique where an electric spark in the air between two conductors produces a light- to light a theatre and storefront in Grand Rapids, Michigan. In 1881, a dynamo connected to a turbine in a flour mill provided street lighting at Niagara Falls, New York. Currently, almost 17 per cent of the world's electricity is produced by hydropower. While Canada produces 60 per cent of its electricity from hydropower, India produces only

13 per cent. Five to ten 10 per cent of electricity in the United States comes from hydropower depending on the supply of water. The hydroelectric power plants at Darjeeling and Shivanasamudram were established as early as in 1898 and 1902, respectively. They were among the first in Asia and India has been a dominant player in global hydroelectric power development.

HARNESSING WATER ENERGY

Hydro energy is available in many forms. It can be categorized as conventional hydropower harnessing the potential energy from high heads of water retained in dams, and kinetic energy from current flow in rivers. The other kind is the oceanic hydropower that harnesses the tidal energy, and kinetic energy from the movement of waves on relatively static water masses (Figure 14.1), thermal gradients and osmotic power. Many ingenious ways have been developed to harness these forms of energy but most involve directing the water flow through a turbine to generate electricity. Besides, it is also harnessed using the movement of the water to drive some other form of hydraulic or pneumatic mechanism to perform the same task.

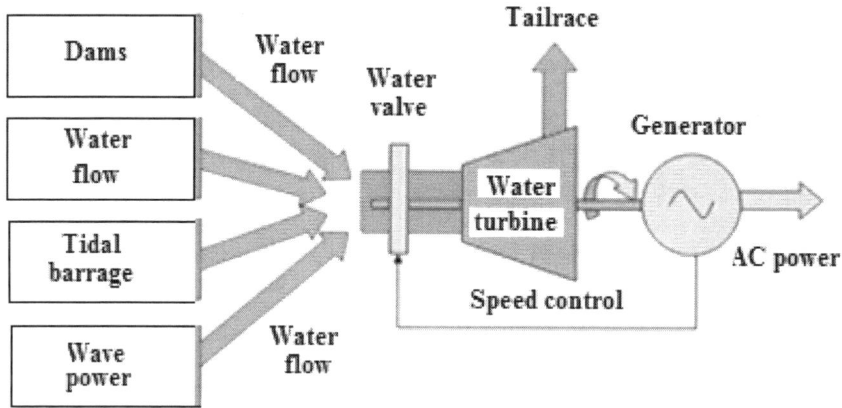

Figure 14.1. Harnessing of water energy for electricity generation.

Gravitational Potential Energy

The water in a reservoir behind a hydropower dam is an example of potential energy. The stored energy in the reservoir is converted into kinetic energy (motion) as the water flows down a large pipe called a penstock and spins a turbine. The turbine spins a shaft inside the generator, where magnets and coils of wire convert the motion energy into electrical energy through a phenomenon called electromagnetism.

Kinetic Energy

It is energy in motion produced through the movement of objects and substances from one place to another. According to Newton's Laws of Motion, objects and substances move when an unbalanced force is acting on them. A river flowing or breeze blowing are examples of motion energy.

CONVENTIOANL HYDROPOWER

As previously stated, hydropower harnesses the power of water available in many forms (NEED, 2014). Hydropower facilities utilizing storage dams are categorized into two main types: conventional and pumped storage. Conventional hydropower plants use the available water from rivers, streams, canal systems, or reservoirs to produce electrical energy. Systems with dams and reservoirs, known as impoundment facilities, store water and use it to generate electricity as per the demand. Projects without reservoirs are known as diversion facilities or run-of-the-river projects. Diversion projects do not require dams; instead, a portion of river water is diverted or channeled through a canal or penstock. Run-of- the -river projects have turbines installed in fast-flowing sections of the rivers, although they do not significantly slow down the rivers' flow. Conventional projects account for more than 80 per cent of hydropower generating capacity in the world. Another type of hydropower plant is known as a pumped storage plant. A pumped storage plant circulates water between two reservoirs- one being at a higher level than the other. When the power demand is low, the plant uses electricity to pump water to the upper reservoir and stores the water there until the power demand is high. At this time water is released from the upper reservoir to flow through turbines and back into the lower reservoir to generate electricity. A pumped storage plant is in many ways like a huge battery that stores the potential energy of the water in the upper reservoir to generate electricity quickly as soon as the demand shoots-up. Pumped storage is likely to constitute about 30 per cent of total new hydropower addition estimated at 230 GW during 2020-2030 (IEA, 2021). The addition from dam based projects is estimated at 56 per cent, run-of-the–river projects at about 13 per cent and others about 2 per cent.

Hydropower Plant

A typical hydropower plant of a dam based system has three main parts namely the reservoir, the dam, and the power plant including turbines and generators (Figure 14.2). The reservoir stores the water (potential energy). The vertical distance between the water surface level in the reservoir to the central axis of the turbine is the head. Openings in the dam control the water flow (kinetic energy). The power plant (turbine and generator) converts the kinetic energy of the moving water into mechanical power and electricity respectively. The generated electricity is sent to the switchyard for voltage transformation and transmission to distant places. The water entering the penstocks returns to the river below the dam and continues its downstream journey. The capacity of a hydroelectric system is given by the maximum rate at which it can produce electricity. It is typically measured in Megawatts (MW) or GW where 1 GW equals 1000 MW. The energy of a hydroelectric system refers to the amount of energy stored as potential energy in the upper reservoir. It is typically measured in Gigawatt-hours (GWh). A reservoir with 10 GWh of storage could operate with power of 1 GW for 10 hr.

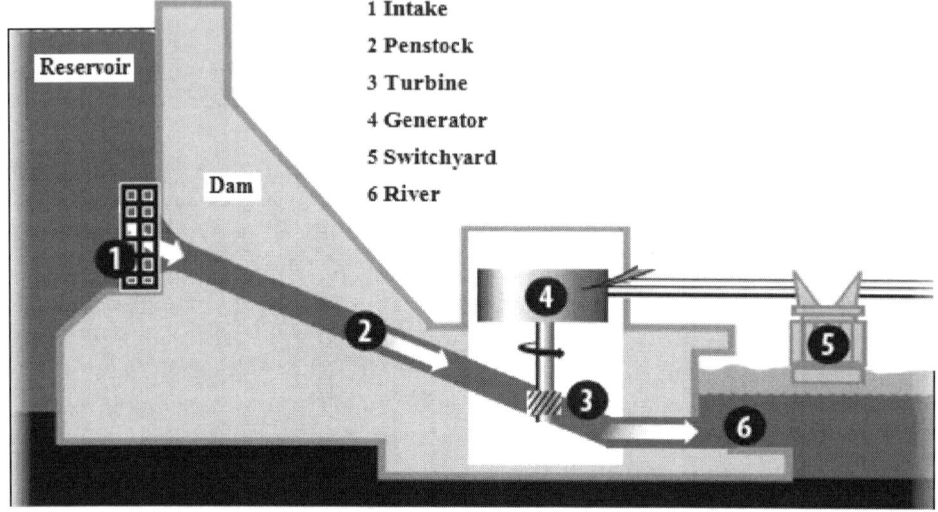

Figure 14.2. Diagram of a dam based hydroelectric facility.

WATER POWER AND ELECTRICITY GENERATION

Hydroelectric power generation is by far the most efficient method of large scale electric power generation. The force of gravity of any object is given by the relation

$$F_g = \text{mass} \times g \tag{14.1}$$

Here g is the gravitational constant, 9.81 m/s². The potential energy is given as this force multiplied by the distance through which the object falls. If H is the height through the object falls, then the potential energy of the object is:

$$PE = F_g \times H \tag{14.2}$$

For water stored in a dam, potential energy per unit volume can be written as ρgH. Here ρ is the density of the water, 1000 kg/m³, and H is the head of water, which is the difference in heights between the surface of the water in the reservoir and the turbine. The potential energy, Watt/s (Joules) of a volume of water v, m³ that falls through a height H, m is given as:

$$PE = 9810 \text{ v H} \tag{14.3}$$

The power P from a dam is given by

$$P = 9.81\ \eta\ H\ v/t = 9.81\ H\ Q\ \eta \tag{14.4}$$

Here Q is the volume of water flowing per second (v/t, the flow rate in m^3/s) and η is the efficiency of the turbine, fraction. For water flowing at one m^3/s from a head of 1 m, the power generated is equivalent to 10 kW assuming an energy conversion efficiency of 100 per cent or just over 9 kW with a turbine efficiency of between 90 to 95 per cent. Since the cost of most components is largely independent of the head, a larger head will generally produce cheaper electricity on a per-unit basis. Typical heads are in the range 100-800 m, although many times even larger and smaller heads are used.

Energy efficiency is the amount of useful energy produced by a system compared to the amount of energy input. A perfect machine would convert all the input energy into useful work. This is nearly impossible as conversion from one form of energy into another form always involves a loss of usable energy. Not all of the potential energy of the water behind the dam is realized because the water running through the pipes encounters drag forces from the pipe walls. The water hitting the turbines generates some heat, as does any type of collision. The water leaving the turbine still has some kinetic energy, which is not given to the turbine. Accounting for all of the energy losses in the system, the hydropower systems are still about 80-90 per cent efficient, which is one of the highest efficiencies for any type of electricity generating facility. For example, a typical coal-fired power plant converts about 35 per cent of the energy in the coal into electricity. The rest of the energy is lost as heat.

Run-of-the-River Systems

Water from a fast flowing river or stream in run-of-the-river scheme is diverted through a turbine, often a Pelton wheel, which drives the electric generator. The local head of water may be essentially zero. The turbine is designed to convert the kinetic energy of the flowing water into the rotational energy of the turbine and the generator. The energy available depends on the quantity of water flowing through the turbine and the square of its velocity. Impulse turbines which are only partially submerged are commonly employed in fast flowing run-of- the-river installations while in deeper, slower flowing rivers with a greater head of water, fully submerged Kaplan reaction turbines may be used to extract the energy from the flowing water. The maximum power output from a turbine used in a run-of- the-river system is equal to the kinetic energy ($\frac{1}{2}mv^2$) of the water impinging on the blades. Taking the efficiency η of the turbine and its installation into account, the maximum output power P_{max} is given by

$$P_{max} = 0.5\,\eta\,\rho\,Q\,V^2 \qquad (14.5)$$

Here V is the velocity of the water flow and Q is the volume of water flowing through the turbine per second. Here, Q is given by A V such that A is the swept area of the turbine blades. Thus

$$P_{max} = 0.5\,\eta\,\rho\,AV^3 \qquad (14.6)$$

The power output is proportional to the cube of the velocity of the water. Thus, the power generated by 1 m^3 of water flowing at 1 m/s through a turbine with 100 per cent efficiency will be 0.5 kW or slightly less taking into account the inefficiencies in the system. This is only one twentieth of the power generated by the same volume of flow from the dam discussed in the previous section. To generate the same power with the same volume of water from a run-of-the-river installation, the speed of the water flow should be $\sqrt{20}$ meters per second (4.5 m/s). Note that this relationship also applies to shrouded turbines used to capture the energy of tidal flows and is directly analogous to the equation for the theoretical power generated by wind turbines.

Micro and Mini Hydel Plants

Hydropower has largely been developed as a major large-scale energy production program. Therefore, designations were relative to very large 'Gigawatt scale' hydropower projects. It is why; there isn't a worldwide agreed definition of small hydro projects. The term mini-hydropower is used in connection with plants having a capacity of over 100 kW, while micro-hydropower units by definition have a capacity of less than 100 kW. Micro and mini hydropower plants are together abbreviated as MHP. The following Table gives generally accepted norms for categorization of hydel projects. Note that numbers of home powered are based on energy use in UK and may vary from one country to another depending upon the energy use pattern.

Hydro Category	*Power Range*	*No. of Homes Powered*
Pico	< 5 kW	< 5
Micro	5 kW – 100 kW	5 – 100
Mini	100 kW – 1 MW	100 – 1,000
Small	1 MW – 10 MW	1,000 – 10,000
Medium	10 MW – 100 MW	10,000 – 100,000
Large	> 100	>100,000

The micro or mini-hydropower plants are mainly run-of- the- rivers schemes located at river streams where water head is insufficient to produce large power. But they can also be installed on piped water supply where pipes provide sufficient head to generate electricity. These hydro power plants operate in a similar way as their larger counterparts. The kinetic energy of water falling from a height (head) is converted into mechanical energy by a turbine which then drives a generator to produce electricity.

Major advantages of micro and mini hydel plants are:

- MHPs use simple technology, are decentralized, renewable, and robust, and typically last for 50 years or more

- ✰ MHPs especially the micro hydropower plants require only a small amount of flow and drop as low as 1 m to generate electricity
- ✰ Electricity can be delivered to the location as far as 1 km away for its use
- ✰ If carefully planned, these hydropower schemes produce a continuous and predictable supply of electrical energy in comparison to other small-scale renewable technologies
- ✰ MHPs have little negative environmental impacts compared to their large counterparts
- ✰ Possibility of local implementation and management minimizes the negative socio-economic impacts to insignificant levels in comparison to large projects
- ✰ Low distribution and running costs
- ✰ It is a cost effective energy solution with minimal initial investments and low maintenance costs

Certain disadvantages of these projects are:

- ✰ All sites are not suitable for MHPs. A suitable site needs to be selected to take full advantage of the electrical potential of small streams
- ✰ The size and flow of small streams may restrict future expansion plans as the power demand in the area increases
- ✰ Electricity generation is highly dependent on stream discharge since there are no reservoirs. In most locations stream discharge fluctuates seasonally. Advanced planning and back-up system may be needed to ensure adequate supply during lean discharge periods

PUMPED HYDROELECTRIC ENERGY STORAGE (PHES)

Pumped hydro-electric energy storage (PHES) has been in use for more than a century. Examples of installed PHES systems as early as 1890 can be found in the Italy and the Switzerland. A pumped-storage plant works much like a conventional hydroelectric station, except that the same water is used over and over again (Figure 14.3). Practically speaking, PHES is a storage mechanism to assist with load balancing in the electricity industry. Both on and off-river systems have been in use. Although basic principle of both is the same, the latter has several advantages over the former as discussed in the next paragraph. In PHES, water stored in an upper reservoir is released into an underground tunnel. The force of the water drives the turbines, which are underground at the base of a reservoir/dam. The spinning turbines are connected to large generators to generate electricity. Most river water passes through the system generating electricity flows down the river except a part of it that is retained in a reservoir. When power demand is low, the turbines spin backward and pump the water from the lower reservoir into the upper reservoir. The head so created is now available to generate electricity whenever needed. Sometimes the pump and the turbine are separate items, but more commonly they are combined. The Francis turbine is capable of reversible operation, utilizing a single unit that acts as a motor-pump or a turbine-generator. PHES systems are highly efficient, capable of reaching and surpassing 80-85 per cent round-trip efficiencies. In other words, about 20 per cent of the energy is lost in a complete pumping/generation cycle.

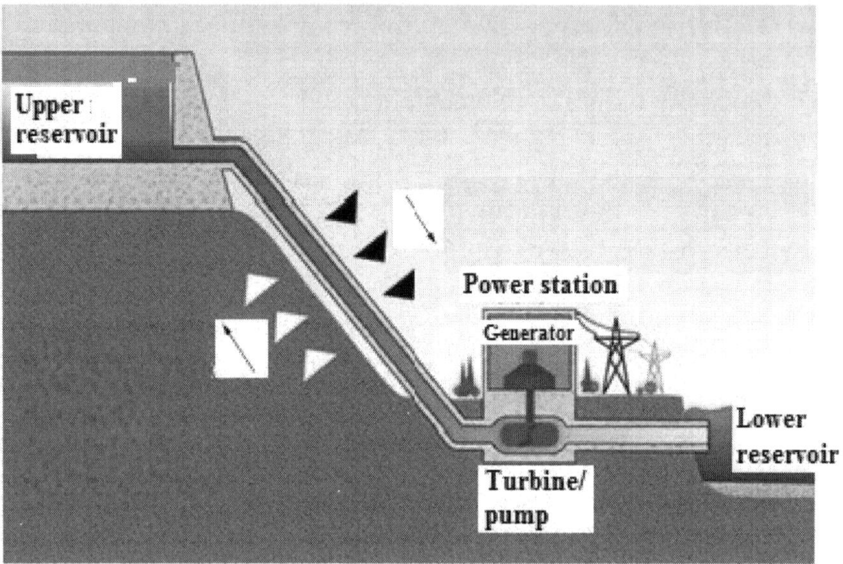

Figure 14.3. A basic schematic diagram of the pumped hydro installation.

Advantages of Hydropower

- Hydropower is a renewable clean energy source, being fueled only by moving water. It does not produce emissions and therefore, does not increase the level of greenhouse gases in the atmosphere
- Hydropower generation is a non-consumptive use of water. The total amount of water in a hydropower system does not change; the moving water is used to generate electricity and is returned to the source from which it is diverted
- Hydropower is usually available when it is needed. It is easy to manage the flow of water through the turbines to produce electricity on demand, and control the amount of electricity generated
- The electricity generated by hydropower facilities is the cheapest source of electricity compared to other sources
- Maintenance costs of hydropower plants are low. Once a plant is established, the water flow that powers the plant is free

- ✰ Hydropower is an efficient way to produce electricity, the average efficiencies being in the range of about 90 per cent
- ✰ Dams based systems create reservoirs that offer a wide variety of non-energy benefits to communities, such as recreational fishing, swimming, and boating
- ✰ The reservoirs can also increase the property value of the adjacent lands
- ✰ Hydropower facilities can help manage the water supply for irrigation, provide flood control and proven to be a reliable source of drinking water during droughts
- ✰ Hydropower dams are very safe and durable-built to last for hundreds of years
- ✰ Hydropower is a flexible energy source to meet varying demand. Most hydropower plants can provide reliable and dependable base load power

Disadvantages of Hydropower

- ✰ Hydropower plants are dependent on water supply. In years of drought, hydropower plants may fail to meet the demand
- ✰ Hydropower dams on rivers permanently change the natural ecology of large areas on the upstream and the downstream
- ✰ Construction of reservoir submerges large areas of land upstream of the dam
- ✰ The natural ecology of the river and adjacent lands downstream is completely changed
- ✰ Hydropower dams can impact water quality and flow. Reservoirs can have low dissolved oxygen levels in the water, a problem that can be harmful to fish and downstream riverbank habitats. Maintaining water flow downstream of a dam is also critical for the survival of habitats
- ✰ Excessive sediments from the catchment can reduce the working life of the dams
- ✰ Some fish populations, such as salmon, migrate upstream to reach spawning grounds before returning downstream. Dams can block fish from completing this natural migration process
- ✰ Development of new hydropower resources can be very expensive because dams have already been built at most economical locations. New sites must compete with other potential uses of the land

OCEANIC HYDROPOWER

Oceans have the potential of a new and significant renewable and environmentally clean energy source. The energy contained in the oceans can be harnessed mainly in the following four forms:

- ☆ Tidal energy
- ☆ Wave energy
- ☆ Ocean's thermal energy
- ☆ Osmotic energy

Tidal Energy

Tides, the daily rise and fall of the Sea's edge, are caused by the gravitational forces between the earth, the moon and to some extent the Sun as well as the spin of the Earth. The moving water has enormous energy as the ocean water rises and falls near its shores as a result of tides. *Thus, tidal energy is a form of renewable energy generated by the tides.* Unlike solar or wind, tidal energy is one of the most reliable sources of renewable energy because two high tides and two low tides are guaranteed every day. Tidal power has been in use as far back as Roman times and the middle Ages to turn water wheels and grind grains. Although the idea of using tidal power for generating electricity is fairly recent, technological advances have shown that it could become a competitive and viable source of renewable energy in near future. Initially, there were two large tidal barrages in the world one being in the France and the other in the Canada. Currently, the Sihwa Lake project located on the west coast of South Korea is the world's largest tidal power station in operation. The 254 MW project consists of 10 water turbine generators having enough power to support the domestic needs of a city with a population of 500,000 people. Tidal energy is harnessed either by harnessing the kinetic or the potential or both the energies of the tides.

Kinetic Energy of the Tidal Currents

The kinetic energy P, W/m^2 of the tidal currents can be assessed by the following relation.

$$P = (1/2)\,\rho V^3 A \tag{14.7}$$

Such that ρ is the density of water usually taken as 1025 kg/m^3, A is the area swept by turbine, m^2, and V is the velocity of water current. Kinetic energy is harnessed by employing a set-up quite similar to a wind turbine. But, the blades of a tidal turbine are smaller than that of wind turbine. It is becuase they can operate at lower velocities as the density of water is much higher than air. Therefore, kinetic energy imparted by water is much higher than air.

Potential Energy

Potential energy during a tide is generated as and when the tide level changes from high to low and vice-versa. Dams or barrages are constructed to tap this potential energy resulting from the differences in water level between a high and a low tide. Consider a rectangular basin behind a barrage having the constant surface area A. If the tidal range is H, then

$$\text{Total volume of basin water} = AH \tag{14.8}$$

$$\text{Mass} = \rho AH$$

$$\text{Potential energy available} = \rho AHg\,(H/2) \tag{14.9}$$

Note that the center of gravity for the mass of water will be at H/2 above the low tide level. If tidal period is given by T, then

$$\text{Average potential energy extracted} = \rho AH^2g/2T \tag{14.10}$$

Apparently, energy available is approximately proportional to the square of the tidal range. Extraction of energy from the tides is practical only at sites where large tides occur and the geography provides suitable sites for the construction of tidal plant. The greatest tidal ranges in India are found in the Arabian Sea, notably at Bhavanagar, in the Gulf of Khambhat (11.6 m), and in the Gulf of Kachchh (Kutch) at Navlakhi (7.8 m). Tidal energy can be exploited by using any one of the following 3 technologies.

Tidal Barrages

A tidal power plant that includes a barrage and other infrastructure is built across an estuary, the area where a river runs into the ocean and the water rises and falls with the tides. A tidal barrage works like a dam with gates and turbines held in a supporting structure, a caisson. In the ebb generation option, gates allow the water to fill the basin to its normal level. The gates are closed at full tide. The water is held back for few hours while the tide begins to recede. At this time sluice gates are opened so that water flows through the sluice gates to run the turbines to generate electricity (Figure 14.4 top). Tidal barrages during ebb generation exploit the potential energy due to the difference in hydraulic head between high and low tides. In the second case called two-way generation, often referred as flood generation, turbines generate power during both flood and ebb tides. In this case, basin is filled through the turbine gates. Sluice gates and turbines are closed until near the end of the high tide when sluice gates are opened and tide water is allowed to flow through the turbines into the basin generating electricity (Figure 14.4 middle). At a point when the hydrostatic head is insufficient for power generation, the sluice gates are closed and kept closed until high tide begins to ebb. When the tide outside the barrage has dropped sufficiently, water is allowed to flow out of the basin through the turbines again generating electricity (Figure 14.4 bottom). Although the energy produced in the latter case is usually less than ebb generation but it is produced over a longer period of time, which can be beneficial depending on the energy requirements of the area. These types of dams not only produce renewable energy, but are also able to reduce flooding especially under global climate change scenario. Tidal power plant of 3.75 MW at Durgaduani in Sunderbans was proposed by the

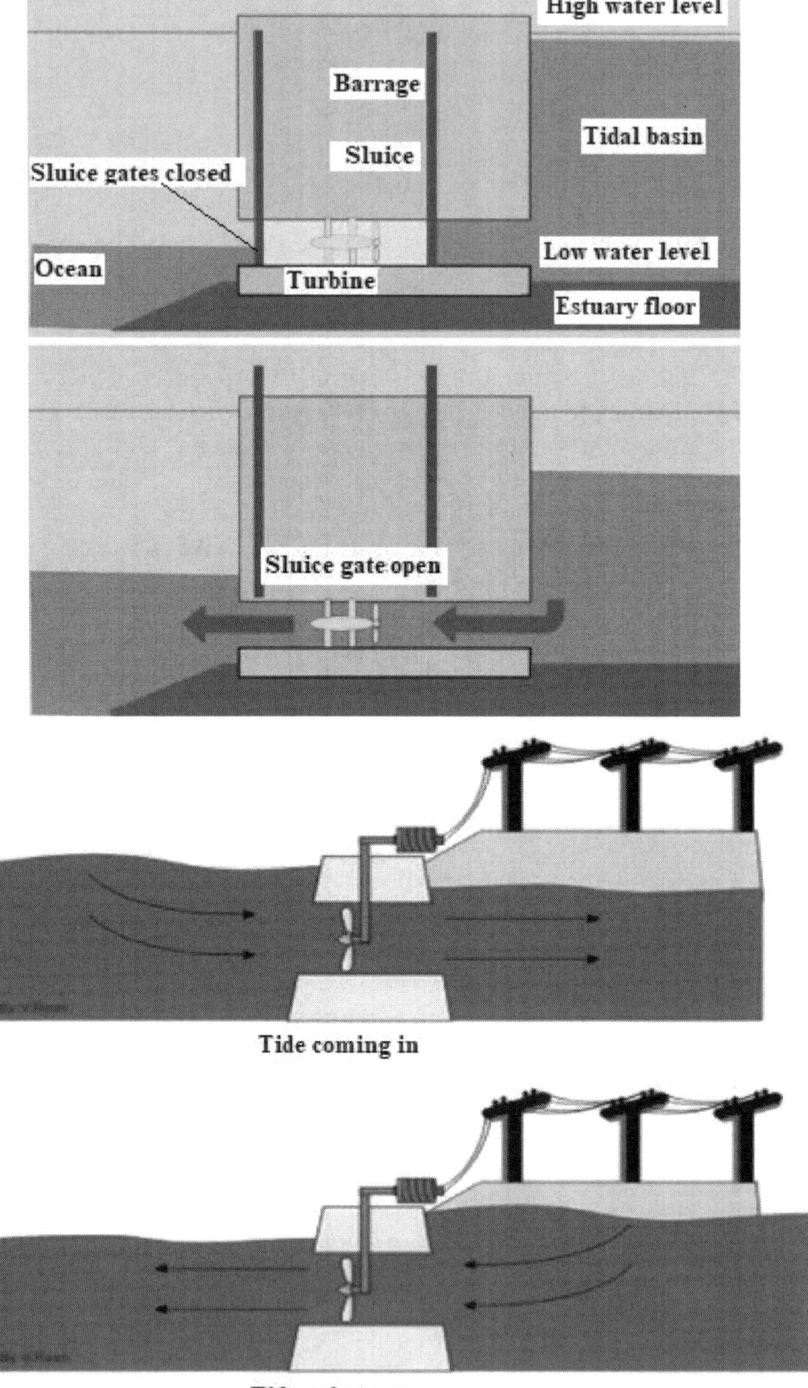

Figure 14.4. Tidal barrage: Ebb generation (top) and flood generation (bottom).

West Bengal Renewable Energy Development Agency. The Gujarat government also proposed to install a 50 MW tidal farm in the Gulf of Kutch. Later on it was decided to increase its capacity to 250 MW. Both these plants were shelved because of their high cost (635 million per MW and Rs 150 million per MW for West Bengal and Gujarat projects respectively) and economic viability.

Tidal Fences/Lagoons

In this case, circular retaining walls embedded with turbines are constructed to capture the potential energy of tides. The size and shape of the lagoons depends on a number of factors such as the tidal range, the water depth, sea conditions, environmental impacts and navigation *etc.* The reservoirs so created are similar to those of tidal barrages, except that they do not obstruct the entire estuary and are smaller structures than a barrage. Numbers of vertical turbines form a fence between two land masses. The system can be in double (or triple) format. When the tide moves in or out, the turbines spin and generate electricity. The turbines in this case are 'bi-directional' and can be flipped during incoming and outgoing tides. These can be combined with pumping to raise the water level in the lagoons. This helps not only in generating more energy but also to flatten out the power output as in the pumped hydro projects.

Tidal Stream Generators/Turbines

Tidal stream generators make use of the kinetic energy of moving water to power turbines, in a similar way as the wind turbines except that these are positioned underwater. These are individual turbines placed wherever there is a strong tidal flow (Figure 14.5 left). These turbines consist of a rotor, gearbox, and a generator and operate both during flood and ebb tides. The turbines are mounted onto some support structure. Different kinds of support structures such as gravity structure embedded in the bed of the Sea, piled structure

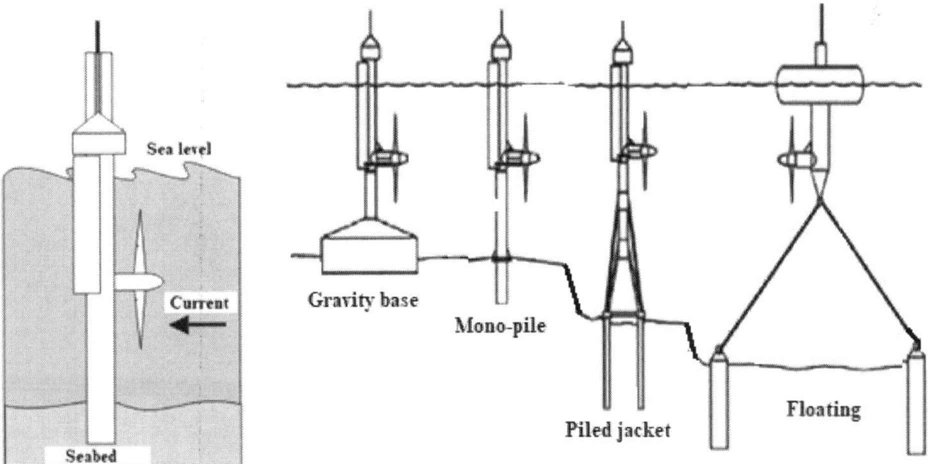

Figure 14.5. A tidal stream turbine (left) and various kinds of support structures (right).

such as mono-pile and piled jacket and floating structure can be used (Figure 14.5 right). Some tidal generators can be built into the structures of existing bridges or can be entirely submerged. In the later case, concerns over impact on the natural landscape are avoided. This method is becoming quite popular than tidal barrages because it is modular in nature and its cost is quite low. Modularity of the system allows for small applications to be expanded once the economic feasibility is settled in a small test project and once environmental impacts are better understood. Moreover, the acoustic properties of the blades in water prevent wildlife from straying into their path helping to minimize the underwater environmental damage.

Dynamic Tidal Power

Basically, dynamic tidal power makes use of the fact that ocean tides don't operate strictly perpendicular to the shore, but also flow in parallel to the shore as well. This feature of tides allows construction of a dam that runs perpendicular to the shore and harvests energy from the tides flowing parallel to the shore (Figure 14.6). The tides flowing in one direction will tend to build-up behind the barrage, and create a level difference that can be used to generate power. Long dams 30 to 60 km long are built perpendicular to the coasts straight out into the Sea or Ocean. The dam looks like a large T. During the movement of tides, water on one side of the dam is at a higher level than the other side. As this water flows through the dam it drives a series of turbines installed within

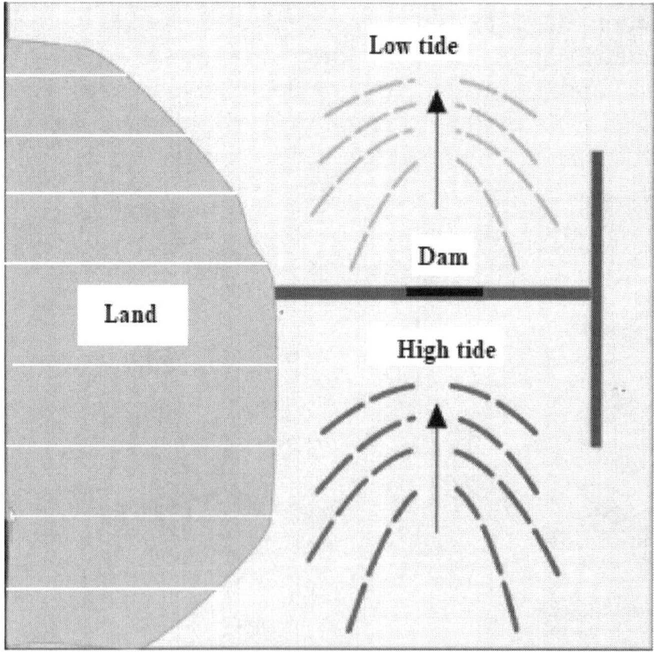

Figure 14.6. A schematic view of dynamic power dam for electric power generation.

the dam and generates electricity. These turbines are also bi-directional, which can be flipped 180° after each tide in order to generate power when the tide comes in and goes out almost doubling the power output. This system does not enclose any area, thus allows more freedom for aquatic life without risk of being trapped. The initial investments are high but the dam can be a multi purpose dam. In that case investments can be easily shared for other benefits such as land reclamation, connections between islands, deep sea and liquefied natural gas ports, mainland coastal protection *etc.*

Wave Power

Ocean waves are the up and down motion of surface water. The waves are caused mainly by the wind. *The highest point of a wave is the crest, while the lowest is the trough.* The size of waves depends on the speed of the wind, wind duration, and the distance of water over which the wind blows. The tremendous energy of Ocean waves can be guessed from the fact that the theoretical annual energy potential of waves off the coasts of the United States is estimated to be as much as 2.64 trillion kWh. Concepts of harnessing wave energy have existed since the 1800s; however modern wave technology began in the 1940s with the experiments of scientist Yoshio Masuda. The first wave power plant in the world opened in 2008 at the Agucadoura in Portugal. Wave power is distinct from tidal power. While in tidal power we capture the energy of the current caused by the gravitational pull of the Sun and Moon, in wave power we capture the energy of wind waves to do useful works such as electricity generation, water desalination, or pumping water. Waves also have both kinetic and potential energies. Kinetic energy of the individual water molecules moving steadily and rather slowly in a circular way can be utilized in different kinds of wave energy converters, either directly via some kind of propeller or indirectly by oscillating water columns wave energy converters. Besides, the individual water molecules during their circular movement are elevated above the still-water line. It represents the potential energy. There are different ways in which wave energy technologies are categorized. Very broad categorization includes oscillating water columns (OWCs), oscillating body converters and overtopping converters. They may be fixed, submerged and floating types. One of the fixed devices is the oscillating water column partially submerged in the water. This device is based on the pressure of the enclosed air in a cavity (column of air), which varies as a function of the level of water in the cavity. The higher the water level greater is the air pressure. The air located in the chamber is drained off through a generator (Wells turbine). As the waves flow in and out of the chamber (ascent and descent of waves), the air inside the chamber is compressed and decompressed. The forced air spins a turbine, which is further attached to a generator to produce electricity.

Another fixed device is a tapered channel (TAPCHAN) system (Figure 14.7). It consists of a channel connected to a reservoir in a cliff. The channel gets narrower as it nears the reservoir, causing the waves to increase in height. When the waves are high enough, they spill over the top of the channel into the reservoir. The stored water flows out of the reservoir through a turbine, generating electricity (Figure 14.7).

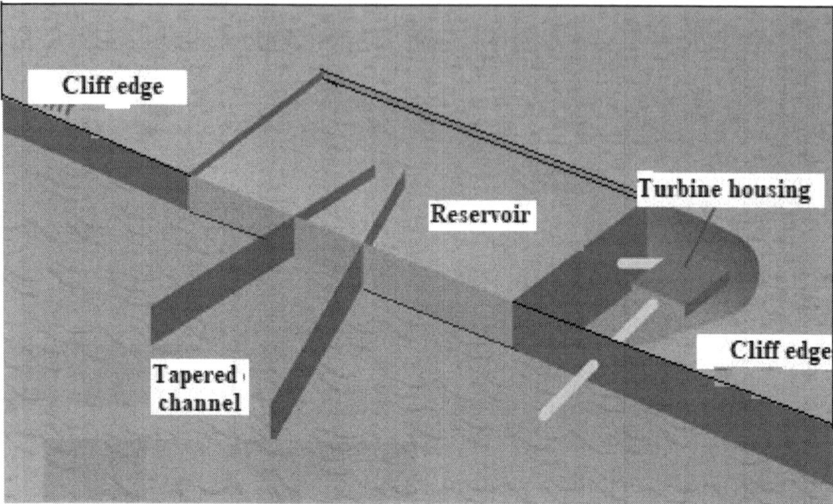

Figure 14.7. View of the tapered channel and turbine housing in the tapered channel system.

Submerged devices include the Archimedes Wave Swing (AWS), a cylindrical buoy moored on the Sea bed. The McCabe Wave Pump having three pontoons linearly hinged together and pointing parallel to the wave direction. The Power Buoy, a device that includes a submerged buoy 1 m below the level of water.

Floating wave devices generate electricity as they are moved by the motion of waves. One of the floating devices is the Pelamis system or an attenuator that looks like a 'Sea snake.' It is a succession of floated empty segments at 50 m or more water depth. The segments are linked to each other by hinged joints. Passing waves cause each tube to rise and fall similar to the giant Sea snake. The motion tugs at the joints linking the tubes. The joints act as a pumping system pushing high pressure oil through a series of motors that drive the generators to produce electricity.

Several other floating wave devices are under development. One such device is the AquaBuOY, a point absorber that is the third generation of two Swedish designs that utilize wave energy to pressurize a fluid which is used to drive a turbine generator. The average cost for wave energy farms (10 MW) is estimated between EUR 330-630/MWh. It is considerably higher than other forms of renewable energies, including offshore wind and tidal current technologies. Funding into wave energy technology has recently increased due to the need for renewable energy sources. In the Indian context, a wave power plant is situated near the breakwaters of vizhinjam port which is about 20 km from Thiruvananthapuram city. The station started its commercial operation in 1991. This oscillating water column (OWC) produces about 150 kW of power.

Ocean Thermal Energy Conversion (OTEC)

Energy from the sun heats the surface water of the ocean but water in lower layers is not heated to a great extent. Surface water in tropical regions can be much warmer than deep water. *The concept of ocean thermal energy conversion makes use of the temperature difference between cooler deep and warmer shallow or surface seawater to run a heat engine to produce useful work, usually in the form of electricity or to desalinate ocean water.* OTEC is firm power (24/7), a clean energy source, environmentally sustainable and capable of providing massive levels of energy. OTEC research began in the United States in 1974 with the establishment of the Natural Energy Laboratory of Hawaii Authority. The laboratory is one of the world's leading test facilities for OTEC technology. The laboratory operated a 250 kilowatt (kW) demonstration OTEC plant for six years in the 1990s. Since, OTEC embraces very large ocean areas; it is quite favourable for islands and many developing countries. It can operate with a very high capacity factor. The main technical challenge in using this technology remains to generate power even with small temperature differences. Current OTEC systems use a temperature difference of at least 20°C to power a turbine to produce electricity. Three kinds of systems are in use to generate electricity using OTEC.

Closed Cycle

This cycle uses a working fluid like ammonia having a low boiling point. The warm water of the surface is injected into a heat exchanger so that the working fluid exchanges the heat to change its state, and is vaporized (Figure 14.8 left). The vapours are directed through the turbine connected to a generator where the electricity is produced. The vapours are condensed back by the deep cold water. The liquid goes back to the heat exchanger and this cycle keeps on repeating. India has built a closed-cycle floating plant of 1 MW rated power. The system's energy conversion efficiency is quite low but it can take advantage of a vast, consistently available resource.

Open Cycle

The open-cycle process does not use any intermediate fluid like the closed-cycle but directly relies upon the sea water. The warm water is placed in a low-pressure container, which carries out the evaporation. Like in the previous cycle, the steam drives the turbine attached to the generator. The offshoot of this process is the production of desalinated water (fresh water) through evaporation (Figure 14.8 right). The steam is condensed into liquid by the deep cold water. The fresh water is captured to supply clean drinking water to the people in nearby locations.

Hybrid Cycle

This cycle combines both open and closed cycles. This system uses the low pressures of the open system and the low boiling point liquids of the closed system to produce energy. No major projects have so far been implemented using the hybrid cycle.

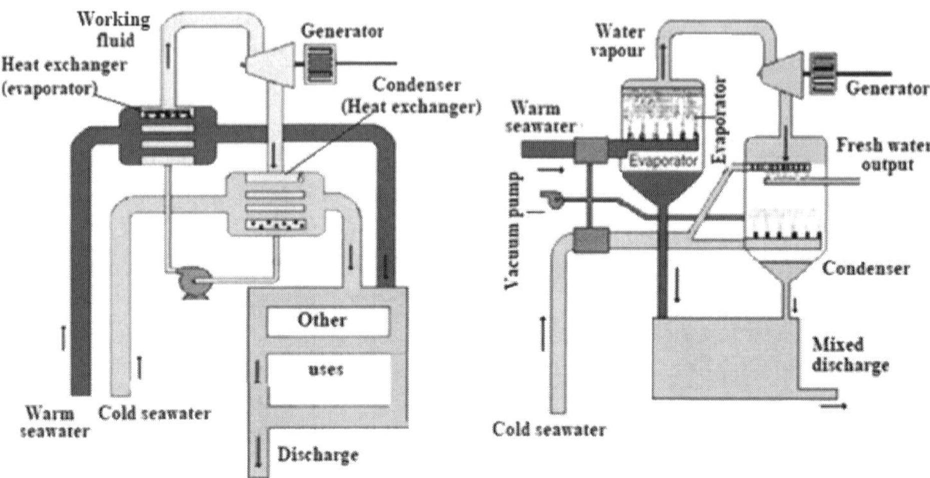

Figure 14.8. OTEC closed cycle (left) and open cycle (right) system.

Osmotic Power

Osmotic power is derived from water flowing between areas of low solute concentration and high solute concentration. Osmotic power is a unique energy source available as a result of difference in salt concentrations of Sea and river waters. When the rivers joins the Sea, two water bodies with different salinity levels merge and release a large amount of energy as these solutions equilibrate their concentrations. A device with a semi-permeable membrane has been developed that transforms this energy into electricity. When two solutions with different salt concentration are placed on two sides of a semi-permeable membrane, the concentrations try to equalize through the process of osmosis at a constant pressure or volume. Thus, the basic principle is to convert pressure differential corresponding to the osmotic pressure that exists between water of high salinity and water of low or no salinity (Andrade, *et al.*, 2016). This represents hydraulic energy that can be transformed into electricity. The energy from osmosis is popularly known as blue energy or pressure retarded osmosis (PRO). Height h shown in Figure 14.9 (left) corresponds to the height reached by the salt water due to osmotic pressure. The height h can be assessed by equalizing the hydrostatic pressure as described below.

$$H_p = A_p + \rho gh \tag{14.11}$$

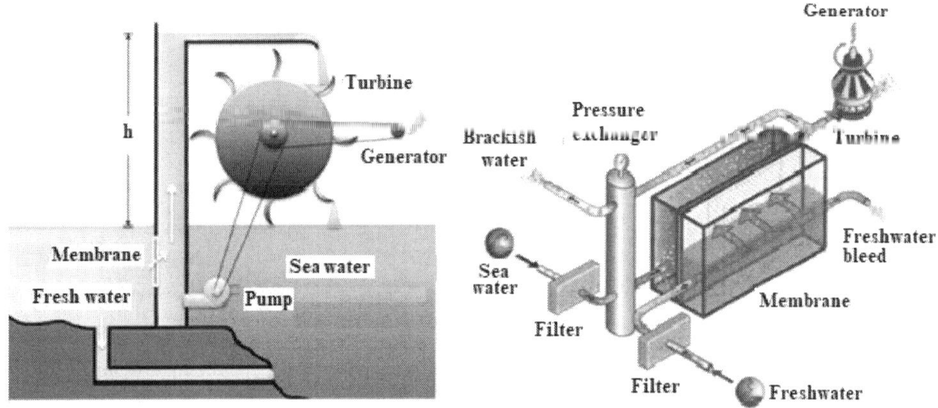

Figure 14.9. Basic diagram showing the principle of pressure retarded osmosis (left) and a modular diagram of the same process (right).

Here H_p is the hydrostatic pressure, Pa, A_p is the atmospheric pressure, Pa, ρ is the water density, kg/m^3, h is the height, m and g is acceleration due to gravity, m/s^2. The equation neglects the salinity of the fresh water assuming it to be negligible. The osmotic pressure, according to Loeb and Norman (1975), varies from 0.6 to 1.1 psi for each 100 ppm of the total dissolved solids. Thus, the osmotic pressure of the Sea water having 33,600 ppm of dissolved solids will be around 336 psi assuming an average value of 1 psi for 100 ppm salt solution. It is nearly equal to 2316639 Pa. Taking atmospheric pressure as 100000 Pa, ρ as 1000 kg/m^3 and g as 9.81 m/s^2, we have

$$h = 225.96 \text{ m}$$

A company named Statkraft has implemented a test plant of 4 kW capacity in Tofte, Hurum in Norway in 2009. A modular diagram of the plant is displayed in Figure 14.9 (right). In the PRO system, seawater is pumped into a pressure chamber where the pressure is lower than the difference between fresh and salt water pressure. Fresh water moves in a semi-permeable membrane chamber and increases its volume in the chamber. As the pressure in the chamber is compensated, a turbine begins to spin which in turn operates a generator to generate electricity. Most appropriate locations of such plants are where freshwater flows into the sea. Besides, the concentration of salt in the seawater at these sites should be sufficiently high to generate the required pressure.

Some characteristics of the osmotic energy are:

- It is a renewable clean energy source depending on an almost inexhaustible resource. It does not produce CO_2 emissions
- It is hydro-generated, due to turbine-generator system used and a marine energy generator, due to seawater use
- It is highly reliable and can be generated continuously

- ☆ Its modular design allows an easy (preventive and corrective) maintenance
- ☆ The system is upgradable

Some limitations of the system are:

- ☆ High risk of clogging and gradual degradation of semi-permeable membranes necessitates pressure-filtering pre-treatment of fresh water and periodic membrane re-placement every 5 to 7 years
- ☆ Freshwater utilization efficiency is only about 40 per cent

Other methods being developed are reverse electro-dialysis and capacitive method. In this method energy can be extracted by mixing saline water and freshwater and cyclically charging up electrodes in contact with saline water, followed by a discharge in freshwater. To avoid the need of membranes, vapour pressure difference methods such as open cycle and absorption refrigeration cycle are also being studied. Since these methods are still in laboratory, these are left out of discussion in this chapter.

Advantages of Ocean Energy

- ☆ It is an inexhaustible source of energy as 71 per cent of Earth's surface is covered by oceans. There is scope to generate renewable energy on large scale
- ☆ Predictable energy output because tides are predictable and constant. Only one need to assess the low or high tides that makes it easier for engineers to design efficient systems
- ☆ Tidal energy is environment friendly as it doesn't produce greenhouse gases
- ☆ Efficiency of tidal power is far greater being around 80 per cent as compared to coal, solar or wind energy
- ☆ Although cost of construction of tidal power is high but maintenance costs are relatively low. Moreover, with technological advancements it will become increasingly cheaper and efficient
- ☆ It can protect coastal flooding due to the stability of the rock armour under different design conditions
- ☆ Equipment and facilities of tidal power can last longer than other renewable technologies, the asset life is assessed at 120 years
- ☆ Wave and tidal energy converters can be located near the coastline. It is easier to install, maintain, capture, and retrieve the energy when they are located close to the coast

Disadvantages of Ocean Energy

- ☆ The main limitation of ocean energy is its cost, construction cost being quite high. For example, a tidal facility with an average annual output of 1050 MW would cost about 1.2 billion dollars excluding maintenance and running costs. This is far more expensive than coal and oil

- There are very few ideal locations for construction of plants and they too are localized to coastal regions only
- Intensity of sea waves is unpredictable and sometimes can damage the power generation units
- The actual generation is for a short period of time. The tides only happen twice a day so electricity can be produced only for that time
- Frozen sea, low or weak tides, straight shorelines, low tidal rise or fall may prove to be some of the hurdles in exploiting ocean power
- Both wave and tidal energy can also have some effect on the environment. Large tidal barrages can affect migration of fishes. Also, spinning turbines can strike or entangle marine organisms. The high water velocities near the units increases the risk of organisms being pulled by these devices
- As with all offshore renewable energies, little is known about how the creation of electromagnetic fields and acoustic outputs will affect the marine organisms

Impediments in Harnessing Ocean Power

- They require high capital investments
- There are concerns on several environmental issues such as habitat change, particularly with tidal barrages
- Maintenance and repair of equipment
- Powerful tides only happen normally 10 hours out of each day. It means some storage capacity needs to be build-up for tidal energy
- The energy produced by the tides is often a long distance from inland where electricity will be used. Therefore, long distance transmission may pose problems

Ocean Energy: Global and Indian Potential

Global potential of energy production with ocean technologies is assessed at 800 TW h for tidal energy, 2000 TW h for osmotic energy, somewhere between 8000 and 80,000 TW h for wave energy and up to 88,000 TW h of OTEC without affecting the ocean's thermal structure. Thus, ocean-based renewable energy can exceed current and likely future global demand for electricity. According to the estimates of the Indian government, India has the potential of 8,000 MW of tidal energy. This includes about 7,000 MW in the Gulf of Cambay in Gujarat, 1,200 MW in the Gulf of Kutch and 100 MW in the Gangetic delta in the Sunderbans region of West Bengal. The Indian Institute of Technology,

Chennai in association with Credit Rating Information Services of India Limited (CRISIL), has conducted a study on "Tidal and Waves Energy in India: Survey on the Potential and Proposition of a Roadmap" in December 2014. As per this report, the estimated theoretical power potentials for tidal and wave energy are 12,455 MW and 41,300 MW respectively. Unfortunately, tidal energy is not included in the 2022 renewable energy target (175 gigawatts) of India. Despite their immense potential to meet global energy demands, ocean technologies still require further research and development for large-scale deployment.

HYDROGEN FUEL CELL

Hydrogen is the simplest and most abundant element in the universe. Despite its abundance, it rarely occurs naturally as a gas on the Earth. It is almost always combined with other elements. It can be generated from oil, natural gas, and biomass or by splitting water using renewable solar or electrical energy. Once hydrogen is produced as molecular hydrogen, the energy present within the molecule can be released by reacting with oxygen to produce water. This can be achieved by either traditional internal combustion engines, or by devices called fuel cells. Hydrogen fuel cell, as the name suggests, uses hydrogen as an energy carrier. In a fuel cell, hydrogen energy is converted directly into electricity with high efficiency and low power losses. Sir William Grove demonstrated the concept of the fuel cell in 1839. However, it was not until 1932 that Francis Bacon developed the first successful fuel cell. Limited progress was made over the next quarter century. The pace of development increased around the time the NASA began to research the technology and used the same in the Gemini program in the 1960's. Today, they use the technology to provide electricity and drinking water for astronauts on Space Shuttle flights.

Construction

A hydrogen fuel cell is a kind of electrochemical cell that converts chemical energy stored in the hydrogen fuel into electricity consuming hydrogen and oxygen. A fuel cell can generate electricity as long as it is supplied with hydrogen and oxygen, and the product water is removed. A fuel cell is a lot like a battery although it differs in many ways. It has two electrodes, one anode and one cathode, where the reactions take place. These are coated with platinum, which acts as a catalyst to break down the hydrogen into protons and electrons and to produce water at the cathode. It has electrolyte that carries the charged particles from one electrode to the other. The electrolyte is contained in a membrane placed between the anode and the cathode. The membrane allows only protons to pass through. It is why; it is called a proton exchange membrane, or PEM. The membrane must be kept moist for its proper functioning. The membrane, catalyst layers (anode and cathode), and diffusion media together form the membrane electrode assembly (MEA). Each MEA in a fuel cell stack is sandwiched between two bipolar plates that help to distribute gases and also serve as current collectors. Therefore, gaskets made of a rubber polymer are added around the edges of the MEA to make a gas-tight seal.

Electrons are released in a hydrogen fuel cell from the hydrogen supplied to the anode. Therefore, the electrodes in hydrogen fuel cells are relatively stable since they act as catalysts in the release or acceptance of electrons and are not chemically changed during this process as in the case of batteries. A typical hydrogen fuel cell produces 0.5 V to 0.8 V per cell. Individual cells are connected in series to increase the voltage. This arrangement is called a fuel cell stack. The cross sectional area of a fuel cell affects its ability to produce current, greater the area-greater the current generated. Therefore, it is possible to generate very large amounts of electrical power by stacking cells in series to build voltage and increasing cell area to boost current (Power = Voltage × Current).

Working

For a hydrogen fuel cell to work, it needs hydrogen (H_2) and oxygen (O_2). Because hydrogen does not occur naturally in the environment, hydrogen fuel is derived from other substances such as methanol, gasoline, natural gas, and water. It is stored in a container for supplying to the cell. The oxygen is typically obtained from the air, though it can be provided directly as pure oxygen. The hydrogen enters the fuel cell at the anode (Figure 14.10). A chemical reaction strips the hydrogen molecules of their electrons and the atoms become ionized to form H^+.

Anode side (an oxidation reaction): $H_2 \Rightarrow 2H^+ + 2e^-$

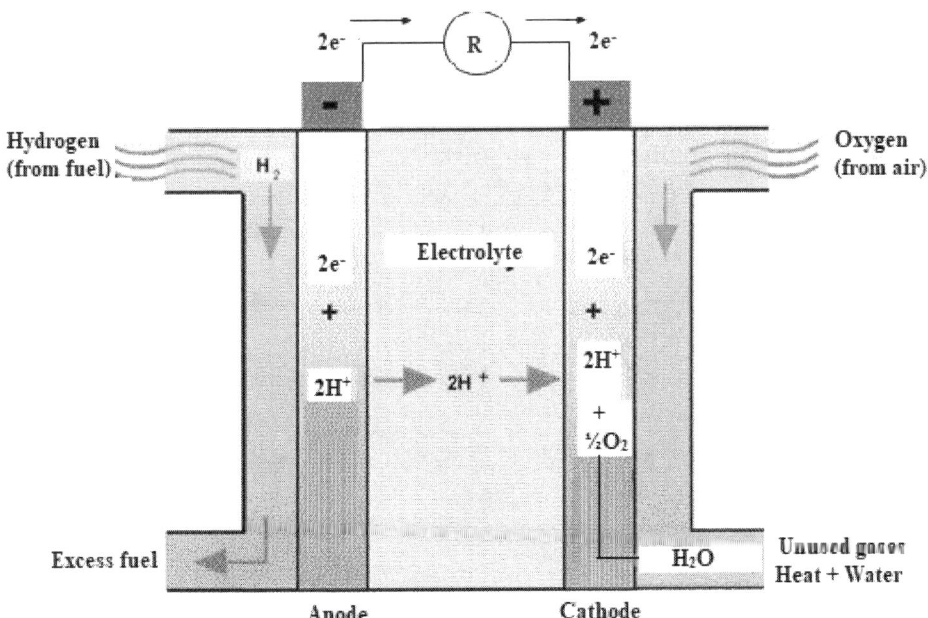

Figure 14.10. Line diagram showing the construction of a hydrogen cell.

The positively charged protons pass through the membrane to the cathode. The negatively charged electrons take a different route. If a circuit is connected between the anode and cathode then the electrons travel through the circuit and provide power to the load connected as part of the circuit. After passing through the circuit and the membrane accordingly, the electrons and protons meet at the cathode. The oxygen enters at the cathode and picks up the electrons and the hydrogen ions (protons) that reach the cathode. Oxygen reacts with these hydrogen ions and electrons to produce water and heat, both of which are removed from the fuel cell.

Cathode side (a reduction reaction): $O_2 + 4H^+ + 4e^- => 2H_2O$

There are several types of fuel cells currently under development, each with its own advantages, limitations, and potential applications. For the purpose of this section, it may suffice to name them as the discussion on each of them is beyond the scope of this book.

- Polymer electrolyte membrane fuel cells
- Direct methanol fuel cells
- Alkaline fuel cells
- Phosphoric acid fuel cells
- Molten carbonate fuel cells
- Solid oxide fuel cells
- Reversible fuel cells

Advantages

- Fuel cells are significantly lighter and more compact
- The fuel cells are quite durable and have long life
- Fuel cells have high efficiency operating at an efficiency of about 40 to 50 per cent compared to fossil fuel fired power plant, which is typically 35 per cent efficient. The internal combustion engine in most vehicles operates at about 15 to 20 per cent efficiency
- The power supply is quite reliable as its quality does not degrade over time
- The working of the fuel cells is much more silent and smooth as there are no moving parts
- It is environmental friendly as it can greatly reduce emissions of CO_2 and other harmful pollutants
- They are modular, meaning they can be stacked to generate required amounts of power
- Fuel cells offer many potential benefits over batteries

Disadvantages

- Expensive to manufacture due to the high cost of catalysts (platinum)
- Lack of infrastructure to support the distribution of hydrogen
- Several currently available fuel cell technologies are in the prototype stage and have not yet been validated
- Hydrogen is not widely available and is expensive to produce
- The flammable nature of hydrogen poses safety concerns for its widespread use
- The extraction of hydrogen for use in fuel cells can consume a lot of energy that may undermine the green benefits of using the fuel cell

FUTURE OF HYDROPOWER

India is at number 5 globally in terms of installed hydroelectric power capacity. As of 31 March 2020, India's installed utility-scale hydroelectric capacity was 46,000 MW. Besides, 4,683 MW is being produced from smaller hydroelectric power units. India's hydroelectric power potential is estimated at 148,700 MW at 60 per cent load factor. An additional 6,780 MW is exploitable from smaller hydro schemes (with capacities of less than 25 MW). Fifty six sites for pumped storage schemes with an aggregate installed capacity of 94,000 MW have been identified. Hydropower will continue to be a major part of the India's and global energy mix for many years to come. Hydropower will play a significant role in dealing with global climate change, being clean, economical source of electricity with minimal carbon dioxide footprints. The future of hydropower offers both challenges and opportunities. New technologies are likely to make power units more efficient to generate more electricity from similar kinds of investments. There will be ample opportunities to add additional units to existing dams increasing their overall capacity with minimal investments. Advances in turbine design and other strategies may help to reduce the impact of conventional hydropower facilities on fish populations, water quality, and the environment.

PRACTICAL EXERCISE

Students are advised to visit their college renewable energy laboratory and understand the hydroelectric power generation model. The working model comprises a plastic container as reservoir, which can be raised or lowered to change the head. The potential energy of water is transferred to the turbine blades. The turbine transfers this kinetic energy to the generator to produce electrical energy. Students may measure the electrical energy output by varying the head. Plot the curve to understand the relation between potential energy and the power generated. Normally, it should result in a linear relation.

Exercise: Study the effect of height of a reservoir on the electrical output of a turbine.

Materials

- ☆ Turbine with benchmark hub
- ☆ Reservoir unit
- ☆ Bucket
- ☆ Multi-meter
- ☆ Alligator clips
- ☆ Water supply
- ☆ Funnel
- ☆ Meter stick
- ☆ Stopwatch or watch with second hand

Preparation

Attach the benchmark hub to the generator. Attach the multi-meter to the ends of the generator wires with alligator clips. Set meter to 200 mV.

Procedure

1. Place the turbine into the water collection bucket
2. Fill the water reservoir unit and position the bottom of the unit 30 cm above the top of the bucket
3. Position the hose at the top of the jug so that the water will flow onto the blades
4. Allow the water to flow for 10 s and record the most consistent output reading
5. Refill the reservoir unit with water from the bucket. Make sure the reservoir unit is completely filled
6. Repeat steps 1-5 two more times. Calculate the average output and record it in the table
7. Repeat steps 1-6 at different reservoir heights

Observations and Data

Height of Reservoir	*Output 1*	*Output 2*	*Output 3*	*Average Output*
Height 1				
Height 2				
Height 3				
Height 4				

Graph your results with height of the reservoir on the x-axis (horizontal) and average output on the y-axis.

Calculate the efficiency of energy transfer in the turbine generator.

Efficiency = (electrical energy output)/(potential energy) (14.12)

Efficiency = (electrical energy output)/[(9800 V H] (14.13)

NOTE: Make sure that both the electrical energy output and the potential energy have the same units of energy. If you measure V in cubic meters and H in meters, then the units for potential energy should be in Joules, which is equal to watt-second.

Conclusion

Write your conclusions in your practical exercise book.

QUESTIONS (THEORY)

1. Briefly discuss the history of hydropower and its future prospectus.
2. Differentiate between dam based and run-off-the river schemes of harnessing water energy.
3. What are the main components of a dam based hydropower plant? Show these components on a neat line diagram.
4. Derive the power equation of a dam based hydropower plant and explain its significance in practical terms.
5. Show that maximum power output from a run-of-the-river scheme is proportional to the cube of the velocity of water.
6. List the advantages and disadvantages of hydropower.
7. What do you understand by Ocean energy? Briefly discuss the various kinds of Ocean energies.
8. Name and describe the various technologies to harness tidal energy to generate electrical energy.
9. Briefly explain the wave energy and discuss various technologies for harnessing the wave energy.
10. What do you understand by Ocean Thermal Energy Conversion? Differentiate between open and closed cycles of harnessing the thermal energy.
11. Describe the osmotic energy of the Oceans. Draw a neat modular diagram of the pressure retarded osmosis and explain its working.

12. List the various impediments in harnessing Ocean power.
13. Briefly explain the construction and working of a hydrogen fuel cell. List its advantages and disadvantages.
14. Write short notes on pumped storage system, and mini and micro hydro plants for hydropower production.
15. Differentiate between kinetic and potential energies of tidal power by giving equations for each of them.

Chapter 15

Wind Energy

INTRODUCTION

Wind energy is basically harnessing of the wind power *i.e.* conversion of the kinetic energy of the wind to mechanical/electrical energy. Harnessing of wind energy is not a new development as it has been in use for centuries. Traditional windmills for grinding flour, pumping water, and sailing ships are some such examples of the use of wind power. Historically, windmills have been commonly used throughout the high medieval (1000 to 1300 AD) and early modern periods (1400 to 1800 AD). The first windmill is traced to be related to the Persians around 500-600 AD. While the horizontal windmill first appeared in Greater Iran during the 9th century, the vertical windmill, an icon of Dutch culture, appeared in northwestern Europe in the 12th century. Prof James Blyth of Scotland built the first version of the current form of wind turbines by the end of the nineteenth century in 1887. Currently, application of wind energy is growing rapidly around the world as most countries favour low-carbon alternatives to reduce GHG emissions to minimize/halt global warming. The technology is being deployed both onshore and offshore. In the latter, either fixed-bottom turbines anchored to the sea floor or floating structures based in deep waters are being employed.

The wind energy over the earth is estimated to be 1.6×10^7 MW. According to the Global Wind Report 2021 prepared by the Global Wind Energy Council (GWEC), the global wind power fleet at the end of 2020 totaled 743 gigawatts (GW). India, with an estimated wind energy potential of 302 GW at 100 m and 695 GW at 120 m hub heights, has been rated as one of the most promising countries for wind power development. Most of this potential *i.e.* 97 per cent at

100 m hub height and 93.8 per cent at 120 m hub height exists in 7 windy states of Gujarat, Karnataka, Maharashtra, Andhra Pradesh, Tamil Nadu, Rajasthan and Madhya Pradesh. The country currently has the fourth highest wind installed capacity in the world with total installed capacity of 38.63 GW (as on 31st March 2020). The three countries at the top are China with just over 288 GW followed by US around 122 GW and Germany with 62.85 GW.

WIND POWER

As the solar radiation enters the earth's atmosphere, different regions of the atmosphere are heated to different degrees because of earth's curvature. The heating is higher at the equator and lowest at the poles. Since air tends to flow from warmer to cooler regions, it sets in global wind patterns as a result of large scale solar heating of different regions of the earth's surface. Wind pattern also change with seasonal variations in solar incidence. Besides, there are localized wind patterns due to the effects of temperature differences between land and seas, or mountains and valleys. Wind speed, a function of height above the ground, generally increases with increasing height. It is because the roughness of ground features such as vegetation and houses cause the wind to slow down. The windmills and wind turbines are used to harness these airflows to produce wind energy. The power in the wind is extracted by allowing it to blow past moving blades that exert torque on a rotor. The rotor turns the drive shaft, which turns an electric generator to produce electricity. The amount of power transferred is dependent on the rotor size and the wind speed. Wind energy, as already stated, is the kinetic energy, KE of the moving air. The kinetic energy of a mass, m with the velocity, V is

$$KE = 0.5\ m\ V^2 \tag{15.1}$$

The air mass m can be determined from the air density, ρ and the air volume, v according to

$$m = \rho v$$

Such that,

$$KE = 0.5\ \rho\ v\ V^2 \tag{15.2}$$

Power is energy divided by time. In a small time, Δt, air particles travel a distance $s = V\ \Delta t$. Multiply the distance with the rotor area of the wind turbine, A, giving the volume of air as:

$$\Delta v = A\ V\ \Delta t$$

Thus, increasing wind velocity increases the amount of air passing the rotor, which increases the output of the wind system. The wind power is then given as:

$$P = KE/\Delta t = 0.5\ \rho\ A\ V\ V^2\ \Delta t/\Delta t \tag{15.3}$$

Simplifying, we have

$$P = 0.5\ \rho\ AV^3 \qquad (15.4)$$

Here, P is power, W (watts), ρ is the air density, kg/m^3, A is the swept rotor area, m^2, and V is the wind speed, m/s. Thus, it can be stated that the power in the wind is proportional to:

- ☆ Area of windmill being swept by the wind
- ☆ Cube of the wind speed
- ☆ Air density - which varies with altitude

If the density of air is taken as 1.2754 kg/m^3, power of the wind per unit area can be given as

$$P = 0.5\ (1.2754\ V^3) = 0.6\ V^3$$

For a wind velocity of 10 m/s, it turns out to be

$$P = 600\ W/m^2$$

The fact that the power is proportional to the cube of the wind speed is quite significant. It means, if the wind speed doubles then the power in the wind increases by a factor of eight. It is why windmills are preferably located at sites having a relatively high mean wind speed. Similarly, doubling the diameter of the turbine's rotor quadruples the swept area and hence the power output from the device in horizontal axis machines. For vertical axis machines the changes in power output with diameter will be determined by the geometry of the rotor.

Two potential wind sites are compared in terms of the specific wind power expressed in watts per square meter of area swept by the rotating blades. It is also referred to as the power density of the site, and is given by the following expression:

$$P = 0.5\ \rho\ V^3 \text{ Watts per m}^2 \text{ rotor swept area} \qquad (15.5)$$

Note that windmills cannot operate at 100 per cent efficiency because the structure itself impedes the flow of the wind. The theoretical maximum efficiency is 59 per cent. This issue is discussed in the next section.

A wind energy system usually requires a minimum average annual wind speed of at least 15 km/h (4.25 m/s). The data presented in Table 15.1 gives some guidelines on wind speeds and their potential to produce electricity. It is worthwhile to note that significant areas of the world have mean annual wind speeds of above 4-5 m/s (14.5 to 18.0 km/h), which makes small-scale wind powered electricity generation an attractive option. Since, wind is an unsteady state phenomenon; it is affected by the time scale *i.e.* hour, month or year, and factors like latitude of the place, altitude of the place, and topography of the place. The time scale is quite important in the design of windmills. The mean hourly velocity (for many years) in fact provides the most reliable data for

Table 15.1. Average wind speed and its suitability for power generation

Average Wind Speed (km/h)	Suitability
Up to 15	Not good
18	Poor
22	Moderate
25	Good
29	Excellent

estimating the potential of a place for tapping wind energy. This data is also useful in mechanical aspects of design of a windmill. The scale of the month is useful to indicate whether it is going to be useful during particular periods of the year and what storage, if necessary, should be provided. In nutshell, it is necessary to obtain accurate wind speed data before any decision is made to establish the windmill or wind park at a particular site. Wind speed data can be obtained from wind maps or from the meteorology office. Unfortunately the general availability and reliability of wind speed data is extremely poor in many regions of the world. Seasonal and site variations may also affect power generation A wind generator, at the same wind speed, will produce lesser power in summer than in winter as air has lower density in summer than in winter (See eq. derived in this section). Similarly, a wind generator will produce lesser power at higher altitudes - as air pressure as well as density is lower -than at lower altitudes.

Some important wind speed limits considered in the design of wind turbines are given as follows.

- Start-up wind speed - the wind speed that will turn an unloaded rotor
- Cut-in wind speed – the wind speed at which rotor can be loaded
- Rated wind speed – the wind speed at which machine is designed to run at optimum tip-speed ratio
- Furling wind speed – the wind speed at which machine will be turned out of the wind to prevent damage
- Maximum design wind speed – the wind speed above which damage could occur to the machine

Obstructions such as trees or hills can interfere with the wind supply to the rotors. It is why; the towers are generally placed 100 m away from the nearest obstacle and the middle of the rotor is placed 10 m above any obstacle within 100 m. The rotors are placed on top of towers to take advantage of the strong winds at higher levels above the ground. These guidelines ensure a relatively consistent wind flow required for an effective wind power system.

MACHINE POWER

Although equation P = 0.5 ρ AV^3 can be used to assess the power in the wind, the actual power extracted is significantly less than suggested by this equation. The actual power extracted by the rotor blades of a horizontal axis wind turbine is the difference between the upstream and the downstream wind powers. Using wind power equation, we have

$$P_o = 0.5 \text{ Mass flow rate per second } (V - V_o)^2 \quad (15.6)$$

Where P_o is the mechanical power extracted by the rotor, *i.e.*, the turbine output power, V is the upstream wind velocity at the entrance of the rotor blades and V_o is the downstream wind velocity at the exit of the rotor blades. The mass flow rate of air through the rotating blades is derived by multiplying the density with the average velocity.

$$\text{Mass flow rate per second} = \rho A [(V + V_o)/2]$$

$$P_o = 0.5 \rho A [(V + V_o)/2] (V - V_o)^2$$

$$P_o = 0.5 \rho A V^3 [(1 + V_o/V)/2] [(1 - (V_o/V)^2] \quad (15.7)$$

Thus, we can say that the power produced by the wind machine can be given by:

$$P_o = 0.5 C_p \rho AV^3 \quad (15.8)$$

Here C_p is the coefficient of performance of the wind machine given as C_p = $[(1 + V_o/V)/2] [(1 - (V_o/V)^2]$. It defines the proportion of the power in the wind that the rotor of the machine can extract. The remaining power is discharged or wasted in the downstream wind. Now if we plot, C_p with respect to V_o/V, it generally varies from 0.25 to 0.45 with a maximum value of 0.59 at V_o/V = 1/3. This theoretical maximum efficiency of a *wind turbine* is termed as the *Betz law or Betz limit. According to Betz's law, no turbine can capture more than 16/27 (59.3 per cent) of the kinetic energy in the wind. The factor 16/27 (0.593) is known as Betz's coefficient and is the upper limit of* C_p *in eq.* (15.2). Practical utility-scale wind turbines achieve 75–80 per cent of the Betz's limit at peak production. The actual power besides wind depends on several factors, such as the type of machine, type of the rotor used, the sophistication of blade design, friction losses, and the losses in the pump or other equipment connected to the wind turbines. Thus, the figure in reality is usually around 45 per cent (maximum) for large electricity producing turbine and around 30 to 40 per cent for a wind pump. It may also be noted that a wind machine operates at its maximum efficiency only for a fraction of the time as a result of variations in wind speed. In the second stage, when mechanical energy captured by wind blades is further converted into electrical energy via wind generators, the converting efficiency also comes into picture. It is determined by several parameters namely gearbox efficiency, generator efficiency, η_{gen}, electric efficiency that encompasses all combined electric power losses in the converter, switches, controls, and cables, η_{ele}. Air density at sea level is 1.225 kg/m^3 and it changes by about 10 per cent with altitude. Taking the density of air as 1.225 kg/m^3, a rough estimate of the output from a wind machine can be obtained from the following equation.

$$P_a = 0.2\ A\ V^3 \tag{15.9}$$

Such that P_a is the average power output in watts over the year and V is the mean annual wind speeds in m/s.

Besides coefficient of performance, two more important parameters determine the performance of a wind turbine. These are torque coefficient and tip speed ratio (TSR). Note that the 3 parameters are well correlated as shown in the following write-up.

The thrust force experienced by the rotor can be expressed as:

$$F = 0.5\ \rho\ A\ V^2 \tag{15.10}$$

Hence we can represent the maximum theoretical torque as:

$$T = 0.5\ \rho\ A\ V^2R \tag{15.11}$$

Such that T is the torque generated, Nm, R is the radius of the rotor, m, A is the rotor swept area, m^2, Since the rotor shaft develops only a fraction of this maximum torque, *a torque coefficient, C_t defined as the ratio between the actual torque to the theoretical torque is used to calculate the actual torque, T_a.*

$$T_a = 0.5\ \rho\ C_t A\ V^2R \tag{15.12}$$

Tip Speed Ratio often abbreviated as TSR is quite important in the design of wind turbine generators. If the rotor of the wind turbine turns too slowly, most of the wind will pass undisturbed through the gap between the rotor blades. Alternatively, if the rotor turns too quickly, the blurring blades will appear like a solid wall to the wind. Therefore, wind turbines are designed with optimal TSR to extract as much power out of the wind as possible. *The tip speed ratio, used to characterize the rotor, is defined as the ratio of the speed of the extremities of a wind turbine rotor to the speed of the free wind.* It is given as V_t/V. if N is the rotation frequency, *i.e.*, rotation per second of a rotor of diameter, D meters, the tip speed, V_t is πND m/s. Thus TSR, λ can be given by the following relation.

$$\lambda = \pi DN/V = Rw/V \tag{15.13}$$

Here w is the angular velocity radians/s. Since radian is dimensionless, TSR unit can be written as 1/s. T_a can also be given as:

$$T_a = P_0/2\pi N = P_0/w \tag{15.14}$$

Incorporating T_a and P_0 in equation (15.14), we have

$$T_a = 0.5\ \rho\ C_t A\ V^2R = 0.5\ \rho\ C_p\ AV^3/2\pi N \tag{15.15}$$

$$C_p/C_t = 2\pi NR/V = \pi ND/V = \text{Tip speed ratio} \quad (15.16)$$

Thus, the relationship between the three factors namely power coefficient, torque coefficient and TSR is given as:

$$C_p = \lambda\, C_t \text{ or } C_t = C_p/\lambda \quad (15.17)$$

Example 14.1

Calculate the theoretical power available for a wind mill having the rotor diameter of 6 m working under a wind speed of 15 km/h. If the coefficient of performance of the wind machine is 40 per cent, what would be the machine power? Assume any value required in the calculations.

Assume the air density at sea level = 1.2 kg/m^3

Wind velocity, $V = 15 \times 1000/(60 \times 60) = 4.17$ m/s

Swept area of the rotor = $(\pi/4) \times 5^2 = 19.63\ m^2$

Using eq. (15.4), theoretical power available is given as:

$P = 0.5\ \rho\ AV^3 = 0.5 \times 1.2 \times (4.17)^3 \times 19.63 = 854.04$ Watts

Machine power using eq. (15.8) is then

$P_o = C_p \times P = 0.4 \times 854.04 = 341.62$ Watts

Advantages of Wind Energy

- It is a renewable infinite source of energy
- It is non-polluting and have little adverse effects on the environment
- No fuel and transportation is required and hence low operating cost
- Wind energy is cost effective, and its prices are still dropping
- Provides employment opportunities in the region
- Wind energy reduces carbon emissions as it replaces fossil fuels

Disadvantages

- Intermittent in nature, varies from time to time and place to place and is diluted and fluctuating in nature
- Unlike water energy, wind energy requires some storage capacity because of its irregularity
- Noise and visual pollution
- Wind power systems have a relatively high overall weight. Estimated weight for large systems is 110 kg/kW
- Large areas are required to establish wind parks
- Wind turbines can damage the habitats of birds and marine life
- The present windmills are neither maintenance free nor practically reliable

WINDMILLS AND WIND TURBINES

In modern times, windmills are also referred to as wind turbines. However the distinction between the two terms must be understood as these are technically different in terms of application principle. A wind turbine is a rotating machine which converts the kinetic energy of wind into mechanical energy. If the mechanical energy is used directly by machinery the machine is usually called a windmill. Windmills have been the first and most effective man-made means of power generation harnessing the power of wind for use in pumping water, saw-milling timber, grinding grain, and so on. On the other hand, if the kinetic energy is converted to mechanical and then electrical energy mostly for commercial applications, the machine is called a wind generator, wind turbine, wind power unit (WPU), wind energy converter (WEC), or aero-generator. In the current section we will use the terms interchangeably although the distinction between the two must be kept in view. *Often times a large number of wind turbines are built close together, which is referred to as a wind project or wind farm or wind park.* It works as a single power plant and supplies power to a given area or utility or feeds the state/national grid.

PRINCIPLE OF WIND TURBINE

We have all seen fans that use electricity to make wind. Contrarily, wind turbines use wind to make electricity. Wind turbines work on a very simple principle of using the aerodynamic force from the rotor blades to turn wind energy into electricity. The blade-element theory is commonly used to understand the movement of the blades by the wind. In this analysis, the cross-section of the blade element looks like an airfoil and is analyzed as such. Actual blade is divided into finite elements of thickness, dr. The first step in calculating the forces is to determine the wind velocities experienced by the blade element. Because the blade is rotating about the hub, the blade element experiences a velocity that is the resultant of the incoming velocity in the axial (perpendicular to the rotor) direction, $V_0(1\text{-}a)$, and other in the tangential (along the rotor plane) direction, wr(1-a′) such that w is the angular velocity at r distance away from the rotor plane (Figure 15.1 top). The coefficients a and a′ take into account the induced rotation of the airflow which goes through the rotor provoked by the sheet of vortices originating at the tip and the hub of the finite blade. Classic airfoil theory states that when a stream of fluid flows past an airfoil, two forces arise due to the difference in air pressure across the two sides of the blade as the fluid is deflected around the wing: a lift force (F_l) and a drag force (F_d) (Figure 15.1 bottom). The basic features that characterize lift and drag are:

- ✰ Drag is in the direction of air flow
- ✰ Lift is perpendicular to the direction of air flow
- ✰ Generation of lift always develops a certain amount of drag

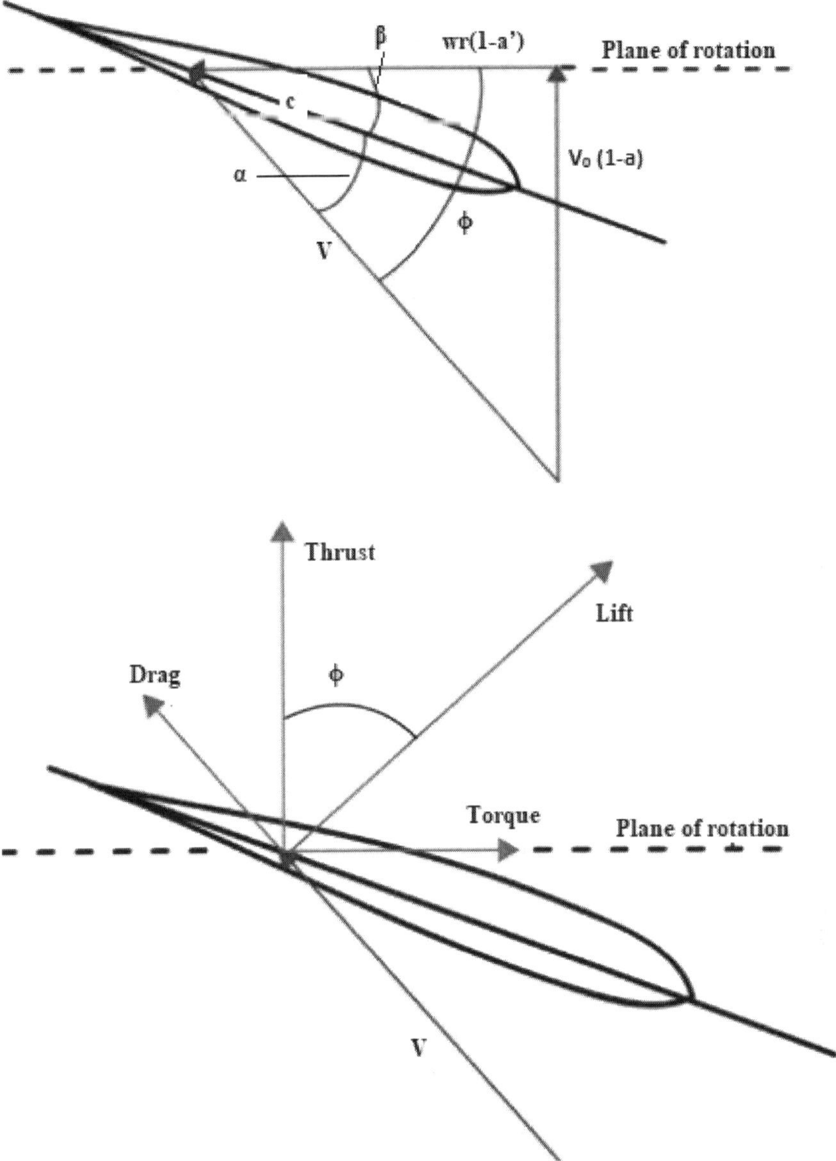

Figure 15.1. Velocities experienced by airfoil (top) and forces acting on blade element (bottom).

- ☆ With a good airfoil, the lift produced can be more than thirty times greater than the drag
- ☆ Lift devices are generally more efficient than drag devices

These forces can be shown to be functions of the geometry of the foil and the flow of the fluid (Buckman, 2015).

$$F_l = 0.5\, c_l \rho V^2\, c \tag{15.18}$$

$$F_d = 0.5\, c_d \rho V^2\, c \tag{15.19}$$

Here F_l and F_d are the lift and drag forces on the airfoil, c_l and c_d are foil specific parameters called the lift and drag coefficients, V is the fluid's velocity, ρ is the density of fluid, and c is the chord length, the length of the line that connects opposite ends of the blade (Figure 15.1top). In the case of turbine blades, air is not flowing directly parallel to the airfoil's chord line and thus we introduce a parameter α, or the angle of attack the angle between the direction of the air and the chord line (Figure 15.1 top) of the blade. The angle of attack is important because the performance coefficients depend on the angle of attack and the Reynolds Number. Moreover, angle of attack is the main parameter of interest in the design of blades because of its influence on drag coefficient and stalling. For small angles of attack, the lift coefficient increases proportionally to the angle of attack at a slope of roughly 2π and the drag coefficient is roughly constant. At a certain angle, c_l begins to drop dramatically in what is known as stalling, *the phenomenon where lift decreases and the drag increases quite substantially*. It introduces higher drag resulting in lower power production.

We can decompose the lift and drag forces into normal and tangential components to the rotor plane using the geometry of the flow and wing. It is because the components of these forces in the tangential direction will cause the drag thrust while components in the normal direction will produce the torque for power output.

$$p_t = F_l \cos\phi + F_d \sin\phi \tag{15.20}$$

$$p_n = F_l \sin\phi - F_d \cos\phi \tag{15.21}$$

Here p_t and p_n are tangential and normal components of the lift and drag forces, F_l and F_d are the lift and drag forces calculated previously and ϕ is relative wind angle being the sum of the angle of attack and pitch angle β, the angle between the chord line and the rotor plane (Figure 15.1 top). Note that pitch angle is given by the blade geometry. Since p_t and p_n are forces per length of the turbine, we integrate these equations along the blade to its tip R from the rotor axis to obtain the total torque and thrust forces.

$$T = \int B p_n \, dr \text{ and} \tag{15.22}$$

$$M = \int r B p_t \, dr \tag{15.23}$$

Here T and M are the torque and thrust forces, dr is the annulus of air at a distance r, the distance of the blade element from the hub and B is the number of blades on the turbine. The forces of the lift/drag cause the rotor to spin. A wind turbine blade for its efficient operation needs to function with as much lift and as little drag as possible because drag dissipates energy.

Golding has derived the expression for aero turbine efficiency, η_a as follows.

$$\eta_a = [1 - K (V_t/V)]/[1 + K (V_t/V)] \qquad (15.24)$$

Here K is F_d/F_l *i.e.* drag to lift ratio and V_t is the tip speed of the blade. Clearly, if there were no drag, *i.e.*, K=0, the efficiency would be unity (η_a=1). In actuality K can be made very small, depending on the airfoil chosen and the angle of attack. Also above equation tells us that the efficiency would be low if V_t/V were very large or again if it were small. It may suffice to say that there exists an optimum ratio of V_t/V at which turbine would operate optimally. Drag devices always have tip-speed ratios less than one and hence turn slowly, whereas lift devices can have high tip-speed ratios (up to 13:1) and hence turn quickly relative to the wind. Most horizontal axis machines work on lift force. The rotor of the turbine is connected to the generator, either directly (if it's a direct drive turbine) or through a shaft and a series of gears (a gearbox) that speed up the rotation. It allows the use of a physically smaller generator. This translation of aerodynamic force to rotational force generates electricity.

Components of Wind Electric Generation System

Basic components of a wind electric system are (Figure 15.2)

- ☆ Tower
- ☆ Blades
- ☆ Nacelle
- ☆ Rotor
- ☆ Shaft
- ☆ Yaw control
- ☆ Transmission/gearbox
- ☆ Generator
- ☆ Braking system
- ☆ Controllers
- ☆ Sensors

Tower

The tower is used to support the structure of the turbine. Commonly, four kinds of supporting towers are used. These are the reinforced concrete tower, the pole tower, the built-up shell-tube tower, and the truss tower. Amongst these, the truss tower is favoured because it is widely adaptable, its cost is low, its parts are readily available, it can be readily transported, it is potentially stiff and has proven its worth. Shell-tube towers also have attractive features and are quite competitive to truss towers. The type of the supporting structure and the tower height is determined by cost considerations and the transmission system used. The tower is designed to withstand the wind load during gusts (even if they occur infrequently and for very short periods). Horizontal axis wind turbines (HAWT) are mounted on towers above the level of turbulence and other ground related effects. In general, for medium and large turbines, the height of the tower is slightly greater than the rotor diameter. Small turbines should have taller towers in comparison with their rotor diameters; otherwise the turbine would be too close to the ground surface and would experience poor wind speeds. Turbines with rated output between 10 kW and 100 kW have tower heights in the range of 20-30 m; 300 kW to 500 kW machines would have towers 35 m to 40 m high. The maximum practical height is estimated to be roughly 60 m.

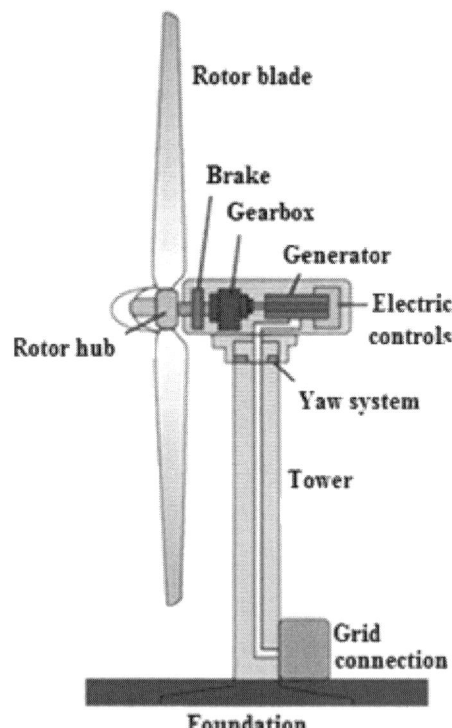

Figure 15.2. Major components of a horizontal axis wind turbine.

The HAWT have been classified as upwind and downwind turbines based on the direction of receiving the wind (Figure 15.3). In the upwind structure the rotor faces the wind directly, while in downwind structure, the rotor is placed on the lee side of the tower.

In the upwind structure, the wind encounters the turbine before reaching the tower. As such, the unsteadiness due to wake passing on the rotor blades causes repeated variations in the wind forces on the tower. Therefore, the tower

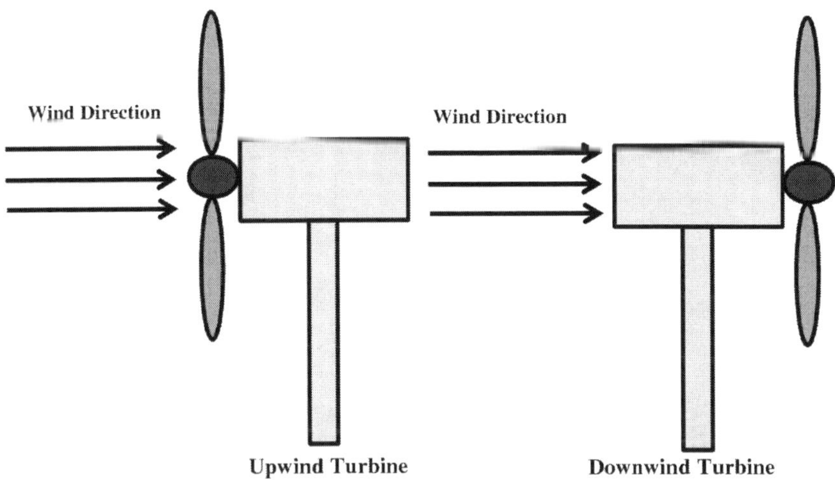

Figure 15.3. An upwind and a downwind turbine.

tends to vibrate and may eventually be damaged. The upwind structure needs a yaw control mechanism to keep the rotor always facing the wind. A HAWT with a simple tail vane type yaw control mechanism is shown in Figure 15.4. On the other hand, the tower vibrations are less in downwind structure, but the blades are now subjected to severe alternating forces as the wind passes through the tower wake. Both upwind and downwind turbines have been used in WEC devices. Downwind rotors are generally preferred for large aero generators. Although other forces acting on the blades of these large machines are significant, tower effects are still important and therefore, tower design is an essential aspect of the overall system design. In any case, it is necessary to avoid amplification of vibration through careful design of the resonant frequencies of the tower, blades, rotor, *etc. viz*-a-viz the wind fluctuation frequencies.

Blades

Blades are physical structures, which are aerodynamically optimized to capture the maximum power from the wind. Each blade is usually 20 m or more in length, depending on the power level. The purpose of pitch control is to maintain the optimum blade angle to achieve certain rotor speeds or power output. Pitch adjustment can be achieved by stalling and furling. By stalling a wind turbine, angle of attack increases causing the flat side of the blade to face further into the wind. Furling decreases the angle of attack, causing the edge of the blade to face the oncoming wind. Pitch angle adjustment is the most effective way to limit output power by changing aerodynamic force on the blade at high wind speeds. This ensures the turbine's safety in the event of high winds, loss of electrical load, or other catastrophic events.

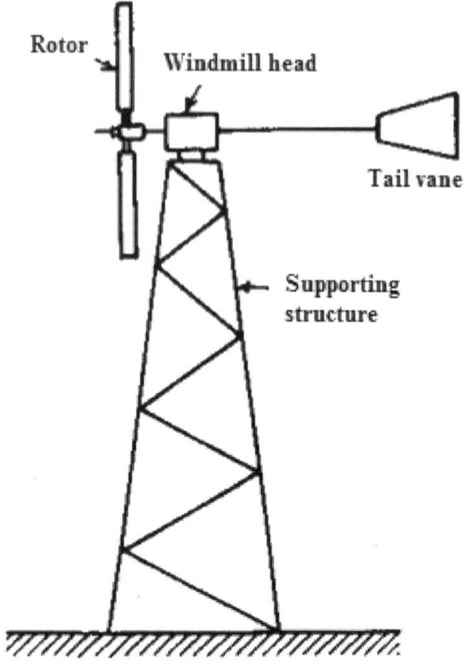

Figure 15.4. Schematic diagram of horizontal axis wind turbine with tail vane.

Nacelle

The nacelle of a wind turbine houses the drive train and other tower-top components. It sits on top of a yaw bearing that allows it to rotate as the wind direction changes. The nacelle must be accessible for maintenance and repair work. Access to the nacelle is via a lift or ladder within the tower. The nacelle is usually assembled at the factory and must be lifted to the top of the tower using specialized lifting equipment. Most modern wind turbines include real-time monitoring systems to monitor the condition of the system. These are also housed in the nacelle. The tower upon which the nacelle is mounted is usually two to three times the rotor blade length although it may be taller to allow the blades to access higher wind speeds.

Rotors

The rotor mounted on the top of the supporting structure on suitable bearings is one of the important components of the wind turbine. It converts energy in the moving air to rotary mechanical energy. All its components need to be properly designed and matched with the rest of the components for its effective utilization. Wind turbines are usually classified into two categories, according to the orientation of the axis of rotation with respect to the direction of wind (see section types of wind turbines). The wind turbine head supports the rotor, and houses the rotor bearings. It also houses any control mechanisms for changing the pitch of the blades, safety devices and tail vane to orient the rotor to face the wind.

Shaft

The shafts are divided into low and high speed shafts. The low-speed shaft transfers mechanical energy from the rotor to the gearbox, while the high-speed shaft transfers mechanical energy from gearbox to generator.

Yaw Control

Vertical axis machines (rotors) do not need a yaw system. It is used only in the horizontal axis wind turbines. The design of a horizontal axis turbine is greatly simplified for localities where prevailing wind blows mostly in one direction. The rotor in such cases can be orientated with the swept area perpendicular to the predominant wind direction. Such a machine is said to be yaw fixed. In other cases, yaw control system rotates the nacelle around the tower axis while pointing the rotor swept area steadily towards the incoming wind direction that helps to increase the energy capture and maximizes the power input to the wind turbine. Yaw systems are also categorized as passive or active types. In the small turbines, *yaw action controlled by tail vane is a passive system* (Figure 15.4). It is similar to that in a typical pumping windmill and utilizes the wind force in order to adjust the orientation of the wind turbine rotor into the wind. *Most large machines are yaw active. Herein a servo-mechanism operated by a wind direction sensor controls the yaw motor to keep the turbine properly oriented.* The entire system is pivoted on strong bearings that allow the rotation of the system around a vertical axis. In order to avoid overshooting in the rotation of the electricity generating system and unnecessary power fluctuations, the yaw break system slows the rotational motion by providing damping. The motor rotates the turbine slowly so that the blades face into the wind as and when there is change in the wind direction.

Transmission

Varying the pitch of the rotor blades can to some extent control the rate of rotation of large wind turbine generator. But in general it is low of about 40 to 50 revolutions per minute (rpm). On the other hand, much greater rates of rotation such as 1800 rpm are required for optimum generator output. Therefore, a mechanical interface consisting of step up gear and a suitable coupling is used to transmit the rotary mechanical energy to the electrical generator. Among the transmission options are mechanical systems involving fixed ratio gears, belts, and chains, singly or in combination or hydraulic systems involving fluid pumps and motors. Fixed ratio gears are commonly used in top mounted equipment because of their high efficiency, low cost, and minimum risk. For bottom mounted equipment which requires a right-angle drive, transmission costs might be reduced substantially by using large diameter bearings with ring gears mounted on the hub. Such a combination offers a high degree of design flexibility and cost saving.

Generator

Electric generators work on the principle of electromagnetic induction. A conductor coil (a copper coil tightly wound onto a metal core) is rotated rapidly between the poles of a horseshoe type magnet. The magnetic field interferes with the electrons in the conductor to induce flow of electric current inside it. Either constant or variable speed generators are used but variable speed units are expensive and/or unproved. Amongst the constant speed generators are synchronous induction and permanent magnet types. Almost all aspects of synchronous generator like construction and working is similar to the synchronous motor but the difference is that generator converts mechanical power into electrical power and motor converts electrical into mechanical power. Amongst the asynchronous generators are the squirrel cage induction and the wound rotor induction generators. Although synchronous generators (large ones at least) are more efficient than asynchronous generators, most wind turbines in the world use three phase asynchronous (cage wound) generators, also called *induction generators* to generate alternating current.

Braking System

A wind turbine is a very sophisticated device equipped with several sensors to detect any major fault. It is also equipped with various safety devices to ensure safe operation during their lifetime. One of the most important safety devices is the braking system. The braking system of a wind turbine ensures that it automatically stops when it detects that one or more of its critical components are not working properly. It may be a result of overheating of the generator or may be for protecting it against high wind speed. It has been assessed that the number of emergency stops during the twenty years of service life of a wind turbine may range from 500 to 1000 stops.

Controllers

Wind turbines require certain control systems. Horizontal-axis wind turbines have to be oriented to face the wind. In high winds it is desirable to reduce the drive train loads and protect the generator and the power electronics equipment from overloading, by limiting the turbine power to the rated value up to the furling speed. At gust speeds, the machine has to be stalled. At low and moderate wind speeds, the aim should be to capture power as efficiently as possible. The purpose of the controller is to sense wind speed, wind direction, shafts speeds and torques at one or more points, output power and generator temperature as may be necessary. They also send appropriate control signals for matching the electrical output to the wind energy input and protect the system from extreme conditions brought upon by strong winds, electrical faults, and the like. The modern large wind turbine generator requires a versatile and reliable control system to perform the following functions:

- ☆ The orientation of the rotor into the wind (azimuth of yaw)
- ☆ Start up and cut-in of the equipment
- ☆ Power control of the rotor by varying the pitch of the blades
- ☆ Generator output monitoring - status, data computation, and storage
- ☆ Shutdown and cut out owing to malfunctioning at very high winds
- ☆ Protecting the generator, and the utility accepting the power and the prime mover
- ☆ Auxiliary and/or emergency power
- ☆ Maintenance mode

Many combinations in terms of the control system are possible involving the following components:

- ☆ Sensor - mechanical, electrical, or pneumatic
- ☆ Decision elements - relays, logic modules, analog circuits, a micro-processor, a fluidics units, or a mechanical unit; and
- ☆ Actuators, which can be hydraulic, electric, or pneumatic. One recommended combination consists of electronic transducers feeding into a micro-processor that signals electrical actuators and provides protection through electronic circuits. Sometimes, a pneumatic slip clutch may also be required.

CLASSIFICATION OF WIND TURBINES

Wind turbines have been classified on the basis of number of factors although the main classification remains on the basis of axis of rotation, which forms the part of this chapter. Nonetheless, the various classification criteria are listed as follows.

Based on Axis of Rotation

- ☆ Horizontal axis machines
- ☆ Vertical axis machines

Based on Aerodynamics

According to this classification turbines are characterized as

- ☆ Drag based
- ☆ Lift based

The force of the wind turbine pushes against a surface like an open sail in drag based wind turbines. In fact, the earliest wind turbines dating back to ancient Persia used this approach. The Savonius rotor is a simple drag based windmill. It works because the drag on the open or concave face of the cylinder is greater than the drag on the closed or convex section. Despite their low efficiencies, these are in use because of their robustness and simplicity to build and maintain. Most horizontal axis machines are lift based turbines. They are capable of extracting more energy from the wind, but require specially shaped airfoil surfaces like those used on airplane wings. The airfoil shape is designed to create a differential pressure between the upper and lower surfaces, leading to a net force in the direction perpendicular to the wind. Rotors of this type must be carefully oriented (the orientation is referred to as the rotor pitch) to enable them to harness the power of the wind with changing wind speed. The Darrieus machines in the vertical axis machines are also lift based machines but are rarely used now. Two examples each of drag and lift types of wind are shown in Figure 15.5.

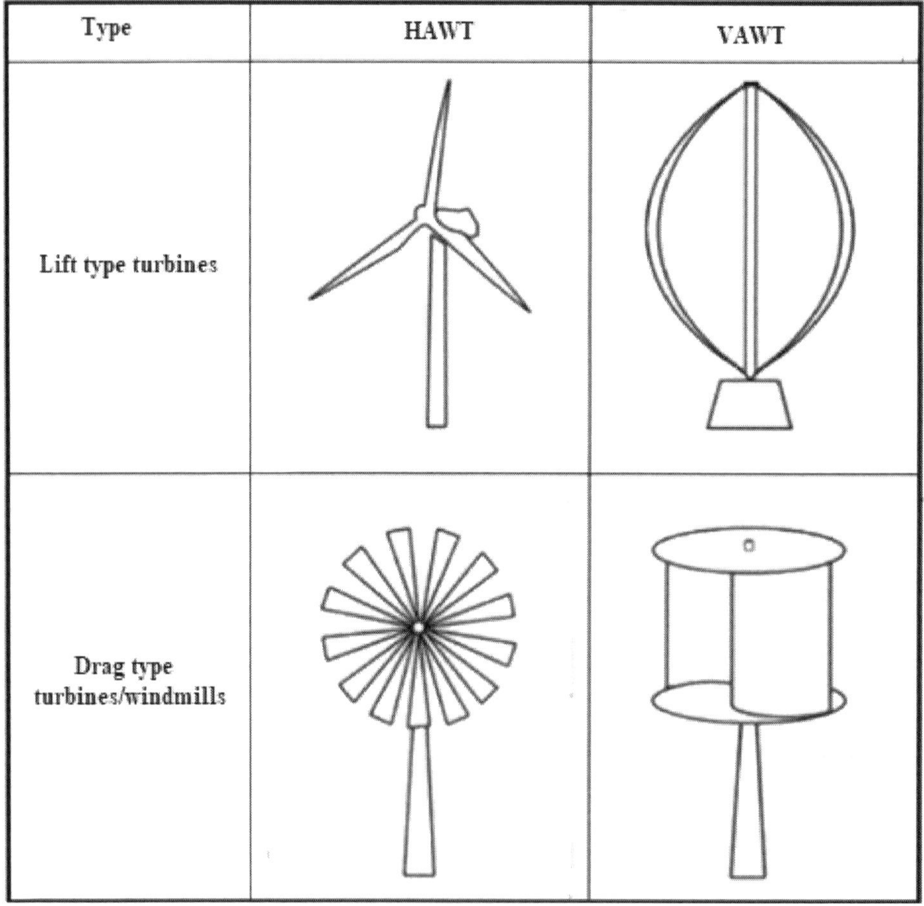

Figure 15.5. Lift and drag type turbines.

Based on Size

- Small size machines (up to 2 kW)
- Medium size machines (2 to 100 kW)
- Large size machines (100 kW and above)
- Single generator at single site
- Multiple generators

Based on Types of Output

- DC output: DC generator or with alternator rectifier
- AC output: Variable frequency, variable or constant voltage AC, constant frequency, variable or constant voltage AC

Based on Rotational Speed of the Turbines

- Constant speed and variable pitch blades
- Nearly constant speed with fixed pitch blades
- Variable speed with fixed pitch blades
- Variable speed constant frequency generating system

Based on Utilization of Output

- Battery storage
- Direct conversion to an electromagnetic energy converter
- Thermal power
- Inter convention with conventional electric utility guides

As previously stated, two basic types of wind turbines on the basis of their axis of rotation, are vertical axis wind turbines and horizontal axis wind turbines (Figure 15.6). Vertical axis wind turbines are further categorized as Savonius or S type wind turbines suitable for areas having low wind velocity and Darrieus wind turbines suited for areas having high wind velocities and H-rotor wind turbines. Horizontal axis wind turbines are categorized as single blade wind turbines, double or multi blade wind turbines and bicycle multi blade type also known as sail type wind turbines.

Vertical-axis Wind Turbines

Vertical-axis wind turbines (VAWT) are relatively simple in design and construction compared to the horizontal axis wind turbines. These are Omni-directional and do not require complicated yaw device. These turbines also have wider operational window because of having a lower cut-in and higher cut-out wind speed. The generator and the gearbox of such systems can be housed at the ground level making the tower design simple and quite economical

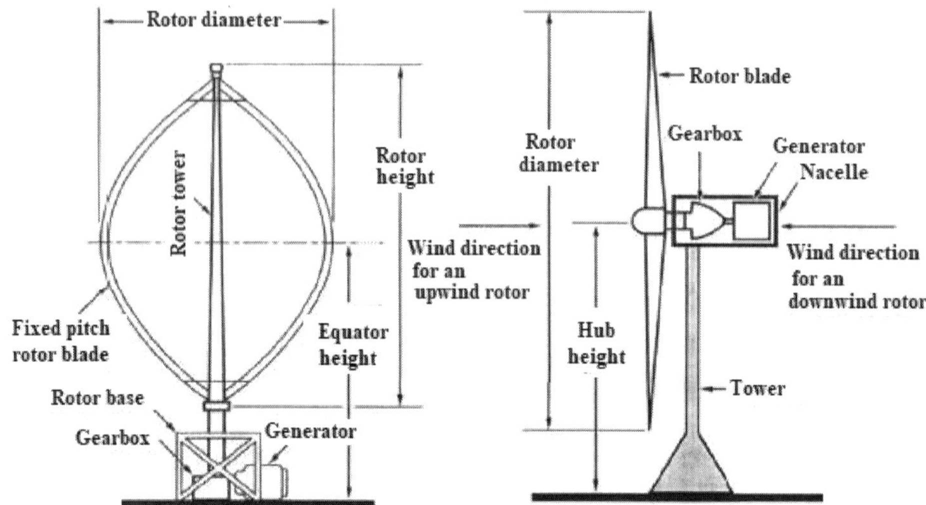

Figure 15.6. Schematic diagram of a vertical (left) and a horizontal-axis wind turbine (right).

(Figure 15.6 left). Moreover, the maintenance of these turbines can be done at the ground level. Vertical-axis wind turbines also need lower-ratio gearboxes, which are less expensive and more efficient than gearboxes needed to operate horizontal-axis wind turbines. Further, vertical axis wind turbines are able to operate at higher wind speeds with relatively lower risk of wind damage. The major disadvantage of vertical-axis machines is that normally they are not self starting. Thus, an additional mechanism may be required to push and start the turbine once it is stopped. Important points to remember regarding VAWT are:

- ✰ Drag is the main force in commonly used Savonius wind turbines
- ✰ Yaw mechanism is not required
- ✰ Lower starting torque
- ✰ Nacelle is placed at the bottom.
- ✰ Difficulty in mounting the turbine
- ✰ Unwanted fluctuations in the power output

Savonius Wind Turbine

The best known vertical-axis rotor is made up of two identical semi cylinders with their axis vertical. Invented by S. J. Savonius in the year 1920, it works on the principle of cup anemometer. Its popularity arises from the fact that it can operate at low wind velocities of about 8 kmph. It consists of two half cylinders facing each other forming an 'S' shaped cross-section. They are mounted on a vertical axis perpendicular to the direction of wind with some gap at the axis between the two cylinders (Figure 15.7 top). The main action of the wind is quite simple. The force of the wind is greater on the cupped face than on the rounded face. A low pressure is created on the convex sides of drums. Torque is produced by the pressure difference between the two sides of the half cylinders facing the wind. Irrespective of the wind direction, the rotor rotates making the convex sides of the buckets head into the wind. Although, the design is quite efficient yet it requires large surface area. The power is tapped from the rotor shaft for uses like water pumping, battery charging, grain winnowing *etc.* Savonius wind energy conversion system having vertical axis eliminates the expensive power transmission system from the rotor to the axis.

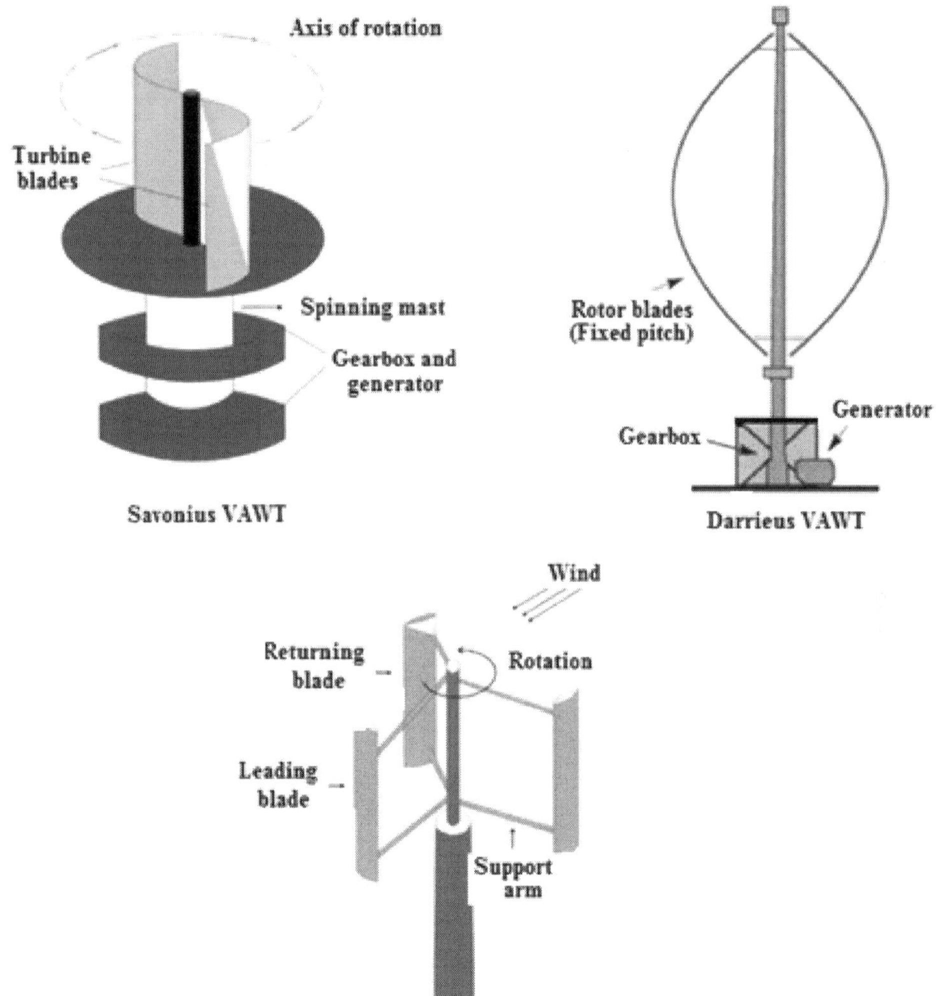

Figure 15.7 Schematic diagram of Savonius (left), Darrieus (middle) and H-rotor windmill (right).

Advantages

- Because of the vertical axis, these turbines continue to work effectively even if the wind changes direction
- There is no need for a tower or other expensive structure to hold it in place because the Savonius design works well even at low wind speeds. It greatly reduces the initial set-up cost
- The device is noise free, easy to build, and relatively small
- Maintenance is easy because the turbine is close to the ground

Disadvantages

- The scoop system used to capture the wind's energy is half as efficient as a conventional turbine resulting in less power generation.

Darrieus Wind Turbine

This machine was invented and patented in 1925 by G.J.M. Darrieus, a French Engineer. Darrieus rotors have three symmetrical curved aerofoil blades, both ends of which are attached to a vertical shaft at both ends to form a cage-like structure (Figure 15.7 top right). The blades look like an egg beater. The force in the blade is pure tension that provides stiffness to withstand the wind forces. As the aerofoil blades rotate, they provide a torque about the central shaft. This shaft torque is transmitted to a generator at the base of the central shaft for power generation. For a given wind speed, Darrieus unit rotates more rapidly and is more efficient than the Savonius rotor. Major advantage of Darrieus wind turbine is that the rotor blades can accept the wind from any point of the compass. The machine can be mounted on the ground eliminating the tower structures. Besides, it requires less surface area than a Savonius type. It supports its blades in such a way that bending stresses are minimal under normal operation. Disadvantages arise from the fact that it experiences lower velocity wind when compared to tower mounted conventional wind energy conversion systems.

The H-rotor turbine is similar to the Darrieus turbine. It has straight and vertical blades instead of curved blades as in Darrieus turbine (Figure 15.7 bottom).

Advantages

- The rotor shaft is vertical. Therefore it is possible to place the load like a generator or a centrifugal pump at ground level
- As the generator housing is not rotating, the cable to the load is not twisted and no brushes are required for large twisting angles
- The rotor can take wind from every direction
- The visual acceptance for placing the windmill on a building might be better than a horizontal axis windmill as these can easily integrate into building

Disadvantages

- Unlike the Savonius wind turbine it is difficult to start. Therefore, a small motor, or another Savonius turbine may be needed to initiate rotation
- Low efficiency

Horizontal-axis Wind Turbines

Most commercial wind turbines belong to this category. Horizontal axis wind turbines (HAWT) have their axis of rotation horizontal to the ground, almost parallel to the wind stream (15.6 right). These machines are highly efficient than their counterpart VAWT having high power coefficient as the blade tips can travel at several times the wind speed. The aerodynamic theory used to design aircrafts is also used to design the shape of the blades of HAWT. The blade tip makes a small angle with the plane of rotation and almost at right angles to the direction of wind. In a modern wind turbine, the velocity of blades can reach as much as six times the wind velocity. Ideally, the blade needs to be twisted, but it is not always the case because of construction difficulties. While the low-speed wind turbines are mainly used for mechanical purposes, like in water pumps, high speed turbines are used for electric power generation. Some important points to remember about HAWTs are:

- Lift is the main force in most currently used wind turbines
- Much lower cyclic stress
- 95 per cent of the existing turbines are HAWTs
- Nacelle is placed at the top of the tower
- Yaw mechanism is required

Horizontal axis machines come with some distinct disadvantages as well.

- They have relatively high cut-in and a low cut-out wind speed leaving only a relatively narrow window of operation beyond which they are prone to damage
- The generator and gearbox of these machines is placed over the tower, which makes its design more complex and expensive

Categorization of HAWT based on Number of Blades

Depending on the number of blades, horizontal axis wind turbines are classified as single blade, two blade, three blade and multi-blade types. In single blade machines, a long blade is mounted on a rigid hub. Induction generator and gear box are arranged at the top of the tower (Figure 15.8). Extremely long blades (60 m) mounted on the hub result in large blade root bending moments due to tower shadow, gravity and sudden shifts in the wind direction. Low cost counter weight is commonly used to centrifugally balance long blade to reduce the rotor cost (Figure 15.8). Single blade turbines are cheaper because of savings on blade materials. Moreover, the drag losses are also minimal. In spite of these, single bladed designs have not become popular due to problems in balancing and visual acceptability.

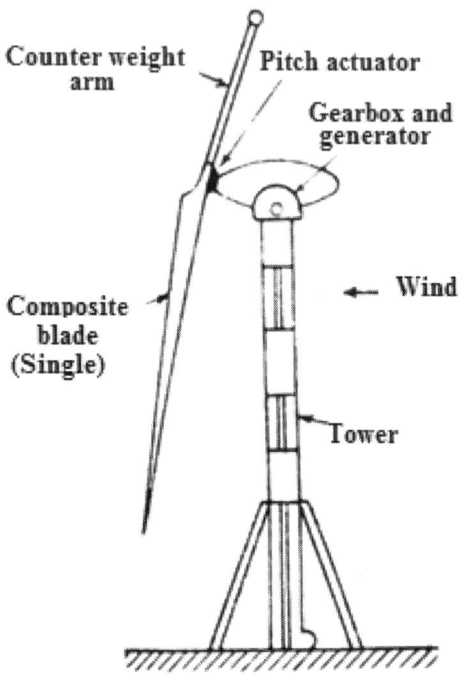

Figure 15.8. Schematic diagram of horizontal axis single blade windmill.

Two blades rotors have similar drawbacks as single blades although the extent is much lesser than the single blade (Figure 15.9 left). The rotor blades are continuously flexed by unsteady aerodynamic, gravitational and inertial loads during operation of the machine. Flexing due to fatigue loading reduces the life of the blades, particularly when these are made of some metal. The tower is also subjected to such loads causing serious damages. It may be quite damaging, if the vibration modes of the rotor happen to coincide with one of the natural mode of vibration of the tower. The rotors with more than two blades are not recommended because of high cost of blades. Yet many commercial turbines for electricity generation have three blades because the performance coefficient of rotors having more than two blades is slightly higher. The choice between the two and three rotors depends on certain factors. The two blade designs have less nacelle weight and are much simpler to erect. Three blade turbines involve 33 per cent more weight and cost, though the power coefficient increases only by 5-10 per cent. On the other hand, the three blade design has smoother power output and a more balance gyroscopic force and therefore less blade fatigue and lesser chances of failure.

Figure 15.9. Schematic diagrams of two blade (left) and multi-blade (right) rotor system in a wind turbine.

The multi-blade design shown in Figure 15.9 (right) consists of blades made from sheet metal or aluminium. Multi-blade rotors have high strength to weight ratios and are strong enough to withstand wind speeds of 60 kmph. These wind turbines (mills) are simple, have good power coefficient, high starting torque, and are low cost.

Sail Type Windmill

Designers have been thinking to design an improved version of vertical-axis wind turbine that captures the inherent advantages of the vertical-axis design, yet minimizes the drawbacks of existing designs. These efforts have lead to a vertical-axis wind turbine with sails that are pivotally attached to parallel and horizontal yardarms. A sail wind turbine system includes a vertical main shaft with a vertical axis of rotation with parallel horizontal yardarms, cross-connected and corner braced to the vertical main shaft. Number of sails used may vary from one design to another. The blades are made by stretching out triangular pieces of canvas cloth or nylon or plastics (Figure 15.10). It runs at 60 to 80 rpm. Moreover, an outer support structure made of vertical columns with lateral cross bracing joins together at the vertical main shaft in the middle and top, and is coupled thereto with self-aligning split bearings. The support columns are anchored to the ground, thereby providing greater support to the structure. This design is an improved version to produce electrical energy at the ground level.

Figure 15.10. Schematic diagram of sail type wind turbine blades.

Relative characteristics of horizontal and vertical axis wind turbines are summarized in tabular form as follows.

Horizontal Axis Wind Turbines	*Vertical Axis Wind Turbines*
Shaft of the rotor is horizontal and is not very long	Shaft of the rotor is vertical and is very long and heavy
Tower is required to support rotor and generator	No such tower may be needed as generator is kept at ground level
Its starting torque is less	Its starting torque is high
Tip to wind speed ratio is more giving more power	Tip to wind speed ratio is less giving less power
Change in the rotor head direction is needed with change in the wind direction	Rotor does not require to change its directions as it can run by the winds coming from any direction
Wind turbine support system has to bear total weight of rotor, blades and generator	Wind turbine support system is made up of guy bars and does not require to bear the weight of rotor and generator
Wind force on the rotor produces stress on every parts of the wind turbine such as rotor, bearings and structures	Wind force on the rotor does not produce stress on all parts of the wind turbine
The support structure required is very strong	Strong support structure is not needed
The wind turbines are very popular for power generation	The wind turbines are not very popular for power generation

RATED CAPACITY AND CAPACITY FACTOR

Worldwide installed capacity of wind energy is calculated using the maximum output of the turbines installed. It is known as the rated capacity. For example, suppose the 60 GW of capacity had a single rated wind speed of 13 m/s. If the wind blew constantly at 13 m/s for the entire year, the output would be 60 × 8760 h/year = 525600 GWh/year, or approximately 525 TWh/year ignoring conversion losses in electricity generation. On the contrary, average wind speeds are significantly lower than the rated wind speed in virtually all locations. Therefore, actual output from the wind system is likely to be substantially lower than the rated capacity. Herein appears the *capacity factor, which for a wind turbine is defined as the actual annual output divided by the annual output that would be obtained if the devices functioned at rated capacity for the entire year.* Given that the capacity factors for wind turbines are typically in the range of 20 to 40 per cent, the actual output will also be 20-40 per cent of the rated capacity. Fossil-fuel-powered plants in general have higher capacity factors. Therefore, for the same installed capacity fossil-fuel-powered plants will on an average produce higher electric output than a wind powered plant of the same capacity.

APPLICATIONS

Windmills are used for quite a number of purposes. Some of the common applications of windmills are for pumping water, saw-milling of timber, milling grains, pumping of drainage water, oil extraction from seeds and machining. It can be used for processing of commodities. The most important application of-course is for producing electricity. Herein we need to define a term solidity as it determines the purpose for which machine can be used. *Solidity is defined as the percentage of the area of the rotor, which contains material rather than air.* High-solidity machines carry a lot of material and have coarse blade angles. They generate much higher starting torque than low-solidity machines but are inherently less efficient than low-solidity machines. High solidity machines will have a low tip-speed ratio and vice versa. The windmills are generally of this type and are used for mechanical tasks. Low-solidity machines run at higher speed and are used for electricity generation.

Water Pumping

Windmills and watermills were probably the original prime movers that replaced animal power as a source of energy. A water-pumping windmill has been used to pump water from wells, ponds, and bore wells for drinking, minor irrigation, salt farming, fish farming, *etc.* Two types of windmills are used for this purpose namely direct drive and gear type. The most common windmill has a horizontal axis rotor of 3–5.5 m diameter, with 12–24 blades with a tail vane mounted on the top of a 10-20 m high mild steel tower. The rotor is coupled with a reciprocating pump of 50-150 mm diameter through a connecting rod. Such windmills become operational even at wind speed of 8-10 kmph. Normally, a windmill is capable of pumping water in the range of 1000 to 8000 L/hr, depending on the wind speed, the depth of water table, and the type of windmill. Windmills can lift water from depths of about 60 m. Today gear type windmills are promoted under the National water pumping windmill program. Pros and cons of windmills for water pumping are as follows.

Pros	*Cons*
One of the alternatives energy source to ensure water for man, animals and land in areas where electricity lines have not yet reached or where the supply is erratic and diesel supplies are expensive	Satisfactory operation only possible in medium wind regimes (12-18 kmph)
Can be installed in remote windy areas	The site should be free from obstacles such as buildings and trees in the surrounding areas
No fuel is required for its operation and hence operational cost is almost nil	The initial cost of the system is high
Easier to operate and maintain, the annual operation and maintenance cost may average about Rs. 500-700	Fluctuating supply of water requires demand and supply management through a storage tank

Wind Turbines for Electricity

Smaller turbines (50-150 W) can be used for individual households. A typical small wind generator produces electricity either at 120/240 volts alternating current for direct domestic use or at 12/24 volts direct current for battery charging. Larger machines generate 3 phase electricity. These machines can be connected to a grid distribution network. In this case electricity is sold to the electricity utility under an agreement between the producer and the grid. When excess electricity is produced, it is released to the grid while it is purchased when the wind is low. The national grid connected system is quite flexible and does away with the need for a back-up system. The third option is to distribute electricity to smaller areas, typically a village or town using a micro-grid. When wind is used for supplying electricity to such a grid, a diesel generator set is often used as a back-up to take care of periods when wind speeds are low. Such a hybrid systems is capable of providing electricity in all weather conditions, although the capital cost of such a system is high but subsequent running costs will be low compared with a pure diesel system.

Battery Charging

Battery charging is an option in areas where households are widely dispersed or where grid costs are prohibitively expensive. For people living in rural areas, a few tens of watts of power are sufficient to meet their lighting requirement and a source of power for a radio/television. Batteries can be recharged at the charging station. An arrangement to replace the batteries can also be made as is the case with LPG cylinders. This will reduce the inconvenience of an intermittent supply resulting from fluctuating wind speeds. Commercial wind generators are also available that produce 12 and 24 volts direct current suitable for battery charging applications.

Wind-solar Hybrid Systems

The wind-solar hybrid system mainly consists of one or two aero-generators along with SPV panels of suitable capacity connected with charge controller, inverter, battery bank, *etc.* to supply AC power. The power generated from both wind and solar components is stored in a battery bank for use whenever required. Since an aero-generator and an SPV system are interfaced, the power generation from these is mutually supplemented providing a reliable and cost-effective electric supply in a decentralized mode where grid power is yet to reach. The cost of the system varies from Rs 2.50 lakhs to Rs 3.50 lakhs per kW depending on the ratio of wind and solar components. The approximate cost of installation including civil works is about Rs 10,000 per kW. Repair and maintenance cost is about Rs 3000 per kW per annum.

FACTORS TO CONSIDER WHILE ESTABLISHING WIND PARKS

Technical, economic, environmental, social and other actors are examined before a decision is made to erect a generating plant on a specific site.

High annual average wind speed: Since the power generated markedly increases with wind velocity, it is obvious to select a site with high wind velocities. A high average wind velocity is the principle fundamental parameter of concern in initially appraising the sites for establishing wind parks.

Availability of anemometry data: An anemometer is used to measure the wind speeds. This data must be available at least over some period for precisely deciding the spots for sitting of wind energy turbines.

Availability of wind velocity- time curve at the proposed site: It is the principal initial controlling factor for predicting the electrical output and hence revenue returns of the WECS machines. It is desirable to have average wind speed 'V' such that V ≥ 12-16 km/hr (3.5 – 4.5 m/s), which is about the lower limit at which present large scale WECS generators 'cut in' *i.e.*, start turning. This curve also determines the reliability of the delivered WECS generator power. WECS reliability will be low if there are long periods of calm, than if the calm periods are short. For precise reliability estimates such curves over a period of about 5 years should be gathered. The ideal case for the WECS would be where V-t curve are flat, *i.e.*, a smooth steady wind blows all the time. A typical site is always less than ideal especially near the ground where wind is turbulent and gusty, and changes rapidly in magnitude and direction. This departure from homogeneous flow is collectively referred to as the structure of the wind that must be properly assessed before deciding the place of sitting.

Altitude of the proposed site: As altitude increases, the amount of gas molecules in the air decreases *i.e.* the air becomes less dense than air nearer to sea level. Contrarily, wind tends to have higher velocities at higher altitudes. Therefore, this contrasting characteristic must be properly matched to assess the wind power.

Terrain and its aerodynamic: One must be fully aware of terrain of the site to be chosen. It should then be possible to make use of hills or mountains which channel the prevailing wind into a pass region, thereby obtaining higher wind power. Moreover, if the WECS is to be placed near the top but not on the top of a not too blunt hill facing the prevailing wind, then it may be possible to obtain a 'speed-up' of the wind velocity over what it would otherwise be. Also the wind here may not flow horizontal making it necessary to tip the axis of the rotor so that the turbine is always perpendicular to the actual wind flow.

Local ecology: If the land surface is base rock then it should be possible to have lower hub heights hence lower structure cost. Presence of trees, grasses and vegetation tend to destructure the wind resulting in higher hub heights and larger system costs than the bare ground case.

Distance to road or railways: Lot of heavy machinery, structure, materials, blades and other apparatus will have to be moved into the chosen site. Therefore, nearness to road and railways will facilitate the transport.

Nearness to local user centers: It is obviously a useful criterion to minimize transmission line lengths and hence losses and cost. After applying all the previous discussed criteria, this criterion may help to narrow down the choice amongst the few appropriate sites.

Nature of ground: A site with stable ground condition should be selected so that the foundations of the structures are secured. Avoid sites that are prone to erosion, as it could possibly wash out the foundations that may damage the whole system.

Favourable land cost: Land cost determines the initial investments and costs of the whole power system. Thus, sites with low land costs such as barren lands not fit for agriculture are best suited for wind power projects that require quite a large area.

Suitable Places for Erection of Wind Turbines

All places where a relatively consistent wind flow is expected are suitable for installing the windmills or construct wind parks. Most suitable places are off-shore and on the seacoast, mountains and plains. An average wind energy value for off-shore and on the seacoast is about 2400 kWh/m^2-year, which reduces to 1600 KWh/m^2-year in the mountains and to 750 KWh/m^2-year in the plains.

Places Unsuitable for Erection of Wind Turbines

Places unsuitable to erect windmills are: humid equatorial region where there is virtually no wind energy and warm, windy countries where wind energy may be unusual because of the frequency of cyclones

ENVIRONMENTAL CONCERNS

Wind power is majorly a clean renewable energy source. However some environmental concerns have been flagged and must be kept in view while planning a wind power scheme. Some of these concerns are as follows.

Electromagnetic interference: Some television frequency bands are susceptible to interference from wind generators.

Acoustic pollution: Wind rotors, gearboxes and generators of the system produce acoustic noise. This needs to be considered while selecting and installing a machine.

Visual impact: Modern wind machines are large objects and have a significant visual impact on their surroundings. Some argue that it is a positive visual impact, while others disagree to this view.

Waste disposal: The lifecycle of a wind turbine is around 20-25 years. It is assessed that by 2023 nearly 140000 wind turbines blades have to be discarded. The disposal of these blades made from glass fiber/epoxy matrix composites will not be easy and may pose special environmental problems.

QUESTIONS (THEORY)

1. Derive the equation to show that wind power increases with cube of the wind velocity.
2. List the advantages and disadvantages of wind energy.
3. Briefly explain the principles of wind turbine.
4. Show that ratio of coefficient of performance and coefficient of torque is equal to tip speed ratio.
5. Use the airfoil theory to derive the tangential and normal components of the lift and drag forces in a wind turbine.
6. Name and describe the various components of wind electric generator system.
7. Discuss various applications of the windmills and wind turbines.
8. Briefly discuss the environmental concerns related to harnessing of wind energy through wind parks.

9. Write short notes on rated capacity, capacity factor, Savonius windmill, Darrieus windmill, horizontal-axis windmill, sail type windmill and grid connected wind turbine system.
10. Differentiate between drag and lift turbines, on-shore and off-shore harnessing of wind energy, an upwind and a downwind turbine, vertical-axis and horizontal-axis wind turbines, and Darrieus windmill and H-rotor Darrieus windmill.
11. Discuss the various factors that must be taken into account while selecting a site to establish a wind park.

Chapter 16

Life Cycle Cost Analysis

Renewable energy projects are being increasingly implemented considering the high cost of fossil fuels and environmental degradation resulting there from. Similarly, old energy projects are being upgraded with new equipments and technologies to improve their efficiencies and lower production costs. Since all these projects are carried out with an intention to receive future benefits, a sound economic analysis is essential to judge whether the investment decision is economically and financially sound. This chapter describes the information that must be assembled to conduct a worthwhile economic or financial analysis. It also describes the procedure(s) to conduct the economic and financial analysis besides discussing some of the social and environmental issues those make a case for the use of these new and green technologies and justify grant of subsidy to the users. It may not be out of place to mention that you have already read cost-benefit analysis in some of the courses such as land drainage, irrigation engineering and protected cultivation. Principles of analysis in some cases especially land drainage (Gupta, 2019) are similar to what we are going to discuss in this Chapter. Some of the terms used may appear to be new as these have been borrowed from the industry.

WHY AN ECONOMIC ANALYSIS?

☆ Life Cycle Cost Analysis (LCCA) is used for project profitability analysis because it takes into account the entire life of a project and time value of money. It plays a dominant role in the decision making whether the management/owner and the financing institutions will invest in the project or not

- It helps to understand the financial implications of various alternatives
- It helps to prioritize various projects and project alternatives
- It helps in fixing the cost of the product taking into account the cost and reasonable profit
- The engineer/energy manager can present the projects in economic terms in order to help the top management/financiers to arrive at sound knowledgeable decisions
- It helps the engineers/energy managers to speak in terms that are better understood by decision makers and the management

Life Cycle Cost Analysis (LCCA)

LCCA or the profitability analysis is an engineering economic analysis tool used to calculate the financial feasibility of a project. It takes into account the entire life of a project and time value of money. It considers all costs, revenues and savings incurred during the service life of the project from cradle to grave. Life-cycle cost is an equivalence method by putting all the costs and benefits, present and future, of a project to calculate net present value and other economic indicators. Similarly, it calculates net present value on the common basis so that an equitable comparison of alternative systems of a project can be made. Thus, it compares the relative merits of competing project alternatives and helps the decision maker to accept or reject a project and prioritize accepted projects.

LCCA PARAMETERS

Discounting

The costs and benefits of a project occur at different times during the project period. The amount available today is worth more than the same money received 10 years after. It means present worth of benefits say received in the 10th year are less than the ones received today. Same is true for costs. Although major costs are incurred in the first year, but there are many projects which span over few years or some costs such as replacement costs and operational and management costs are incurred after the project has been completed. These costs and benefits are taken into account by following the discounted cash flow method of analysis. *Discount rate is the rate at which the future benefits (losses) can be converted to their present value. The discounted cost shows the value of money/project at a specific time in the future that is equivalent in value to a specified sum today.* Therefore, assigning an appropriate discount rate is the key to properly valuing future cash flows. Discount rate is calculated by using one of the three parameters namely risk free interest rate, inflation and risk premium. The discount rate for an investment also depends on the type of financing, equity or loan. In the case of pure equity financing, the discount rate can be the best possible return from other similar projects. In the case of a loan, the discount rate equals the interest rate at which loan has been taken. If there is a mix of equity and loan, then the project's discount rate is the weighted average of these two rates. Usually discount rates vary from a low of 10 per cent to about 15 per cent. Current bank rates on fixed deposits are good indicators of the discount rates to be adopted in economic evaluation. Since bank deposits rates fluctuate, it may be worthwhile to make the analysis using 3-4 discount rates (Less and more than the average) to derive most appropriate information. Once the discount rate is fixed, it can be used to calculate the discount factors (DF) as per the following relation:

$$DF_t = 1/(1+ D_r/100)^t \qquad (16.1)$$

Where D_r is the discount rate (per cent) and t is the year say 1, 2, 3 ....30. DF_1 for t=1 and D_r = 6 per cent would be:

$$DF_1 = 1/(1+ D_r/100) = 1/(1+0.06) = 0.9434$$

Discount factors for a discount rate of 8 per cent over a period of 30 years are given in Table 16.1 (Gupta, 2019).

Another term to understand the time value of money is the Future Value (FV). *It shows the value of money/project at a specific time in the future that is equivalent in value to a specified sum today.* We use compounding to calculate FV. For compounding, the interest amount for each interest period (year) is calculated on the principal plus the total amount of interest accumulated in all previous periods. The total sum available after a number of interest periods can be calculated as follows:

$$S_n = P\,(1+ I/100)^n \qquad (16.2)$$

Where S_n is the sum available at the end of n years, P is the principal amount, I is the interest rate expressed in percentage (annual interest rate), and n is the number of years. The factor $(1+ I/100)^n$ is called the future worth (FW) factor.

Thus, present worth or discount factor = 1/Future worth factor, provided discount and interest rates are the same.

Table 16.1. Discount factors at a discount rate of 8 per cent

Year	*Discount Factor*	*Year*	*Discount Factor*	*Year*	*Discount Factor*	*Year*	*Discount Factor*
1	0.926	9	0.501	17	0.270	25	0.146
2	0.857	10	0.464	18	0.251	26	0.135
3	0.784	11	0.429	19	0.232	27	0.125
4	0.735	12	0.397	20	0.215	28	0.116
5	0.607	13	0.368	21	0.199	29	0.108
6	0.681	14	0.341	22	0.184	30	0.099
7	0.584	15	0.316	23	0.171		
8	0.541	16	0.292	24	0.158		

Inflation

Inflation is the rate of increase in prices over a given period of time. It occurs due to downward change in the value of the money. Therefore, Inflation can be understood as the amount of money required to purchase the same amount of a product will be higher due to inflation in prices. If there is no inflation in prices then the amount required will be the same but higher in case inflation in prices. its effect on the project should be taken into account because the future costs and benefits are calculated in terms of present prices. Although effect of inflation in most calculations is neglected yet it should be possible to account for it by valuing the costs and benefits for an assumed inflation rate. Let us assume that C is the cash in hand today for buying an energy efficient device and f is the inflation rate in fraction. Assuming uniform inflation over the years, the investment for the same device after n years would be $C(1 + f)^n$. The present value of such money with interest/discount component is given by:

$$P_f = C (1 + f)^n/(1+i)^n \tag{16.3}$$

Where P_f is the present worth with inflation taken into account and i is the interest or discount rate in fraction. There may be several interesting cases depending upon the values of i and f. If $i = f$, no change in worth, year after year. If $i > f$, it may be better to save and not to buy the device now. If $i < f$, it may be better to buy the device now and not to save.

Analysis Period

One must distinguish between the actual life and the economic life of a project while considering time period of a project. The actual life of a well maintained project may be quite large. In economic terms, future costs and benefits occurring say after 30 years are insignificant. For example; the discounted benefit for the 12 per cent discount rate in the 30th year will be 0.033 times the actual benefits. It will further go down thereafter. As such, the economic life of a project is generally taken as 30 years. Another important issue in an economic evaluation is the financial life of the project often imposed by the financing organizations. A stakeholder having obtained the loan from a financial institution must pay back the loan in the pre-decided duration say in 10 years. In that event, cash flow must exceed the costs during the financial life. Although the project has a terminal value in this case, yet it has little meaning to the stakeholder who is concerned with cash flow during the financial life of the project. Short analysis period (*e.g.*, 10-15 years) is normally used for energy projects under unstable economic situation with high interest rates. Savings and expenses beyond 15 years become trivial due to heavy discounting.

Residual or Salvage Value

Salvage value is the value of the investment at the end of the analysis period normally given by the worth of the project at the end of analysis period. It is the expected market price at the end of the project. A rule of thumb is to assign a residual value @ 10 per cent of purchase price. It acts as a credit to the project in the final year. It is normally assigned to the costs with negative sign.

LCCA INDICATORS

Net Present Value (NPV), Savings-to-Investment Ratio (SIR) or what we call benefit-cost ratio (B:C ratio), and Internal Rate of Return (IRR) are the three major indicators that demonstrate the economic feasibility of a project. Besides payback period (PBP) and breakeven benefit (BEB) are also calculated to get greater insight of the financial feasibility. In most projects, four indices namely NPV, SIR, IRR and PBP are determined to assess the economic of financial viability of the project.

Net Present Value

Net Present Value is also called Net Present Worth (NPW). Basic rule of net present value is that the money in hand today is worth more than in the future. *NPV shows how much value will be generated during the life cycle of the project. It is the difference between the present values of benefits over the present values of costs.* Two assumptions are made to make things simple (a) all recurring costs are expressed as annual expenses incurred at the end of each year and (b) One time costs are incurred at the end of the year in which they are incurred. Discounting is used to calculate NPV. To estimate the NPV, cost involved and benefits realized on all future dates are assigned equivalent present values using discount factors. Higher the discount rate, the lower the present value of the future cash flows. The difference in net present value of benefits and net present value of costs gives the NPV of the project as follows.

$$\text{NPV of benefits} = \sum_{t=1}^{n} \frac{B_t}{(1+i)^t} = \sum_{t=1}^{n} B_t \times DF_t \qquad (16.4)$$

Here B_t is the benefit received in t^{th} year; n is the number of years; DF_t is the discount factor in t^{th} year (Eq. 16.1). Other factors are as described previously.

$$\text{NPV of costs} = \sum_{t=1}^{n} \frac{C_t}{(1+i)^t} = \sum_{t=1}^{n} C_t \times DF_t \qquad (16.5)$$

Here C_t is the costs incurred in t^{th} year. Other factors are the same as described previously.

Subtracting NPV of costs from NPV of benefits, we get net NPV as:

$$NPV = \sum_{t=1}^{n} \frac{B_t - C_t}{(1+i)^t} = \sum_{t=1}^{n} (B_t - C_t) \times DF_t \qquad (16.6)$$

A positive NPV shows the net amount a project is likely to generate in its lifetime. A negative NPV gives the net amount that is likely to be lost, if the project is implemented. As such, it brings out a project's absolute feasibility in terms of cash flow.

Savings-to-Investment Ratio (SIR)

The savings-to-investment ratio (SIR) is the same as the benefit to cost ratio. *It is the ratio of the present value of benefits* (*Eq. 16.4*) *divided by the present value of costs* (*Eq. 16.5*). It shows the value of the project relative to the investment. SIR as previously stated also considers time value of money both for the investments and net cash flows generated during the project life. If SIR > 1.0 then the project makes more money than its cost. If SIR < 1.0 then the project costs more than the money it makes.

Internal Rate of Return (IRR)

IRR is the rate of return which a project is likely to earn over its useful life. Practically, IRR is the discount rate at which NPV is zero and SIR or B-C ratio is one.

$$NPV = \sum_{t=1}^{n} \frac{B_t - C_t}{(1+IRR)^t} = 0 \tag{16.7}$$

It means it is the rate at which project just breaks even. The term internal refers to the fact that it is independent of economic environmental factors. To determine IRR, a trial and error procedure is used. Initially, a DR is assumed. If the NPV of the project is positive at this discount rate, the IRR of the project is more than the assumed discount rate. If the NPV is negative, the IRR is less than the assumed discount rate. Since it is difficult to pinpoint the rate at which NPV is zero, a guess is made once two discount rates at which NPV is positive and negative are known. IRR is then approximated by the following relation (Tripathi, 2007):

$$IRR = D_l + (D_h - D_l) \times \frac{N_l}{|(N_h - N_l)|} \tag{16.8}$$

Here, D_l is the lower discount rate; D_h is higher discount rate; and N_l and N_h are the NPVs at the lower and higher discount rate. A check on NPV is made with the assessed IRR using eq. (16.6). Minor adjustments might be needed in the assessed value. It is not possible to give a fixed value of IRR at which the project is viable. Microsoft excel sheet can be used to easily calculate the IRR rather than following the trial and error procedure. The calculated IRR is compared with the minimum acceptable rate of return (say prevailing market rate) or the project's cost of capital. Any project with an IRR greater than the project's cost of capital is profitable (and will have positive NPV). Companies often set *hurdle rates for IRR i.e. minimum IRR that the company will accept for implementing projects*. Hurdle rates are normally higher than the company's cost of capital. Therefore, only projects with higher profits than the capital cost are selected.

Payback Period (PBP)

All projects require some kind of investments and in turn generate some returns spread over the life of the project. *Payback period is the time period for an investment to generate sufficient incremental cash to recover its full initial capital outlay.* Thus, the PBP measures number of years it will take for the net undiscounted benefits to repay the investment. If the cash flow is uniform over the years, the payback period, P is given by

$$P = (I_i/E) \quad (16.9)$$

Where I_i is the initial investments and E is the annual net benefits.

If annual cash flows vary over the years, the payback period is determined when the accrued savings equal the initial investment costs. This happens when the cumulative cash flow balance becomes zero. The payback period is also categorized as simple and discounted. It is quite easy to compute the simple payback period but its major weakness is that it ignores the time value of money, *i.e.* it does not take into account the interest rate, inflation rate, escalation in cost, maintenance costs, *etc.* Secondly, it ignores net benefits that occur after the payback period and therefore, does not measure profitability. Nevertheless, simple payback period gives a quick estimate of how long it will take to recover the investment. Simple PBP is useful for projects where returns are quite quick. For example, if projects can payback in a year, there is little need to calculate discounted future values. For longer paybacks, simple PBP becomes inaccurate. In that case one should calculate discounted PBP in which discounted rather than simple cash flows are used in the calculations. *Thus, discounted PBP is time required for the cumulative savings to recover its initial investment cost and other accrued costs, taking into account the time value of money.*

Break Even Benefits (BEB)

Break even benefit is the minimal benefit required to meet the costs. It is calculated as follows (Smedema and Rycroft, 1983):

$$\text{BEB} = \text{NPV of costs} / \sum_{i=1}^{n} \text{discount factors} \quad (16.10)$$

For a project to be viable, the real undiscounted average benefits per year should exceed the BEB.

STEP-WISE PROCEDURE TO CALCULATE LCC

1. Determine costs of various items
2. Choose discount rate
3. Choose analysis period
4. Estimate residual value of equipment at end of service life
5. Determine benefits from main and by-products

6. Calculate present value of investments
7. Calculate life cycle cost and annual life cycle cost
8. Calculate present value of benefits
9. Calculate net present value (NPV)
10. Calculate savings-to-investment ratio
11. Calculate internal rate of return (IRR)
12. Calculate simple or discounted PBP
13. Calculate break even benefits

The following sections briefly discuss some of the important steps.

Life Cycle Costs

There are numerous costs associated with any project. The costs of a project can be categorized as tangible costs and intangible costs. *Tangible or direct costs are those costs that can be measured in financial terms and can be assigned a numeric value.* Some of these are land acquisition, purchase of equipment, civil construction and installation, operation, maintenance, and disposal at the end of the project. *Intangible costs are the costs that cannot be quantified directly in economic terms, but may still have a very significant business impact.* For example one may find it difficult to assign a numeric value to reduced green house gas emissions in implementing green technologies especially for small scale projects. For large scale projects, carbon credit earned can be assigned some value. Normally both intangible costs and benefits are left out of financial analysis, as it is often difficult to assign numbers. Nonetheless, it may be desirable to list all kinds of costs. Life cycle cost can be described in the form of an equation as follows.

$$\text{LCC} = \text{Capital} + \text{OM and R cost} + \text{Energy cost} + \text{Water cost} + \text{Other costs} - \text{Salvage value}$$

All costs have to be discounted. In that respect, *LCC is the total LCC in present-value (PV)*. Capital costs that include purchase, acquisition and construction costs need not be discounted, if incurred in the base year (zero). But, if these are spread over few years, even those costs incurred following the base year are discounted and added to cost made in the base year. Operation, maintenance and replacement (OM and R) costs include costs incurred on the operation, maintenance and equipment replacement, if any. For example, storage batteries for a solar PV system may need replacement say after every five years or so. All these costs need to be discounted as per year of incurring the cost. Other costs include contract costs and miscellaneous costs. Salvage value anticipated as receipt has to be converted to present worth for calculating LCC and assigned a negative value. LCC of various items of cost for a 5 MW solar based power generation project in Barmer (Rajasthan) is shown in Figure 16.1. It shows that PV panels and electric conversion and storage cost about 70 per cent of the total cost (Ranganath and Sarkar, 2021).

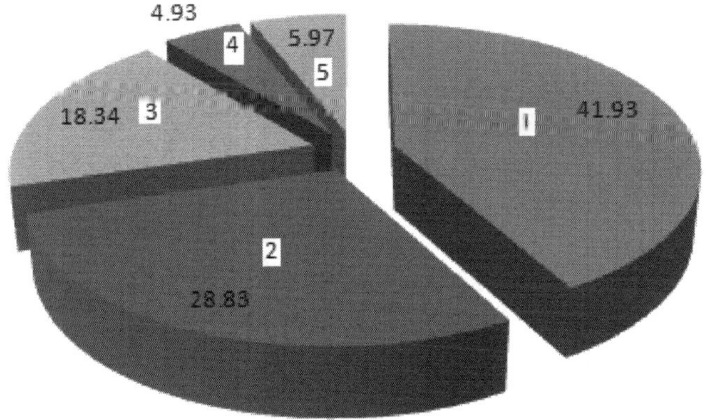

1 PV panels

2 AC and DC electric system

3 Civil/mech design/installation

4 Peripheral works, commissioning and testing

5 Pre project activities

Data indicate relative cost in percent for each item of cost

Figure 16.1. Distribution of cost for various items in life cycle cost of a 5 MW solar plant in Rajasthan.

Once we have calculated the LCC, we can easily find the Annual Life Cycle Cost (ALCC). If inflation is taken into account then ALCC is calculated by the following relation.

$$ALCC = LCC/[(1+f)/(i-f)][1-\{(1+f)/(1+i)\}^n] \quad (16.11)$$

Here f is the annual inflation rate in fraction. If inflation is neglected, it can be taken as zero such that

$$ALCC = LCC/[1/i][1-\{1/(1+i)\}^n] \quad (16.12)$$

Cost per unit product can be determined by the following relation.

$$\text{Cost per unit of product} = \text{ALCC/Annual production in desired units} \quad (16.13)$$

Desired units for electricity is kWh, for gas and water pumped, it is m^3 and so on.

Benefits

Similar to costs, all tangible and intangible benefits should be listed. In renewable energy programs, the benefits mainly arise from selling the energy and any by-product that is generated in the process. As an example, slurry, besides gas as the main product, is one of the by-products in biogas plants. Slurry can be sold as manure. Similarly benefits arising from carbon credits earned can be included in the benefits. Present value of benefits needs to be calculated as the benefits arise over the years. If the benefits arise uniformly from year one, the present value of benefits can be obtained by the following equation.

$$\text{Present value of benefits} = \text{benefits per annum} \times [(1/i)\{1\text{-}1/(1+i)^t\}] \quad (16.14)$$

Calculate Net Present Value (NPV)

The NPV is calculated by subtracting present value of total investments and costs from the present value of the total savings/revenues.

$$\text{NPV} = \text{NPV of benefits} - \text{LCC} \quad (16.15)$$

For a project to be profitable or financially feasible the NPV should be more than 0.

Calculate Savings-to-Investment Ratio (SIR)

The relative value, SIR (Savings-to-investment ratio) of the project is calculated by dividing present value of savings by present value of investments.

$$\text{SIR} = \text{NPV of benefits/LCC} \quad (16.16)$$

For the project to be profitable, SIR should exceed 1.0 although some projects which generate employment or otherwise take forward any societal need can be implements even if SIR is equal or less than 1.0.

Calculate Internal Rate of Return (IRR)

IRR is a hypothetical discount rate that results in SIR = 1.0 or NPV = 0. If the IRR ≥ the discount rate used in the analysis, the investment is worthwhile (economically feasible). A high IRR reveals more profit per unit of investment made.

The whole LCC analysis procedure is illustrated with an example of a home PV system.

Example 16.1

A PV system comprising panels, batteries to store electricity, and inverter to convert DC into AC is proposed to be installed. The information available is: cost of panels (C_p) = Rs. 50000/kW, cost of inverter (C_i) = Rs. 20000/kW, cost of batteries (C_b) = Rs. 25000 per kWh of storage, solar power installation kit including installation charges= 15000, annual O and M including miscellaneous cost (A_c) = 1000 per year from year 1 to year 25. Life of the panels, inverter and batteries is 25, 25 and 5 years respectively. If the interest rate I is 10 per cent, calculate the LCC, ALCC and the cost of electricity generated. Assume any other value(s) as per guidelines of analysis and calculate per unit cost of the electricity generated.

Capital cost in the base year or zero year = Rs. (50000 + 20000 + 25000 +15000) = Rs. 110000

Replacement cost of batteries = $[25000/(1+0.1)^5] + [25000/(1+0.1)^{10}] + [25000/(1+0.1)^{15}] +$

$[25000/(1+0.1)^{20}] = 34862$

Present value of annual cost (Eq.16.14) = $1000[(1/0.1)\{1-1/(1+0.1)^{25}\}] = 9080$

As per guideline, salvage value is taken as 10 per cent of purchase price of panels, inverter and batteries = 9500

Present value of salvage value in 25th year = $9080/(1+0.1)^{25} = 838$

Life cycle cost = Rs. (110000 + 34862+9080 - 838) = Rs. 153104

Annual life cycle cost (Eq.16.11) = $153104/[(1/0.1)\{1-1/(1+0.1)^{25}\}] = 16862$

Assuming that total kWh produced from 1 kW PV system is around 4-5 units say on an average = 4.5 units

Total kWh produced in 1 year = 300 × 4.5 = 1350 (Assuming PV system works for 300 days in a year)

Therefore, cost per kWh = Annual life cycle cost/(300× 4.5) =16862/(300× 4.5) = 12.50

Example 16.2

Calculate the NPV, SIR and payback period for the example 16.1 if annual saving in electricity bill is Rs. 20000 per annum.

We can proceed in a similar fashion as in example 16.1 but to illustrate that such kind of analysis can be made in excel or a spread sheet, we follow the tabular method of analysis for this example. Following step-wise procedure is followed.

- Column 1 shows the number of years varying from 1 onwards. First year is written as zero, the base year (See Table 16.2)
- In Column 2, all the costs incurred are entered year wise. The initial investment is shown against zero year while battery replacement cost is written in the respective years. Similarly the salvage value is written against year 25 with a negative sign. O and M including miscellaneous costs are entered in Column 3. The sum of the costs (sum of values in columns 2 and 3) are written in column 4
- The discount rate (DR) is entered in column 5. Year wise discount factors (DF = $1/(1+i)^n$) are calculated and entered in column 6
- Discounted cost for each year is calculated by multiplying total costs (column 4) by DF (column 6) and entered in column 7
- All benefits accruing due to projects are entered in column 8.

- Discounted benefits in column 9 are the values obtained by multiplying values in column 8 with DF in column 6
- Cash flow for each year is calculated by subtracting the total costs (Column 4) from net benefits (column 8). Note the first value is negative since initial investments in first year are higher than the net benefits, which is zero in this case

Table 16.2. Calculations of B:C ratio, NPW, and payback period for a home solar system

Year	*Cost*	*O and M cost*	*Total Cost*	*DR*	*DF*	*DC*	*Net Benefits*	*DB*	*Cash Flow*
1	*2*	*3*	*4*	*5*	*6*	*7*	*8*	*9*	*10*
0	110000	0	110000	10		110000			-110000
1	0	1000	1000	10	0.9091	909	20000	18182	**19000**
2	0	1000	1000	10	0.8264	826	20000	16528	**19000**
3	0	1000	1000	10	0.7513	751	20000	15026	**19000**
4	0	1000	1000	10	0.6830	683	20000	13660	**19000**
5	25000	1000	26000	10	0.6209	15523 + 621	20000	12418	**-600**
6	0	1000	1000	10	0.5645	565	20000	11290	**19000**
7	0	1000	1000	10	0.5132	513	20000	10264	**19000**
8	0	1000	1000	10	0.4665	467	20000	9330	19000
9	0	1000	1000	10	0.4241	424	20000	8484	19000
10	25000	1000	26000	10	0.3855	9638 + 386	20000	7710	-600
11	0	1000	1000	10	0.3505	351	20000	7010	19000
12	0	1000	1000	10	0.3186	319	20000	6372	19000
13	0	1000	1000	10	0.2897	290	20000	5794	19000
14	0	1000	1000	10	0.2633	263	20000	5266	19000
15	25000	1000	26000	10	0.2394	5985 + 239	20000	4788	-600
16	0	1000	1000	10	0.2176	218	20000	4352	19000
17	0	1000	1000	10	0.1978	198	20000	3956	19000
18	0	1000	1000	10	0.1799	180	20000	3598	19000
19	0	1000	1000	10	0.1635	164	20000	3270	19000
20	25000	1000	26000	10	0.1486	3716 + 149	20000	2972	-600
21	0	1000	1000	10	0.1352	135	20000	2704	19000
22	0	1000	1000	10	0.1228	123	20000	2456	19000

Year	Cost	O and M cost	Total Cost	DR	DF	DC	Net Benefits	DB	Cash Flow
1	2	3	4	5	6	7	8	9	10
23	0	1000	1000	10	0.1117	112	20000	2234	19000
24	0	1000	1000	10	0.1015	102	20000	2030	19000
25	-9080	1000	-8080	10	0.0923	-838 + 92	20000	1846	19000
			226920				500000		286600
						153104		181540	

O and M: Operation and maintenance, DR, DF, DC and DB are the discount rate, discount factors discounted costs and discounted benefits.

NPV = Discounted benefits – Discounted costs = ΣColumn 9 - Σcolumn 7 = 181540 – 153104 = 28436

SIR or Benefit: Cost ratio = ΣDiscounted benefits/ΣDiscounted costs = ΣColumn 9/Σcolumn 7 = 181540/153104 = 1.19

Payback period = 7 years (The initial investments are recouped in 7 years as shown in bold)

The average annual break even benefit is calculated as follows.

Present value of costs for 25 years = 153104

$$\sum_{i=1}^{25} \text{Discount factor} = 9.08$$

Annual break even benefit (Eq. 16.10) = 153104/9.08 = 16861

The real average benefits during the 25 years are 20000 (column 8/25), more than the break even benefits.

Looking at all the parameters, the project is financially viable and can be implemented.

OTHER APPLICATIONS

Life cycle cost analysis can also be used to calculate and prioritize the various alternative options or projects. In this case, LCC analysis for all options/projects is carried out in the same way as for any project discussed in the previous sections. It can also be used to decide whether to rehabilitate any old project or improve it with new more efficient technology. In this case PV values of the difference in investments, re-investments, operation and maintenance and other items of cost and benefits are calculated. The alternative that saves the most in terms of the total life-cycle savings will be the most favorable option.

INDIRECT BENEFITS

On-site renewable energy generation can produce significant benefits in terms of energy use along with many environmental, and economic benefits to local governments and communities. Some of the indirect socio-economic impacts of the renewable projects are given as under.

Use of wastelands: The farmers owning wastelands or less productive lands can offer their lands on lease to establish solar and wind parks. It will provide them an assured regular source of income.

Diversification of energy sources: the renewable energy sources help to diversify energy supplies reducing dependence on imported fossil fuels adversely impacting the national economy.

Employment generation: Renewable energy projects support economic growth through job creation. Skilled and unskilled human resource required for construction, installation, and maintenance of on-site renewable energy generation systems can create local jobs. According to NREL (2012) average jobs created per year (Full time equivalent (FTE)/year) for wind and solar energy projects is given as under.

During construction	
Large wind projects	44000-66000
Photovoltaic projects	8300-9700
During operation (Annual for lifetime of the system)	
Large wind projects	4500-4900
Photovoltaic projects	610-630

Inter-sectoral terms of trade: Resource mobilization from one sector to another is an indication of strengthening inter sectoral terms of trade and economic development. Although share of the industrial sector dominates the total cost, on-site renewable energy generation systems also require considerable amount of raw materials. Purchasing these materials from local businesses can increase local manufacturing and trading opportunities. Thus, investing in on-site renewable energy generation helps stimulate local, state, and regional economies by boosting ancillary industries. In the long run it may even strengthen the agriculture sector. As such, one can anticipate infrastructure development and boost to regional economy.

Environmental improvement: Thermal energy is one of the great polluter affecting both the human and animal population nearby besides affecting agricultural crops and water resources. Renewable energy reduces greenhouse gas emissions or other pollutants. The relative environmental pollution resulting from various sources of power is schematically shown in Figure 16.2. It shows that coal is the greatest polluter while green power (solar, wind and biomass) is the least polluter.

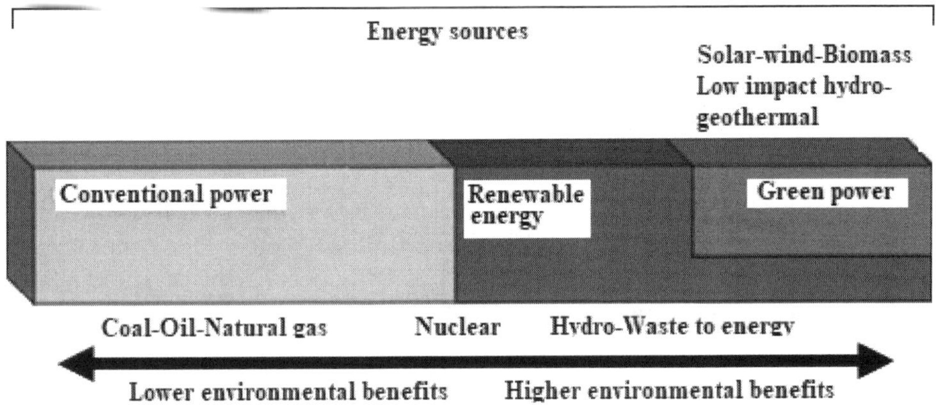

Figure 16.2. Energy sources and environment pollution potential.

Improved power quality and supply reliability: Quality electricity is characterized by delivery of required amount of energy in a consistent manner without much variation. Sags or spikes in voltage are considered poor power quality. The complex network of interconnections involved in generating, transmitting, and delivering grid-based electricity causes temporal variations in the characteristics of delivered power. Because on-site renewable energy generation systems have fewer interconnections (*e.g.*, transmission substations), electricity from these sources is likely to have higher power quality than electricity delivered through the grid from fossil fuel sources.

CASE FOR SUBSIDY

Not few years ago, the cost of solar systems was unaffordable to the common man. Even today it is on the higher side. Cost of 1 kW panel ranges from Rs. 45,000 to Rs. 70,000. Although, it will save money in the longer run, yet the cost of a solar system especially for pumping irrigation water amply justify some kind of subsidy to minimize the burden of high initial investments. Moreover, the following factors justify the grant of subsidy:

- People living in disadvantaged areas where grid electricity has not reached yet or is unreliable or farmers only getting supply during night can use renewable sources for making their life comfortable
- Fossil fuel is in short supply and India has to spend lot of foreign exchange in importing the crude oil, which puts a major drain on our economy

- India has planned to use half of the total energy requirement by renewable sources by 2050
- India is committed to achieve the target of carbon neutral country by 2070
- The renewable energy installations have some indirect benefits discussed in the previous section. Reduced load on state and national grids may ultimately ensure energy security of the nation, the commitment to supply electricity to all 24×7

It is noteworthy that Haryana provides solar subsidy of 40 per cent on systems in the capacity range of 1 kW to 3 kW and 20 per cent for systems more than 3 kW and up to 10 kW. Probably the quantum needs to be revisited at least for the higher range commonly needed by farmers to pump water.

FEASIBILITY REPORTING

Life cycle cost analysis (LCCA) is always presented in a report format. It has the following components.

- An executive summary in the beginning of the report
- Summary table showing the investment, LCC indicators and other indicators, if any
- A brief description of the project
- A brief explanation about the feasibility of the project
- Conceptual or schematic drawing showing the design intent of the project
- A site plan or layout of the proposed project or integration to the facility whichever is applicable
- Detailed LCCA of the project
- Recommendations

QUESTIONS (THEORY)

1. What do you understand by life cycle cost (LCC) analysis? Why one should carry out for the LCC analysis?
2. Name and describe various parameters used in LCC analysis.
3. Name and briefly describe various LCC indicators.
4. What do you understand by the time value of money? Differentiate between discounted and future value of money.
5. Briefly describe the steps-wise procedure of LCC analysis.

Glossary of Terms

Acid hydrolysis: The treatment of cellulosic, starch, or hemicellulosic materials using acid solutions (usually mineral acids) to break down the polysaccharides to simple sugars.

Acidogenic: Acid-forming- used to describe micro-organisms that break down organic matter to acids during anaerobic digestion process.

Acoustic pollution: Pollution resulting from acoustic noise by wind rotors, gearboxes and generators of the wind energy system.

Aerobic: Able to live, grow, or take place only where free oxygen is present.

Aerobic fermentation: Fermentation processes that require presence of oxygen.

Agricultural residue: Agricultural crop residues are the plant parts, primarily stalks and leaves, not removed from the fields with the primary food or fiber product. Examples include corn stover (stalks, leaves, husks, and cobs), wheat straw, and rice straw *etc.*

Alcohol: An organic compound with a carbon bound to a hydroxyl (hydrogen and oxygen, or –OH) group. Examples are methanol (CH_3OH), and ethanol (CH_3CH_2OH).

Alkaline hydrolysis: The use of solutions of sodium hydroxide (or other alkali) in the treatment of cellulosic material (wood) to break down cellulose to simple sugars.

Ampere (amp): A measurement of electric current.

Anaerobic: Living or active in an environment without oxygen.

Anaerobic digestion: Degradation of organic matter by microbes that produces a gas comprised mostly of methane and carbon dioxide in the absence of oxygen and usually under wet conditions.

Anhydrous: A material that does not contain water, either adsorbed on its surface or as water of crystallization.

Aquatic plants: The aquatic biomass resources, such as algae, giant kelp, other seaweed, and water hyacinth. Certain microalgae can produce hydrogen and oxygen while others manufacture hydrocarbons and a host of other products.

Ash content: Residue remaining after ignition of a sample determined by a definite prescribed procedure.

Ba: A mixture of A per cent biodiesel and (100- A) per cent petroleum diesel based on volume.

Bacteria: A small single-cell organism that ferment sugars to ethanol.

Bagasse: Residue remaining after extracting juice from sugar-containing plants like sugar cane.

Batch distillation: A process in which liquid feed is placed in a single container and the entire volume is heated (see batch fermentation and batch process).

Batch fermentation: Fermentation conducted from start to finish in a single vessel without addition to, or removal of, a major substrate or product stream until the process is complete (see batch distillation and batch process).

Batch process: Unit operation where one cycle of feedstock preparation, cooking, fermentation, and distillation is completed before the next cycle is started (see batch distillation and batch fermentation).

Beam radiation: The solar radiation propagating along the line joining the receiving surface and the Sun. It is also referred to as direct radiation. In other words solar radiation that has not been absorbed or scattered and reaches the ground directly from the sun is called beam radiation or direct radiation.

Betz's law: According to this law, any turbine can capture no more than 16/27 (59.3 per cent) of the kinetic energy in the wind.

Binder: The binders are materials that are added to biomass to enhance bonding and impart stable properties to the briquettes.

Bioconversion (or biochemical conversion): A general term describing the use of biological systems to transform one compound into another. Examples are digestion of organic wastes or sewage by micro-organisms to produce methane and the synthesis of organic compounds from carbon dioxide and water by plants.

Biodiesel: A biodegradable fuel for use in diesel engines that is produced through the transesterification of organically derived oils or fats. It may be used either as a replacement for or as a component of diesel fuel.

Bioenergy: Energy produced from biomass.

Biofuel intermediate: A biomass based feedstock that serves as a petroleum replacement in downstream refining, (*i.e.*, sugars, intermediate chemical building blocks, bio-oils, and gaseous mixtures). Algal biofuel intermediates include extracted lipids, lipid-extracted biomass, or bio-oil resulting from hydrothermal liquefaction.

Biofuels: Any fuel derived from biomass that is plant or algae material or animal waste is known as biofuel. In its most general sense, biofuels are all types of solid, gaseous and liquid fuels that can be derived from such biomass.

Biogas: A gaseous mixture of carbon dioxide and methane produced by the anaerobic digestion of organic matter.

Biogas upgrading: A process whereby a significant portion of the carbon dioxide, water, hydrogen sulphide and other impurities are removed from raw biogas (digester gas) primarily leaving methane.

Biomass: Biomass has been defined as living or recently dead organisms and any by-products of those organisms, plants or animals. It is an energy resource derived from organic matter. These include wood, agricultural waste, and other living-cell material, algae, sewage, and other organic substances that may be used to make energy through chemical processes.

Biopower: The use of biomass feedstock to produce heat through direct combustion of the feedstock, gasification and combustion of the resultant gas, or using other thermal conversion processes. Electric power is generated with engines, turbines, fuel cells, or other equipment.

Break even benefits (BEB): Break even benefit is the minimal benefit required to meet the costs.

Burner: A device that positions a flame in the desired location by delivering gas and air to the location so that controlled, continuous combustion is achieved.

Burner head: The part of a burner that incorporates the burner ports.

By-product: Leftover material or material generated as a result of an industrial process or as a breakdown product in a living system. In general, something formed during the making of something else.

Calorific value: It represents the energy content of a fuel, expressed in units, such as kilocalorie or Joule.

Capacity factor: Capacity factor for a wind turbine is defined as the actual annual output divided by the annual output that would be obtained if the device functioned at rated capacity for the entire year.

Carbohydrate: A class of organic compounds made up of carbon, hydrogen, and oxygen and having approximate formula $(CH_2O)_n$; includes cellulosics, starches, and sugars.

Carbon dioxide: Represented as CO_2, it is a colourless, odourless gas produced by the respiration and combustion of carbon-containing fuels. It is used by plants as food in the photosynthesis process.

Carbon monoxide: Represented as CO, it is a colourless, odourless, poisonous gas produced during incomplete combustion.

Catalyst: A substance that increases the rate of a chemical reaction without being consumed or produced by the reaction. Enzymes are catalysts for many biochemical reactions.

Cellulase: A family of enzymes that break down cellulose into glucose molecules.

Cellulose: Cellulose, a structural component of a green plant's cell walls, comprises carbon, hydrogen, and oxygen in the form of starches, proteins, and sugars. Also defined as a carbohydrate made of linked glucose molecules (a six-carbon sugar) that strengthen the cell walls of most plants.

Cetane (also called Hexadecane): An alkane hydrocarbon with the chemical formula $C_{16}H_{34}$. Consists of a chain of 16 carbon atoms, with 3 hydrogen atoms bonded to the 2 carbon atoms, and 2 hydrogen atoms bonded to each of the 14 other carbon atoms. Cetane is often used as a short term for cetane number, a measure of the detonation of diesel fuel. Cetane ignites very easily under compression. It is assigned a cetane number of 100, which serves as a reference for other fuel mixtures.

Cetane number: Often abbreviated as CN, it is a measurement of the combustion quality of diesel fuel during compression ignition. It serves as an expression of diesel fuel quality amongst of other measurements that determine overall diesel fuel quality.

Chemical energy: Chemical energy is the energy stored in the bonds of atoms and molecules. It is the energy that holds these particles together.

Cloud point (CP): The temperature at which oil starts to solidify.

Co-combustion: Co-combustion also known as co-firing or co-utilization is the burning of more than one fuel to produce power.

Co-firing: The use of a mixture of two fuels within the same combustion chamber.

Co-gasification: It is defined as the gasification of a mixture of waste/biomass and coal.

Co-generation: The technology of producing electric energy or any other form of useful energy (usually thermal) for industrial, commercial, or domestic heating or cooling purposes through the sequential use of the energy source.

Combustion: Combustion is a high-temperature exothermic redox chemical reaction between a fuel (the reductant) and an oxidant, usually atmospheric oxygen that produces oxidized, often gaseous products, in a mixture termed as smoke.

Combustion air: The air fed to a fire to provide oxygen for combustion of fuel.

Combustion efficiency: It is the efficiency of converting available chemical energy in the fuel to heat. In other words, it can also be stated as ratio of actual heat produced by combustion to total heat potential of the fuel consumed.

Compressed natural gas: Abbreviated as CNG, it is natural gas that has been compressed to 3,000 to 3,600 pounds per square inch, gauge (psig), usually for purposes of onboard fuel storage for natural gas vehicles.

Concentration ratio: It is used to determine the effectiveness of the concentrator and is defined as the ratio of the concentrator aperture to energy absorbing area of the receiver.

Conduction: Conduction is a process of heat transfer from a high temperature region to a low temperature region within a body or between different bodies which are in direct physical contact.

Convection: Heat transfer under the combined action of heat conduction and mixing motion is known as heat convection.

Conventional hydropower plant: A facility that uses available water from rivers, streams, canals, and reservoirs to produce electricity.

Cracking: Reduction of molecular weight by breaking bonds, which may be done by thermal, catalytic, or hydrocracking. Heavy hydrocarbons such as fuel oils are broken up into lighter hydrocarbons such as gasoline.

Crest: The highest point of a wave.

Current (Ocean): A continuous movement of ocean water in s specific direction. Currents have been likened to rivers within the larger ocean itself.

Dam: A barrier constructed across a waterway to control the flow or raise the level of water.

Degumming: Degumming is the process of removing phospholipids or gums from the crude vegetable oils.

Dehydration: The removal of the water from any substance.

Dehydrogenation: The removal of hydrogen from a chemical compound.

Dendro energy: Wood preferably grown in a sustainable manner used in electrical generation or thermal energy refers to dendro energy. Note that Dendro is a Greek word for wood but used for biomass in its general sense.

Densification: The densification has been defined as the process of compaction of biomass into a product of higher density than the original raw material. Sometimes, it is also referred as briquetting.

Diffuse radiation: The solar radiation scattered by aerosols, dust and molecules. It does not have any unique direction.

Digester: A biochemical reactor in which anaerobic bacteria are used to decompose biomass or organic wastes into methane and carbon dioxide.

Direct radiation: See beam radiation.

Disaccharides: The class of compound sugars that yields two monosaccharide units upon hydrolysis; examples are sucrose, maltose, and lactose.

Discount rate: Discount rate is the rate at which the future costs and benefits (losses) can be converted to their present value.

Discounted cost: The discounted cost shows the value of costs incurred at a specific time in the future that is equivalent in value to a specified sum today.

Distillers dried grains (DDG): The dried grain, a by-product of the grain fermentation process, which may be used as a high-protein animal feed.

Diversion project: A hydropower facility that does not require a dam but instead diverts river water from its course.

Downdraft gasifier: A gasifier in which the product gases pass through a combustion zone at the bottom of the gasifier.

Drop-in fuel: A substitute for conventional fuel that is completely interchangeable and compatible with conventional fuel. A drop-in fuel does not require adaptation of the engine, fuel system, or the fuel distribution network and can be used "as is" in currently available engines in pure form and/or blended in any amount with other fuels.

Dryer: An equipment to dry, normally used to dry agricultural products.

E10: A mixture of 10 per cent ethanol and 90 per cent gasoline based on volume.

Elastic energy: Elastic energy is a form of potential energy that is stored in an elastic object - such as a coiled spring or a stretched elastic band

Electrical energy: Electrical energy is the movement of electrons (the tiny particles that makeup atoms, along with protons and neutrons). Electricity and lightning are examples of electrical energy.

Elemental analysis: The determination of the mass fraction of carbon, hydrogen, nitrogen, oxygen, and additional heteroatoms (sulphur, chlorine, ash, *etc.*) in a sample. See ultimate analysis.

Emissions: Waste substances released into the air or water.

Endothermic: Chemical reactions in which reactants absorb heat energy from the surroundings to form products. Melting of ice is an example of an endothermic process because it absorbs heat from its surroundings.

Energy: Energy has been defined as the capacity to do work or capability to produce an effort. It is expressed in N-m or Joules.

Energy crop: A commodity (crop) grown specifically for its fuel value. These include food crops such as corn and sugarcane and nonfood crops such as poplar trees and switch grass.

Energy security: Energy security is defined as the uninterrupted availability of energy at an affordable price with easy access to all as per the following equation.

Energy security = Availability + Affordability + Accessibility + Acceptability

Enzymatic hydrolysis: Use of an enzyme to promote the conversion of a complex substance into two or more smaller molecules by reaction with water.

Enzyme: A protein or protein-based molecule that speeds up chemical reactions occurring in living things. Enzymes act as catalysts for a single reaction, converting a specific set of reactants into specific products.

Ester: A compound formed from the reaction between an acid and an alcohol. In esters of carboxylic acids, the -COOH group of the acid and the -OH group of the alcohol lose a water and become a -COO linkage.

Esterification: Esterification is any reaction typically between a fatty acid and an alcohol that results in the production of an ester.

Estuary: The area of water at the wide mouth of a river as it joins the sea.

Ethanol (CH_3CH_2OH): A colourless, flammable liquid produced by fermentation of sugars. Ethanol is used as a fuel oxygenate; also the alcohol found in alcoholic beverages.

Excess air ratio: Excess air ratio is the ratio between the locally available and the stoichiometric amount of combustion air.

Exothermic: A process or reaction that releases heat. For example, wood burning in the presence of oxygen is an example of an exothermic reaction.

Fast pyrolysis: Pyrolysis in which reaction times are short resulting in higher yields of certain fuel products ranging from primary oils to olefins and aromatics depending on the severity of conditions.

Fatty acid: A fatty acid is a carboxylic acid (an acid with a -COOH group) with long hydrocarbon side chains.

Feedstock: Any material used directly as a fuel, or converted to another form of fuel or energy product. Bioenergy feed stocks are the original sources of biomass. Examples of bioenergy feed stocks include corn, crop residue, and woody plants.

Fermentation: A biochemical reaction that breaks down complex organic molecules (such as carbohydrates) into simpler materials (such as ethanol, carbon dioxide, and water). Bacteria or yeasts can ferment sugars to ethanol.

Fixed bed: A collection of closely spaced particles through which gases move up or down in gasification or combustion.

Flash point: The temperature at which a combustible liquid will ignite when a flame is held over the liquid; anhydrous ethanol will flash at 51°F.

Flocculation: The process by which individual particles of algae aggregate into clot like masses or precipitate into small lumps. Flocculation occurs as a result of a chemical reaction between the algae and another substance, usually salt water. The term is also used in water industry where flocculants are used in the water purifying process.

Fluidized bed: A gasifier or combustor design in which feedstock particles are kept in suspension by a bed of solids kept in motion by a rising column of gas. The fluidized bed produces approximately isothermal conditions with high heat transfer between the particles and gases.

Fly ash: Small ash particles carried in suspension in combustion products.

Fossil fuels: Any material from which hydrocarbons may be extracted for energy applications. Few examples are coal, petroleum, natural gases, oils from shales and tar sands, methane hydrates *etc.*

Fuels (nuclear breeder reactors): These include U-238 and Th-232, which may be converted to fissile isotopes (*e.g.*, U-233, U-235, Pu-239, and Pu-241) as a result of neutron capture.

Fuels (nuclear fission reactors): The naturally occurring fissile isotope U-235, as well as Pu-239 and Pu-241 produced by neutron capture in U-238 and U-233 from Th-232.

Fuel-air ratio (FAR): It is the ratio between mass of fuel and the mass of air in the fuel-air mixture at any given moment.

Fuel wood: Wood primarily used for conversion to some form of energy especially in residences.

Fungi: Plant-like organisms having cells with distinct nuclei surrounded by nuclear membranes, incapable of photosynthesis. Fungi are decomposers of waste organisms and exist as yeast, mold, or mildew.

Furnace: An enclosed chamber or container used to burn biomass in a controlled manner to produce heat for space or process heating.

Gasification: Any chemical or heat process used to convert a feedstock to a gaseous fuel. Generally refers to thermo-chemical processes.

Gasifier: A device that converts solid fuel to gas.

Gas turbine: Gas turbine converts the energy of hot compressed gases (produced by burning fuel in compressed air) into mechanical power, which can be used to generate electricity. Sometimes it is also called a combustion turbine.

Geothermal energy: It refers to the use of the energy stored as heat (associated with volcanoes, hot springs, geysers, *etc.*) beneath the Earth's surface.

Global warming: A term used to describe the increase in average global temperatures due to the greenhouse effect.

Glycerin ($C_3H_8O_3$): A liquid by-product of biodiesel production. Used in the manufacture of dynamite, cosmetics, liquid soaps, inks, and lubricants.

Gravitational energy: It is the energy of place or position, a kind of potential energy.

Green diesel: A diesel fuel substitute made from renewable feedstocks by using traditional distillation methods. It is also known as renewable diesel.

Green gasoline: A liquid identical to petroleum-based gasoline, but is produced from biomass such as switch grass and poplar trees. It is also known as renewable gasoline.

Greenhouse effect: The heat effect due to the trapping of radiant energy in the earth's atmosphere by greenhouse gases due to which sun's radiant energy cannot be reradiated.

Greenhouse gas: A gas, such as water vapour, carbon dioxide, tropospheric ozone, methane, nitrous oxide, and low level ozone, which contributes to the greenhouse effect.

Head: Vertical change in elevation between the head water level and the tail water level expressed in either meters or feet.

Heating value: Heating value or heat of combustion is the amount of heat energy released by the combustion of a unit value of fuel.

Heat rate: The amount of fuel energy required by a power plant to produce one kilowatt-hour of electrical output. A measure of station thermal efficiency generally expressed in Btu per net kwh.

Hemicellulose: Hemicellulose is a polymer of five different sugars consisting of short, highly branched chains of sugars. In contrast cellulose is a polymer of only glucose.

Herbaceous biomass: It refers to all types of biomass with a grassy nature, such as cereal and hay.

Herbaceous plants: Non-woody species of vegetation, usually having low lignin content such as grasses.

Herbaceous energy crops: Perennial non-woody crops that are harvested annually, though they may take two to three years to reach full productivity.

Examples include: Switchgrass (*Panicum irgatum*), Reed canary grass (*Phalaris arundinacea*), Miscanthus (*Miscanthus × giganteus*), and Giant reed (*Arundo donax*).

Higher heating value (HHV): In contrast to lower heating value, If the water vapor in the flue gas produced from fuel hydrogen during oxidization is condensed within the conversion device, additional energy is released. The higher heating value (gross calorific value) includes this energy as well. See its contrast with lower heating value.

Hydrocarbon (HC): An organic compound molecule that only contains hydrogen and carbon.

Hydrocarbon emissions: In vehicle emissions, these are usually vapours of hydrogen-carbon compounds created from incomplete combustion or from vaporization of liquid gasoline. Emissions of hydrocarbons contribute to ground-level ozone.

Hydro cracking: A process in which hydrogen is added to organic molecules at high pressures and moderate temperatures; usually used as an adjunct to catalytic cracking.

Hydrogenation: Treatment of substances with hydrogen and suitable catalysts at high temperature and pressure to saturate double bonds.

Hydraulic retention time (HRT): HRT is the average time a 'volume element' of fluid resides in a reactor. It is computed from liquid-filled volume of an anaerobic digester divided by the volumetric flow rate of liquid medium.

Hydrolysis: A chemical reaction that releases sugars that are normally linked together in complex chains. Hydrolysis reactions break down the cellulose and hemicellulose in the biomass during ethanol production.

Hydropower: The use of water to generate electricity.

Hydrothermal liquefaction: In the hydrothermal liquefaction pathway, bio-oils are separated from water via high heat and pressure so they can be catalytically hydro-treated and converted to advanced hydrocarbon fuels. This pathway is especially suited for conversion of wet feed stocks, such as algae, because no drying step is required, and the high-quality oil produced is lower in oxygen than bio-oil derived from other conversion processes.

Inclined grate: A type of furnace in which fuel enters at the top part of a grate in a continuous ribbon. Then it passes over the upper drying section where moisture is removed, and descends into the lower burning section. Ash is removed at the lower part of the grate.

Indirect liquefaction: Indirect liquefaction is a process carried out in the presence of catalyst to convert non-condensable products and gases from pyrolysis or gasification into liquid products.

Inflation: Inflation is the rate of increase in prices over a given period of time.

Injector jet: A jet removable and/or adjustable by means of which a calibrated amount of gas is allowed to pass through an orifice.

Insolation: The term applies specifically to solar energy irradiation. The symbol H is used for insolation for a day. The symbol I is used for insolation for an hour (or other period if specified).

Intangible benefits: Intangible or indirect benefits are the benefits that cannot be quantified directly in economic terms, but may still have a very significant business impact.

Intangible costs: Intangible or indirect costs are the costs that cannot be quantified directly in economic terms, but may still have a very significant business impact.

Internal rate of return (IRR): IRR is the rate of return which a project is likely to earn over its useful life. Practically, IRR is the discount rate at which NPV is zero and SIR or B-C ratio is one.

Iodine value (IV): The iodine value is a measure of the total unsaturation of vegetable oils. It is also an indicator of their susceptibility to oxidation.

Irradiance: The rate at which radiant energy is incident on a surface per unit area of surface. The symbol G is used for solar irradiance, with appropriate subscripts for beam, diffuse, or spectral radiation. Its units are, W/m^2.

Irradiation: The incident energy per unit area on a surface is found by integration of irradiance over a specified time, usually an hour or a day. Also known as radiant exposure expressed in J/m^2.

Joule: Metric unit of energy, equivalent to the work done by a force of one Newton applied over a distance of one meter (= 1 kg m^2/s^2). One joule (J) = 0.239 calories (1 calorie = 4.187 J).

Kilowatt (kW): A measure of electrical power equal to 1,000 Watts. 1 kW = 3412 Btu/hr = 1.341 horsepower.

Kilowatt hour (kWh): A measure of energy equivalent to the expenditure of one kilowatt for one hour. For example, 1 kWh will light a 100-Watt light bulb for 10 hours. 1 kWh = 3,412 Btu.

Kinetic energy (KE): Kinetic energy is the energy of motion given by 0.5 mV^2 where m is the mass of the moving object and V is its velocity. Observable as the movement of an object, particle, or set of particles such as electromagnetic and radio waves, electrons, atoms, molecules, substances, and objects.

Lagoon: A shallow pond that can connect with or communicate with a larger body of water. A tidal lagoon is constructed of a 360 degree barrage in a tidal basin or estuary.

Landfill gas: Biogas produced from the natural degradation of organic material in landfills.

Life cycle cost analysis (LCCA): LCCA or the profitability analysis is an engineering economic analysis tool used to calculate the financial feasibility of a project.

Light energy: Light energy is a type of wave motion and is a form of electromagnetic radiation. It comprises photons produced when atoms of an object's heat up.

Lignin: A complex polymer that serves as a component of wood and vascular plants, making them firm and rigid.

Lipids: A diverse group of naturally occurring molecules containing hydrocarbons which make up the building blocks of the structure and function of living cells. Examples of lipids include fats, oils, waxes, and most of the non-protein membrane of cells.

Lower heating value (LHV): The LHV (net calorific value) characterizes the energy released during complete oxidation of dry biomass without counting the latent heat of the water produced.

Megawatt (MW): A measure of electrical power equal to one million Watts (1,000 kW). See Watt.

Melt point or pour point: Melt or pour point refers to the temperature at which the oil in solid form starts to melt or pour.

Metabolism: The sum of the physical and chemical processes involved in the maintenance of life and by which energy is made available to the organism.

Methane (CH_4): The major component of natural gas. It can be formed by anaerobic digestion of biomass or gasification of coal or biomass.

Methanol (wood alcohol) (CH_3OH): An alcohol formed by catalytically combining carbon monoxide with hydrogen under high temperature and pressure in the ratio of 1:2.

Micro-emulsion: Micro-emulsion is defined as a mixture of two immiscible liquids to form homogenous and stable solution stabilized by surfactant(s) or combination of surfactant(s).

Micro-organism: Any microscopic organism such as yeast, bacteria, fungi, *etc.*

Mtoe: It is a unit of energy defined as the amount of energy released by burning one tonne of crude oil.

Municipal solid waste (MSW): Any organic matter, including sewage, industrial, and commercial wastes, from municipal waste collection systems. Municipal waste does not include agricultural and wood wastes or residues.

Net present value (NPV): NPV is the value likely to be generated during the life cycle of the project or the present (today's) value of the project. It is the difference between the present values of benefits over the present values of costs.

Nitrogen fixation: The transformation of atmospheric nitrogen into nitrogen compounds that can be used by growing plants.

Nitrogen oxides (NO_x): A product of photochemical reactions of nitric oxide in ambient air; the major component of photochemical smog.

Nitrous oxide: N_2O, a greenhouse gas having 310 times the global warming potential of carbon dioxide.

Non-renewable resource: A resource that cannot be replaced after use. Although fossil fuels like coal and oil are in fact fossilized biomass resources, they form at such a slow rate, that in practice, they are non-renewable.

Nuclear energy: Nuclear energy is the energy stored in the nucleus of an atom - the energy that holds the nucleus together.

Ocean thermal power: Power derived by exploiting differences in water temperature between the surface and deeper parts of the Ocean.

Octane rating (octane number): A measure of a fuel's resistance to self-ignition, hence a measure of the antiknock properties of the fuel. Diesel fuel has a low octane rating, while gasoline and alcohol have high octane ratings, suitable for spark-ignition engines.

Oils and fats: Oils and fats are water insoluble, hydrophobic substance primarily composed of fatty esters of glycerol (triacylglycerol).

Organic compound: Compound containing carbon chemically bound to hydrogen. Often contain other elements (particularly O, N, halogens, or S).

Organic residuals: Surplus or waste organic materials derived from biomass, including, but not necessarily limited to products, production and processing wastes and by-products, remnants, bodily fluids, excrement or effluents derived from plants, animals, or micro-organisms that have been previously used, processed, preprocessed, or remains after their primary production or use has occurred.

Osmotic power: Power derived from water flowing between areas of low solute concentration and high solute concentration.

Oxygenate: A compound which contains oxygen in its molecular structure. Ethanol and biodiesel act as oxygenates when they are blended with conventional fuels. Oxygenated fuel improves combustion efficiency and reduces tailpipe emissions of CO.

Particulates: Fine liquid or solid particles such as dust, smoke, mist, fumes, or smog, found in air or emissions.

Payback Period (PBP): Payback period is the time period for an investment to generate sufficient incremental cash to recover its full initial capital outlay.

Penstock: A closed conduit or pipe for conducting water to a water wheel, turbine, or powerhouse.

Petroleum: Substance comprising a complex blend of hydrocarbons derived from crude oil through the process of separation, conversion, upgrading, and finishing, including motor fuel, jet oil, lubricants, petroleum solvents, and used oil.

Photo-bioreactor: A closed system of plastic or glass containers which allows for the input of light, nutrients, and carbon dioxide to cultivate phototrophic micro-organisms, such as microalgae.

Photo conversion: Conversion of light into other forms of energy by chemical, biological, or physical processes.

Photoelectric effect: The physical process by which a PV cell converts sunlight into electricity.

Photosynthesis: A complex process used by many plants and bacteria to build carbohydrates from carbon dioxide and water, using energy derived from light. Photosynthesis is the key initial step in the growth of biomass and is depicted by the equation: $CO_2 + H_2O + \text{light} + \text{chlorophyll} = (CH_2O) + O_2$.

Physico-chemical conversion: This may be used to produce liquid fuel by physical (*e.g.*, pressing) or chemical (*e.g.*, esterification) means. The important processes with a certain market potential is vegetable oil production from oil seed and esterification of this vegetable oil to fatty acid methyl ester as a substitute for diesel fuel.

Polymerization: It is any process in which relatively small molecules, called monomers, combine chemically to produce a very large chainlike or network molecule, called a polymer.

Polysaccharide: A carbohydrate consisting of a large number of linked simple sugar, or monosaccharide units. Examples of polysaccharides are cellulose and starch.

Potential energy: It is stored energy or the energy of position.

Pour point or melt point: See melt or pour point.

Power: Power is how fast energy is used or transmitted. Thus, power is the amount of energy divided by the time it takes to use the energy. Its unit is Watt, which is one joule per second of energy used.

Pre-combustion phase: The brief period after injection of the fuel into the combustion chamber but before the onset of ignition is termed as pre-combustion phase.

Process heat: Energy, usually in the form of hot air or steam, needed in the manufacturing operations of an industrial plant.

Producer gas: Fuel gas high in carbon monoxide (CO) and hydrogen (H_2) produced by burning a solid fuel with insufficient air or by passing a mixture of air and steam through a burning bed of solid fuel.

Proof: It is a unit to measure per cent of ethylene in the mixture during its production. 1 per cent is equal to 2 proof.

Proximate analysis: The determination, by prescribed methods, of moisture, volatile matter, fixed carbon (by difference), and ash. It does not include determinations of chemical elements or determinations other than those named.

Protein: A chain of up to several hundred amino acids, folded into a more or less compact structure. Amino acids are used by living matter in making proteins. In their biologically active state, proteins function as catalysts in metabolism and to some extent as structural elements of cells and tissues. Protein content in biomass (in mass per cent) can be estimated by multiplying the mass per cent nitrogen of the sample by 6.25.

Pumped storage plant: A hydropower facility with two reservoirs (one higher than the other) such that water from the lower reservoir is pumped into the higher reservoir and kept their until the power demand is high.

Pyrolysis: Pyrolysis is thermal decomposition of biomass occurring in the absence of oxygen producing a range of useful products. It involves breaking down of large complex hydrocarbons into relatively simpler molecules of gas, liquid and char.

Radiant energy: Radiant energy is electromagnetic energy that travels in transverse waves. Solar energy is an example of radiant energy.

Radiation: Heat exchange between two bodies or mediums, which are separated and are not in contact with each other, is called radiation heat transfer.

Reaction: A dissociation, recombination, or rearrangement of atoms.

Refractory lining: A lining capable of resisting and maintaining high temperatures. Usually ceramic.

Refuse-derived fuel: Fuel prepared from combustible portion of the municipal solid waste after removing noncombustible materials such as rocks, glass, and metals.

Renewable energy resource: An energy resource that can be replaced as it is used, including solar, wind, geothermal, hydro, and biomass. Municipal solid waste (MSW) is also considered to be a renewable energy resource.

Residues, biomass: By-products from processing all forms of biomass having significant energy potential.

Run-of-river project: A hydropower facility with turbines placed in fast flowing sections of rivers to generate power without impeding the river's natural flow.

Saccharide: A simple sugar or a more complex compound that can be hydrolyzed to simple sugar units.

Saccharification: A conversion process using acids, bases, or enzymes in which long-chain carbohydrates are broken down into their component fermentable sugars.

Salvage value: The value of the investment at the end of the analysis period normally given by the worth of the project at the end of analysis period.

Savings-to-investment ratio (SIR): It is the ratio of the present value of benefits divided by the present value of costs. It is same as benefit-cost ratio.

Scrubber: An air pollution control device that uses a liquid or solid to remove pollutants from a gas stream by adsorption or chemical reaction.

Silviculture: Theory and practice of manipulating forest vegetation to enhance fresh growth and establish new forest stand and wildlife habitat.

Slow pyrolysis: Thermal conversion of biomass to fuel by slow heating to less than 842°F (450°C), in the absence of oxygen.

Sludge: Bio-solids separated from liquids during processing. Sludge may contain up to 97 per cent water by volume.

Solar collector: A device that collects and/or concentrates solar radiations from the Sun, used to convert solar energy to thermal energy.

Solar cell: A solar cell also known as a photovoltaic (PV) is a semiconductor device that directly generates electricity using the sun light.

Solar constant: The solar constant (SC) is defined as the average radiation intensity falling on an imaginary surface perpendicular to the Sun's rays at the edge of the Earth's atmosphere.

Solar dryer: A solar dryer is a food dehydrating system that uses solar energy to remove the moisture from the produce.

Solar insolation: The total amount of solar energy received at a particular location during a specified time period, often in units of kWh/m^2- day.

Solar radiance: The solar radiance is the rate at which radiant energy is incident per unit area of the surface. It is an instantaneous power density in units of kW/m^2.

Solidity: Solidity is defined as the percentage of the area of the rotor, which contains material rather than air.

Sound energy: Sound is the movement of energy through substances in longitudinal (compression/rarefaction) waves. Sound has much less energy compared to any other form of energy.

Stalling: This term is commonly used in relation to wind turbines. When the lift force decreases and the drag force increases quite substantially, the phenomenon is known as stalling.

Starch: A molecule composed of long chains of a-glucose molecules linked together (repeating unit $C_{12}H_{16}O_5$). These linkages occur in chains of a-1,4 linkages with branches formed as a result of a-1,6 linkages. This polysaccharide is widely distributed in the vegetable kingdom and is stored in all grains and tubers.

Steam turbine: A device for converting energy of high-pressure steam produced in a boiler into mechanical power. This power can then be used to generate electricity.

Stoichiometric mixture: Mixture of gas and air in proportions determined by the theoretical air requirement.

Stover: The dried stalks and leaves of a crop that remain after the grain has been harvested.

Substrate: The base on which an organism lives or a substance acts upon (as by an enzyme).

Sunshine hours: It is the sum of the time intervals (in hours) during which the direct normal solar irradiance exceeds a threshold of 120 W/m^2.

TAG: An ester formed from glycerol and three fatty acid groups. Triglycerides are the main constituents of natural fats and oils.

Tangible benefits: Tangible or direct benefits are the benefits that can be measured in financial terms and can be assigned a numeric value.

Tangible costs: Tangible or direct costs are the costs that can be measured in financial terms and can be assigned a numeric value.

Tar: A liquid product of thermal processing of carbonaceous materials.

Thermal energy: Thermal energy (or heat) is the internal energy in substances- the vibration and movement of atoms and molecules within substances. Geothermal energy is an example of thermal energy.

Thermo-chemical conversion: The use of heat to chemically change substances to produce solid, liquid, or gaseous energy products under various conditions with some access to air. Gasification, pyrolysis, and charcoal production are the main examples.

Tidal barrage: A facility built like a dam that allows the tides to power turbines and generate electricity.

Tidal energy: A kind of ocean energy obtained from rising and falling tides that can be converted into useful forms of power, mainly electricity.

Tidal stream power: Hydropower derived from swift, steady ocean currents.

Tides: Tides are defined as the rise and fall of sea level caused by the gravitational pull of the moon and to some extent the sun on the Earth.

Tip speed ratio: It is defined as the ratio of the speed of the extremities of a wind turbine rotor to the speed of the free wind.

Total radiation: The sum of the beam and diffuse radiation, sometimes known as global radiation.

Total solids: The amount of solids remaining after all volatile matter has been removed from a biomass sample by heating at 105ºC to constant weight.

Transesterification: A process that includes chemical reactions of alcohols and triglycerides contained in vegetable oils and animal fats to produce biodiesel and glycerin.

Triglyceride: A combination of glycerol and three fatty acids. Most animal fats comprise primarily of triglycerides.

Trough: The lowest point of a wave.

Turbine: A machine with a series of curved blades or buckets that converts the kinetic energy of a moving fluid to mechanical power.

Ultimate analysis: The determination of the elemental composition of the organic portion of carbonaceous materials.

Vacuum distillation: The separation of two or more liquids under reduced vapour pressure; reduces the boiling points of the liquids being separated.

Volatile: A solid or liquid material that easily vaporizes.

Volatile matter: Those products, exclusive of moisture, given off by a material as a gas or vapour during combustion, determined by definite prescribed methods that may vary according to the nature of the material.

Waste streams: Unused solid or liquid by-products of a process.

Watt: The Watt (abbreviated W) is the International System of Units' (SI) standard unit of power (energy per unit time), the equivalent of one joule per second.

Wave crest: The highest point of a wave.

Wave energy: Kinetic and potential energy that results from the oscillation of water (waves) that can be converted to useful forms of power, mainly electricity.

Wave trough: The lowest point of the wave.

Wet scrubber: An air pollution control device used to remove pollutants by bringing a gas stream into contact with a liquid.

Wood: Wood is a heterogeneous, hygroscopic, cellular and anisotropic material. It consists of cells, and the cell walls are composed of micro-fibrils of cellulose (40–50 per cent) and hemicellulose (15–25 per cent) impregnated with lignin (15–30 per cent).

Xylose ($C_5H_{10}O_5$): A five-carbon sugar. A product of hydrolysis of xylan found in the hemicellulose fraction of the biomass.

Yeast: Any of various single-cell fungi capable of fermenting carbohydrates.

Objective Type Questions and Answers

QUESTION 1: FILL IN THE BLANKS

Chapter 1

1. Kilowatt is a unit of _________ and a kilowatt-hour is the unit of _________.
2. The process of release of nuclear energy when two or more than two smaller atoms are combined into a larger one is known as _________.
3. Electromagnetic energy that travels in transverse waves is known as _________ energy.
4. ___________ energy is due to the vibration and movement of atoms and molecules within the substances.
5. The energy output of the sun is assessed at _________ kW/y
6. 1 gigawatt is equal to ________ megawatts.
7. Approximately ______ per cent of the solar energy striking the Earth's surface is converted to kinetic energy of the wind.
8. ________ power is a form of hydropower that converts the energy obtained from tides into useful forms of power.
9. __________ is a kind of organic material made of plants and animals (micro-organisms).

10. Hydro power plants with generation capacity exceeding 30 MW are categorized as ________ scale power plants.

Chapter 2

11. Cellulose, the structural component of a green plant's cell walls comprises carbon, hydrogen, and ________.
12. Molecular weight of hemicellulose is ________ than cellulose.
13. Lignin is a major contributor to ________ yield during pyrolysis of biomass.
14. The amount of heat available in a fuel (kJ/kg) is known as its ________ value.
15. Direct compaction of biomass can be achieved through ________ briquetting with or without binder and ________ briquetting.
16. Non-glucosidic organic solutions such as resins can be used as a ______ in briquetting of biomass.
17. High pressure machines are used for residues having high ______ content.
18. Briquettes with __________ structure helps in efficient burning because of better aeration during burning of the fuels.
19. In a roller machine two closely rotating rollers rotate at the same speed but in ________ directions.
20. Charcoal briquettes produce ________ smoke compared to non-carbonized briquettes.

Chapter 3

21. _________ conservation is maintained throughout the whole process of combustion.
22. In stoichiometric combustion, all the carbon in the fuel forms __________ and all the hydrogen forms ___________ vapours.
23. Excess air ratio is the ratio between the locally available and the _________ amount of combustion air.
24. Incomplete combustion occurs when there is ___________ oxygen to completely convert the fuel to carbon dioxide and water.
25. __________ combustion methods have been devised to reduce emissions of nitrogen oxides (NO_x).
26. ________ furnaces are used for biomass materials such as wood chips or those containing relatively low ash content.
27. Burning of blended biomass with coal in the furnace of a conventional coal-fired plant is known as ___________.
28. Indirect co-firing is a combination of _________ and combustion.

29. AFBC stands for atmospheric __________ bed combustor.
30. PFBC stands for __________ fluidized bed combustor.

Chapter 4

31. Pyrolysis is thermal decomposition of biomass occurring in the absence of ________.
32. Pyrolysis at high temperatures greater than 800°C and at high heating rates mainly produces _________.
33. Slow pyrolysis takes several hours to complete and gives ________ as the main product.
34. Liquid bio-oil yield in flash pyrolysis may exceed ________ per cent
35. The optimum moisture content of the biomass for pyrolysis should be around _________ per cent.
36. In plasma pyrolysis, cold plasmas are generated at vacuum pressure, while thermal plasmas are generated at ____________ pressure.
37. Beehive oven is a kind of __________bed reactor.
38. Plenum chamber in a bubbling fluidized bed reactor is provided with _________ for uniform distribution of the gas.
39. Increase in pressure affects the temperature profile in the _________ flow reactor.
40. Biomass in an ___________ pyrolysis reactor is pressed against a hot surface either by centrifugal or mechanical forces.

Chapter 5

41. Syngas is strictly defined having only CO, _________, CO_2 and H_2O as its constituents.
42. Gasification is carried out in a ____________ environment at relatively high temperature
43. Equivalence ratio (ER) is defined as the actual air (supplied in the gasification) to the ______________ air required for combustion.
44. The process of pyrolysis products reacting with CO_2 as an oxidant to form CO, CO_2 and H_2 is known as _______ reforming
45. ____________ analysis is carried out to quantify chemical constituents of biomass namely carbon, hydrogen, and oxygen.
46. Syngas produced with air as the gasifying agent has _________ heating value.
47. In an updraft gasifier the __________ zone is at its bottom.
48. Minimum fluidization velocity results in drag forces on the particles, which is _________ to the weight of the particles in the bed that gives them a fluid like behaviour.

49. The calorific value of the high calorific value gas is more than ______ MJ/m^3.
50. Two major processes through which tar can be cracked into lower molecular weight compounds are __________ and catalytic processes.
51. Cooling of gas is necessary for directly using it in an internal combustion engine as well as to __________ its energy density.
52. Thermal cracking requires some additional energy as it takes place at ________ temperature than the gasification process.
53. ________ based catalysts are widely applied in catalytic cracking because of price advanatage.
54. Cyclone filter is capable of removing particles equal to or greater than _____ µm.
55. On an average use of producer gas in a compression ignition (diesel) engine can save about ________ per cent of diesel.

Chapter 6

56. Digestion is a biological process that occurs in the _________ of oxygen and in the presence of __________ organisms at temperatures ranging from 35 to 70°C and atmospheric pressure.
57. The biogas normally is a mixture having methane (CH_4) in the range of ______ to ______ per cent.
58. Calorific value of biogas on an average is about ________ MJ/m^3.
59. In the process of biogas formation, process of rupturing of chemical bonds using water is known as ________.
60. Acetogenesis is accomplished by facultative and anaerobic group of organisms collectively known as __________ formers.
61. Acid formers and _________ fermentors must remain in a state of dynamic equilibrium for efficient digestion.
62. An overloaded biogas reactor accumulates excessive _______ that eventually affects gas production.
63. Survival of bacteria is severely affected at pH above ________.
64. Addition of _________ increases gas production for materials low in C:N ratio.
65. Micro-organisms that don't require oxygen in order to survive or proliferate are known as ____________.

Chapter 7

66. Floating-drum type biogas plant has been developed by Khadi and _________ Industries Commission.

67. _________ biogas plant model has been developed by AFPRO (Action for Food Production)
68. Transportable biogas plant has a _________ digester.
69. Digester of a moveable biogas plant is buried below the ground in an ________ position to ensure two phase system.
70. HRT stands for ________ retention time.
71. Density of the feeding water + dung slurry on an average is taken as ______ kg/m^3.
72. As per general guidelines, 1 kg of dry cattle dung produces approximately _____ m^3 of biogas.
73. Biogas is about ________ per cent lighter than air.
74. Biogas production is anticipated to reduce significantly when the ambient temperature goes below ________ °C.
75. As far as possible, the pH of the contents in the digester tank should be close to ________.
76. Too high or variable feeding rate can _______ the pH of the digester contents.
77. Use of LPG stoves for biogas results in _________ burning efficiency.
78. The amount of energy released on complete combustion of biogas is about ________ MJ/m^3.
79. Lighting is excellent if the gas pressure is about ______ mm WG.
80. *Eucalyptus* leaves should be soaked in water for about _____ days prior to feeding them into the biogas plant.

Chapter 8

81. High viscosity of virgin vegetable oils may lead to poor atomization of the fuel, incomplete combustion, and __________ of the fuel injectors.
82. Triacylglycerols that are liquids at room temperature are generally called as ______ and those that are solids are called as _______.
83. Molecules of virgin oils crack because of their higher ________ weight.
84. The amount of heating energy released by the combustion of a unit value is known as its _______ value.
85. The temperature at which oil starts to solidify is known as _____ point.
86. Polyaromatic hydrocarbons (PAHs) emissions from neat vegetable oils are generally _________ than conventional diesel fuel emissions
87. The polymerization of the vegetable oil in the ring belt area results in _______ of the piston rings.
88. Viscosity of oil is a function of its _________.

89. The process of blending two non-miscible phases using a substance known as surfactant is known as __________.
90. Esterification is any reaction typically between a fatty acid and an _______ that produces an ester.

Chapter 9

91. Complex sugars comprise between two and ten __________ linked together.
92. Biodiesel is a generic name for fuels obtained by _________ of a vegetable oil.
93. FAME stands for fatty acid ________ ester.
94. FAEE stands for fatty acid _________ ester.
95. Surfactant(s) are added to a liquid to reduce its _________ tension, thereby increasing the liquid's spreading and wetting properties.
96. A by-product of the transesterification process is_______.
97. The stoichiometric ratio requires 1 mol of oil to react with ______ mol of alcohol to produce 3 mol of fatty acids methyl esters.
98. Oil and methanol are in single phase in the ___________ state.
99. The flash point of biodiesel is significantly __________ than diesel or gasoline.
100. Chemical formula for gasoline is _______.
101. About ____ kg of wheat is required to produce 1 L of absolute ethanol.
102. The rate of fermentation in starchy feed stocks considerably reduces at alcohol concentration of about _____ per cent.
103. _________ are used to break-up the complex cellulose-hemicellulose-lignin structure of cellulosic material before the fermentation process can begin to produce ethanol.
104. Mileage per liter covered with E85 (85 per cent ethanol mix) is ________ than with petro-diesel alone.
105. A ________ kilo liter per day second generation bio-refinery can create about 10 jobs in plant operations alone.

Chapter 10

106. The diameter of the sun is __________ km.
107. Wavelength of ultraviolet radiation is less than _______ nm.
108. Solar radiation received from the sun after its direction has been changed by reflection and scattering by the atmosphere is known as ___________ radiation.
109. A thermopile pyranometer consists of _______ concentric hemispheres.

110. Sunshine hours are the sum of the time intervals (in hours) during which the direct normal solar irradiance exceeds a threshold of ____ W/m^2.
111. Campbell-Stokes recorder is used to measures the ______ hours/day above the threshold value.
112. The units of thermal conductivity of the material is ________/(m-K).
113. Radiation heat transfer takes place most effectively in a ________.
114. The energy emitted by non-black bodies is _________ as compared to a black body when both are maintained at the same temperature.
115. Maximum energy emitted as radiation by a body at a particular temperature is governed by ____________ -Boltzmann law

Chapter 11

116. Flat plate collectors with ________ as the heat transfer medium are mostly used for agricultural drying.
117. Water-in glass evacuated tube solar collector is also known as open _________ tube system.
118. The vacuum in the evacuated heat pipe solar collector enables the liquid to boil at a __________ temperature than at normal atmospheric pressure.
119. ___________collectors are capable of increasing the intensity of energy flux from a low value to high values of the order of 10,000.
120. A parabolic dish collector in the shape of a parabolic dish concentrates solar radiations on a receiver placed at its _______ point.
121. In a power tower collector (PTC) large number of flat, sun-tracking mirrors known as _________ are used.
122. Solar chimney (or thermal chimney) is a form of __________ heating.
123. A solar chimney is generally placed on the __________ facing wall in the northern hemisphere.
124. The whole unit consisting of the chimney (storage) wall with vents and the glazing is known as _____________ wall named after its designer.
125. The vent at the top of the thermal chimney remains __________ during heating of the space.
126. An array of air type collectors is arranged on the roof/side of the building in ___________ space heating.
127. A common thumb rule to design the capacity of storage tank in domestic water heaters is to use _________ liters of storage for every m^2 of collector area.
128. Normally a __________ L capacity domestic water heating system is considered optimum for a family of 4 adult members.

129. The __________ of the plane mirror reflector on the underside of the lid is adjusted in such a way that the strong beam of sunlight falls on the glass cover.
130. Convection currents in non-convective solar ponds are controlled by dissolving __________ in the bottom layer of the pond making it too heavy to rise even after heating.

Chapter 12

131. Natural convection solar drying process is much ________ than the forced convection solar drying process.
132. Chimney in a dryer is used to create the draft force to improve natural ________.
133. An electric fan or blower is used in a ____________ circulation drying to force the air into or out of the dryer.
134. Solar tunnel dryer is based on __________ house effect.
135. Solar heat is stored in the form of specific and ________ heat in the thermal energy storage (TES) materials/devices.

Chapter 13

136. A photovoltaic (PV) cell is also known as _________ cell.
137. The physical process by which a photovoltaic cell converts sunlight into electricity is known as the ____________ effect.
138. Silicon has a total of ______ electrons.
139. Phosphorus and arsenic have __________ electrons in their outer energy level compared to four in the silicon.
140. Boron and gallium have ________ electrons in their outer energy level compared to four in the silicon.
141. When n-type and p-type semiconductors are placed in contact with each other to form a solar cell, a kind of _________ zone is created in an area around the junction.
142. Current produced by a solar cell is ___________ current.
143. Power output of a photovoltaic solar cell under load is approximately equal to _______ Watt DC
144. Solar panels are normally made by using the cells in a ________ combination.
145. Magnitudes of both the voltage and current ___________ in a series-parallel combination of solar cells.
146. The combination of the solar _________ together is known as the solar panel/array.

147. To absorb maximum sunlight, the panel movement along with the sun is controlled by the maximum power point __________.
148. PM-KUSUM stands for Pradhan-Mantri Kisan _______ Suraksha evam Utthaan Mahabhiyaan.
149. All __________ wire fencing systems are best suited for wet and greener regions where earth conductivity is good.
150. __________ return wire system is used in areas where ground is unable to conduct enough power because of its dryness or for some other reasons.

Chapter 14

151. The water stored in a reservoir behind a hydropower dam has ________ energy.
152. Run-of-the-river projects utilize ________ energy of the flowing water to generate electricity.
153. Pumped storage systems of generating electricity have _____ reservoirs as against one in the conventional dam based plants.
154. In a conventional dam based project potential energy of the water is converted into kinetic energy (motion) as the water flows down a large pipe called _________.
155. In a pumped hydroelectric energy storage schemes, about ______per cent of the energy is lost in a complete pumping/generation cycle.
156. The available potential energy from the tides is approximately proportional to the _________ of the tidal range.
157. The energy produced in a two way generation of electricity using a tidal barrage is usually _________ than one way generation but electricity is produced over a longer period of time.
158. Tidal power plant proposed by the West Bengal Renewable Energy Development Agency and the government of Gujarat were shelved because of their _______ investment costs.
159. The reservoirs created in tidal fences are similar to those of tidal barrages; except that they do not obstruct the entire ___________.
160. ___________ tidal power makes use of the fact that ocean tides don't operate strictly perpendicular to the shore, but also flow in parallel to the shore as well.
161. The highest point of a wave is the ________, while the lowest is the ______.
162. _________channel system is one of the ways to harness wave power of the Oceans

163. Floating device known as Pelamis system or an attenuator to generate electricity from the waves looks like a ________ snake.
164. Current Ocean Thermal Energy Conversion systems use a temperature difference of at least ________ °C to power a turbine to produce electricity.
165. ___________ power is derived from differences in salt concentrations between sea and river waters.
166. A device with a __________ membrane is used to harness the osmotic energy of the Ocean.
167. Atmospheric pressure is approximately equal to ________ Pa.
168. The energy from osmosis is popularly known as ________ energy.
169. As per revised estimates of IIT, Chennai, the theoretical power potentials of India for tidal and wave energy are ________ MW and ________ MW respectively.
170. A floating device named AquaBuOY is under development to harness ________ energy of the ocean.

Chapter 15

171. Wind energy basically converts the _________ energy of the wind to ___________ energy.
172. A wind energy system usually requires a minimum average annual wind speed of at least _______ km/h.
173. Other factors remaining the same, a wind generator will produce lesser power at higher altitudes because air pressure and air ________ is lower than at lower altitudes.
174. _______ wind power is expressed in watts per square meter of area swept by the rotating blades.
175. A wind turbine tower should generally be placed at ________ m away from the nearest obstacle.
176. The law that states that no turbine can capture more than 59.3 per cent of the kinetic energy in the wind is known as _______ law.
177. Relative wind angle is the sum of the angle of ________ and the ______ angle.
178. The phenomenon where lift decreases and the drag increases quite substantially is known as ________.
179. Vertical axis machines (rotors) do not need a _________ system.
180. The _______ turbine is similar to the Darrieus turbine except that it has straight and vertical blades instead of curved blades of Darrieus turbine.

Chapter 16

181. _________ rate is the rate used to convert future costs/benefits to their present value.
182. If discount and interest rates are same, it is possible to write: Discount factor = 1/_______ worth factor
183. Salvage value is usually taken as ______ per cent of purchase price.
184. Saving to investment ratio is same as _________ to cost ratio.
185. Minimum internal rate of return fixed by a company to undertake any project is known as _________ rate of return.

ANSWERS TO QUESTION 1

1. Power, Energy
2. Fusion
3. Radiant
4. Thermal
5. 2.8×10^{23}
6. 1000
7. Two
8. Tidal
9. Biomass
10. Large
11. Oxygen
12. Lower/Less
13. Char
14. Calorific/heat
15. Cold, Hot
16. Binder
17. Lignin
18. Honeycomb
19. Opposite
20. Less
21. Mass
22. Carbon dioxide, Water
23. Stoichiometric
24. Insufficient or not enough
25. Staged
26. Understoker
27. Co-combustion
28. Gasification
29. Fluidized
30. Pressurized
31. Oxygen
32. Gases
33. Biochar
34. 75
35. 10
36. Atmospheric
37. Fixed
38. Distributors
39. Entrained
40. Ablative
41. H_2
42. Reducing
43. Stoichiometric
44. Dry
45. Ultimate
46. Low
47. Combustion
48. Equal
49. 28
50. Thermal
51. Increase
52. Higher
53. Nickel
54. 5
55. 70-75
56. Absence, Anaerobic
57. 55, 65
58. 20
59. Hydrolysis
60. Acid
61. Methane
62. Acid
63. 8.5
64. Urea
65. Anaerobes
66. Village

67. Deenbandhu
68. Tubular
69. Inclined
70. Hydraulic
71. 1090
72. 0.04
73. 20
74. 20
75. 7.0
76. Reduce
77. Low/Reduced
78. 24
79. 150
80. 5
81. Coking
82. Oils, Fats
83. Molecular
84. Calorific
85. Cloud
86. Lower
87. Seizing
88. Temperature
89. Emulsification
90. Alcohol
91. Monosaccharides
92. Transesterification
93. Methyl
94. Ethyl
95. Surface
96. Glycerol
97. 3
98. Supercritical
99. Higher
100. C_8H_{15}
101. 3
102. 15
103. Enzymes
104. Lower/Less
105. 100
106. 1.39×10^6
107. 0.286
108. Diffuse
109. 2 (Two)
110. 120
111. Sunshine
112. W (Watt)
113. Vacuum
114. Less
115. Stefan
116. Air
117. Vacuum
118. Lower
119. Focusing
120. Focal
121. Heliostats
122. Passive
123. South
124. Trombe
125. Closed
126. Active
127. 50
128. 100
129. Angle
130. Salt
131. Slower
132. Convection
133. Forced
134. Green
135. Latent
136. Solar
137. Photoelectric
138. 14

139. 5
140. 3
141. Depletion
142. Direct
143. 1.38
144. Series-parallel
145. Increases
146. Modules
147. Tracker
148. Urja
149. Live
150. Earth
151. Potential
152. Kinetic
153. 2 (Two)
154. Penstock
155. 20
156. Square
157. Less
158. High
159. Estuary
160. Dynamic
161. Crest, Trough
162. Tapered
163. Sea
164. 20
165. Osmotic
166. Semi-permeable
167. 100000
168. Blue
169. 12455, 41300
170. Wave
171. Kinetic Mechanical/ Electrical
172. 15
173. Density
174. Specific
175. 100
176. Betz
177. Attack, Pitch
178. Stalling
179. Yaw
180. H-rotor
181. Discount
182. Future
183. 10
184. Benefits
185. Hurdle

QUESTION 2: SELECT THE CORRECT OR MOST APPROPRIATE ANSWER (A, B, C OR D) FROM THE FOLLOWING

Chapter 1

1. **Energy is found to exist as**
   (a) Light (b) Heat
   (c) Motion (d) All of these

2. **The process of a larger atom splitting into two or more smaller ones in nuclear energy is known as**
   (a) Fission (b) Fusion
   (c) Both (a) and (b) (d) None of these

3. **A rock sitting on top of a hill contains**
   (a) Kinetic energy (b) Potential energy
   (c) Chemical energy (d) None of these

4. **Normally, sound has much ________ energy than any other form of energy**
   (a) More (b) Equal
   (c) Less (d) Any of these

5. **Sunshine hours over India range from _____ hours per-annum**
   (a) 1000-1200 (b) 1500-2000
   (c) 2000-2500 (d) None of these

6. **The windmill might get damaged if wind speed exceeds**
   (a) 90 km/h (b) 50-60 km/h
   (c) 12-14 km/h (d) None of these

7. **The temperature of the sea water _________ with depth**
   (a) Increases (b) Equal throughout
   (c) Decreases (d) None of these

8. **The most appropriate statement that describes the characteristic(s) of non-conventional energy sources**
   (a) Inexhaustible (b) Do not cause environmental degradation
   (c) Are locally available (d) All of these

9. **Limitations of geothermal power plants include**
   (a) High initial cost to build
   (b) Waste water released can damage soil/aquifers
   (c) Can be a cause of mini tremors in the areas
   (d) All of these

10. **Which is not a kind of hydropower plant**
   (a) Fission of atoms
   (b) Based on harnessing run-of-the-river energy
   (c) Based on pumped storage schemes with two reservoirs at different levels
   (d) Harnessing of potential energy of water stored behind a dam

Chapter 2

11. **Which is not an agricultural residue?**
   (a) Corn stover (b) wheat straw
   (c) Cattle dung (d) rice straw

12. **Main structural cell constituent(s) in plant-derived biomass**
   (a) Cellulose (b) Hemicellulose
   (c) Lignin (d) All of these

13. **To characterize the biomass, bulk density and particle density of the biomass are used to calculate**
   (a) Porosity (b) Void ratio
   (c) Both (a) and (b) (d) None of these

14. **Which of the macromolecular substance of the biomass serves the most important role in binding during densification**
   (a) Lignin (b) Cellulose
   (c) Hemicellulose (d) None of these

15. **A binder is always needed for briquetting of biomass**
   (a) True (b) False
   (c) Neither (a) nor (b) (d) None of these

16. **The preparation of raw materials before briquetting comprises**
   (a) Drying
   (b) Size reduction
   (c) Mixing of raw materials/binder in correct proportions
   (d) All or whichever is required

17. **Which one of them is not a binder used in biomass densification**
    (a) Glucosidic materials (b) Sodium silicate
    (c) Straw (d) Bentonite
18. **Two rollers in a roller machine are**
    (a) One driving- one driven (b) Both driven
    (c) Both driving (d) None of these
19. **The quality of briquettes is affected by**
    (a) Moisture content of the feed material
    (b) Size range of the feed material
    (c) Shape of the feed material
    (d) All of these
20. **Charcoal briquettes are preferred because of**
    (a) Low smoke emissions compared to non-carbonized briquettes
    (b) Low ash content
    (c) Easy marketability (d) All of these

Chapter 3

21. **Which is the highest level of conversion**
    (a) Pyrolysis (b) Gasification
    (c) Combustion (d) None of these
22. **Primary air for combustion in air staging is normally kept _______ stoichiometric level**
    (a) Below/less than (b) More/higher than
    (c) Equal (d) None of these
23. **Energy released during incomplete combustion is _______ than during complete combustion**
    (a) More (b) Less
    (c) Equal (d) None of these
24. **NO_x abatement can be achieved by**
    (a) Selective non-catalytic reduction (SNCR)
    (b) Selective catalytic reduction (SCR)
    (c) Any of (a) and (b)
    (d) None of these

25. Direct co-firing of biomass may results in

(a) Corrosion
(b) Fouling
(c) Slagging
(d) All of these

Chapter 4

26. Pyrolysis at low temperatures less than 450°C and at slow heating rates mainly produces

(a) Biochar
(b) Gases
(c) Bio-oil
(d) None of these

27. Fast pyrolysis takes seconds to complete the process and is a preferred technology as it yields

(a) Equal or more than 60 per cent bio-oil
(b) Up to 20 per cent biochar
(c) Up to 20 per cent syngas
(d) All of these

28. In flash pyrolysis one must ensure

(a) Very high heating and heat transfer rates that requires finely ground feed
(b) Residence time of pyrolysis vapours in the reactor less than 1 s
(c) Quenching (rapid cooling) of the pyrolysis vapours to give bio-oil
(d) All of these

29. Chemically carbon monoxide is

(a) CH_4
(b) CO
(c) CO_2
(d) C

30. Chemical formula for cellulose is

(a) $CH_{10}O_5$
(b) $C_{10}H_5O_5$
(c) $C_6H_{10}O_5$
(d) None of these

31. On the basis of biomass movement (gas-solid contacting mode) pyrolyzers are classified as

(a) Fixed bed
(b) Moving (fluidized) bed
(c) Entrained bed
(d) All of these

32. Bubbling fluidized bed reactor comprises

(a) A plenum chamber
(b) Bubbling zone
(c) Disengagement zone
(d) All of these

33. Which is not the disadvantage of circulating fluidized bed reactor

(a) High heat transfer rates from sand

(b) Erosion problems inside the reactor

(c) Complex operation with high operating costs

(d) Need of large quantities of heat carriers (sand)

34. Cyclone reactor is another version of

(a) Rotating cone reactor (b) Ablative pyrolysis reactor

(c) Vacuum furnace reactor (d) None of these

35. Biomass pyrolysis is ________ in a double screw auger than a single screw auger pyrolyzer

(a) Equal (b) Slower

(c) Faster (d) None of these

Chapter 5

36. Producer gas from a gasifier is also known as

(a) Syngas (b) Generator gas

(c) Wood gas (d) All of these

37. The reaction of pyrolysis products with water as an oxidant to form CO, CO_2, and H_2 is known as

(a) Dry reforming (b) Steam reforming

(c) Both (a) and (b) (d) None of these

38. The reaction $C + CO_2 \leftrightarrow 2CO$ is known as

(a) Water-gas shift reaction

(b) Heterogeneous water-gas shift

(c) Boudouard equilibrium

(d) Steam reforming reaction

39. Which determination is not a part of proximate analysis of the biomass

(a) Hydrogen (b) Moisture

(c) Volatile matter (d) Ash

40. In which gasifier feed is introduced from the top and air from the bottom through a grate

(a) Downdraft (b) Updraft

(c) Cross draft (d) All of these

41. Imbert gasifier is a ________ gasifier

(a) Updraft
(b) Cross draft
(c) Downdraft
(d) None of these

42. Fluidizing medium in a fluidizd bed gasifier comprises

(a) Silica sand
(b) CaO
(c) olivine
(d) Any of these

43. The velocity at which a gasifying agent is introduced is faster in

(a) Bubbling bed gasifier
(b) Circulating bed gasifier
(c) Updraft gasifier
(d) Downdraft gasifier

44. Which of the following organic compound in the product gas is not a part of tar

(a) Phenolic ethers
(b) Gaseous hydrocarbons
(c) Poly-aromatic hydrocarbons
(d) Alkyl phenolics

45. Secondary tar removal can be achieved through

(a) Wet cleaning
(b) Dry cleaning
(c) Both (a) and (b)
(d) Activated carbon based catalysts

Chapter 6

46. Biogas can be naturally captured from

(a) Marshes
(b) Landfills
(c) Sewage
(d) All of these

47. The chemical formula for acetic acid is

(a) CH_3COOH
(b) $C_2H_4O_2$
(c) Any of (a) and (b)
(d) None of these

48. The raw cow dung is mixed with water usually in the proportion of

(a) 1:1
(b) 1:1.25
(c) 1:3
(d) Any of (a) and (b)

49. Retention time is given as

(a) Volume of digester divided by the feedstock added per day
(b) The feedstock added per day divided by the volume of digester
(c) Volume of digester multiplied by the feedstock added per day
(d) None of these

50. Carbon to nitrogen ratio is highest in

(a) Cow dung (b) Wheat straw

(c) Grass (d) Potato tops

Chapter 7

51. Which is not a fixed dome type biogas model

(a) Janata model (b) Deenbandhu model

(c) KVIC model (d) None of these

52. Characteristic(s) of bag type biogas plant approved by Bureau of Indian Standard

(a) Portable

(b) Made of UV resistant plastic

(c) Has a life of about 5 years

(d) All of these

53. Capacity of the plant is described by

(a) The quantity of gas produced in a day

(b) The capacity of the gas holder

(c) Capacity of the digester

(d) Sum of the capacity of gas holder and digester

54. On an average which of the following animals produces maximum dung per day

(a) Cow (b) Ox

(c) Buffalo (d) Pig

55. HRT stands for

(a) Hydrologic retention time (b) Hydraulic retention time

(c) Hydraulic removal time (d) Hydraulic regression time

56. Volume of gas holder is kept as

(a) 50 per cent of the capacity of the plant

(b) 80 per cent of the capacity of the plant

(c) 100 per cent of the capacity of the plant

(d) 1 per cent of the capacity of the plant

57. Bio-slurry helps in

(a) Stimulating the primary production of algae and micro-organisms in a fish pond

(b) Reducing the toxicity of aluminium in acid soils

(c) The production of bio fertilizers like Azolla and aquatic biomass Spirulina

(d) All of these

58. Reduced biogas yield can be due to

(a) Reduction in feedstock quality

(b) Temperature drop in the digester

(c) Inhibition by foreign compounds that might have entered the digester

(d) All of these

59. If the solid content of the slurry manure is about 5 to 10 per cent then it is categorized as

(a) Liquid manure (b) Slurry manure

(c) Semi-solid manure (d) Solid manure

60. Highest hydraulic retention time is recommended for which states

(a) Haryana and Punjab

(b) Andhra Pradesh and Kerala

(c) North Eastern states

(d) Equal for all states

Chapter 8

61. Which one is not a disadvantages associated with vegetable oils?

(a) Higher viscosity

(b) Lower volatility

(c) Gum formation due to oxidation and polymerization

(d) Seldom contain sulphur

62. Which one is not a non-edible vegetable oil?

(a) *Jatropha curcas* (b) Karanjia (*Pongamia glabra*)

(c) Palm oil (d) *Pongamia pinnata*

63. Most engine manufacturers designate a range of required CNs for their engines. In most cases, the range is around

(a) 45-50 (b) 35-40

(c) 100-120 (d) None of these

64. Iodine value is measured by the amount of iodine in grams that reacts with

(a) 100 gm of given oil
(b) 100 mL of given oil
(c) 1 kg of given oil
(d) 1000 mL of given oil

65. Flash point of neat biodiesel compared to petroleum based diesel fuel is

(a) Below the diesel fuel
(b) Nearly equal to the diesel fuel
(c) Above the diesel fuel
(d) Can be either below or above it

66. High viscosity alters the structure of the fuel spray coming out of the injectors in the form of

(a) Increased droplet size
(b) Decreased spray angle
(c) Increased spray penetration
(d) All of these

67. Viscosity of vegetable oils can be managed to a greater extent by

(a) Heating the oil
(b) Dilution with diesel fuel
(c) Emulsification
(d) Any of these

68. Transesterification is the reaction of

(a) An ester with an alcohol
(b) A fatty acid and an alcohol
(c) Any of (a) and (b)
(d) None of these

69. Modification in the engine to increase the injector valve opening pressure (IVOP) helps in

(a) Decreasing fuel spray droplet diameter
(b) Increasing spray velocity
(c) Increasing penetration distance
(d) All of these

70. Blend B10 means

(a) 10 per cent vegetable oil in diesel
(b) 10 per cent diesel in the vegetable oil
(c) 10 per cent vegetable oil, 10 per cent biodiesel in diesel
(d) None of these

Chapter 9

71. Cellulosic biomass comprises

(a) Cellulose (b) Hemicelluloses

(c) Lignin (d) All of these

72. Emission resulting from pure biodiesel compared to petroleum diesel have high

(a) Carbon-monoxide (b) Hydrocarbons

(c) NO_x (d) Particulate matter

73. The chemical formula for methanol is

(a) CH_3OH (b) C_2H_5OH

(c) C_2H_3OH (d) None of these

74. About ______ kg of glycerol is produced for every tonne of biodiesel manufactured

(a) 10 (b) 1000

(c) 100 (d) Nil

75. Fatty acid alcohol esters have ________ density than glycerol

(a) Higher (b) Lesser

(c) Equal (d) None of these

76. Bagasse, a waste material from sugarcane, can be used

(a) As a energy source

(b) To produce second generation ethanol

(c) Both (a) and (b)

(d) None of these

77. India is second largest producer of bagasse after

(a) China (b) Brazil

(c) USA (d) Mexico

78. The syngas (a mixture of carbon monoxide and hydrogen) is reacted over certain catalyst at elevated temperatures and pressures to form

(a) Methanol (b) Ethanol

(c) Both (a) and (b) (d) None of these

79. *Jatropha* oil seed cake can be used as

(a) Animal feed (b) Manure

(c) Both (a) and (b) (d) None of these

80. **Biodiesel has a density of**
   (a) ~ 0.87 g/cm^3
   (b) ~ 1.0 g/cm^3
   (c) ~ 1.2 g/cm^3
   (d) ~ 0.60 g/cm^3

Chapter 10

81. **The radiation coming from the Sun as viewed from the earth appears to be essentially equivalent to that coming from a __________ surface at 5762 K.**
   (a) Black
   (b) White
   (c) Grey
   (d) Blue

82. **The diameter of the earth is**
   (a) 1.39×10^6 km
   (b) 1.27×10^4 km
   (c) 1.50×10^8 km
   (d) None of these

83. **Nearly all the ultraviolet radiation is absorbed by**
   (a) Earth surface
   (b) Water vapours and carbon dioxide in the atmosphere
   (c) Ozone layer in stratosphere
   (d) None of these

84. **The value of the Sun's constant is ________ W/m^2.**
   (a) More than 1350 but less than 1400
   (b) Less than 1300 but more than 1000
   (c) Less than 1000 but more than 500
   (d) None of these

85. **Which one is not a greenhouse gas**
   (a) Carbon dioxide
   (b) Nitrous oxide
   (c) Methane
   (d) Oxygen

86. **A pyranometer limits the spectral response from 300 to ________ nanometers.**
   (a) 10
   (b) 1500
   (c) 2800
   (d) 3500

87. **National Institute of Solar Energy assessed the potential solar power in the country to be around 750 GWp, here GWp stands for**
   (a) Gigawatts probable
   (b) Gigawatts peak
   (c) Gigawatts projected
   (d) Gigawatts potential

88. **Transfer of heat between two regions/bodies maintained at different temperatures can occur through**
   (a) Conduction
   (b) Convection
   (c) Radiation
   (d) Any one or a combination of these

89. **Transfer of heat in radiation mode from two bodies kept at two different temperatures takes place (identify the most appropriate statement)**
   (a) Only from the body kept at higher temperature
   (b) Only from the body kept at lower temperature
   (c) From both the bodies towards each other but net balance is from the body at higher temperature
   (d) From both the bodies towards each other

90. **Units of convective heat transfer coefficient is given as**
   (a) W/m^2-K
   (b) W/m-K
   (c) WK/m^2
   (d) None of these

Chapter 11

91. **Solar collector converts solar energy to __________ energy**
   (a) Direct electric
   (b) Thermal
   (c) Both (a) and (b)
   (d) None of these

92. **Parabolic trough collector (PTC), is a kind of**
   (a) Line focus collector
   (b) Point focus collector
   (c) Contact focus collector
   (d) None of these

93. **Solar space heating can be achieved through**
   (a) Passive heating
   (b) Active heating
   (c) Both (a) and (b)
   (d) None of these

94. **The vent at the top of the chimney remains open only during**
   (a) Heating
   (b) Cooling and ventilation
   (c) Both (a) and (b)
   (d) None of these

95. **In a box type solar cooker a reflecting mirror is fitted on the __________ of the lid of the box.**
   (a) Top
   (b) Both on the top and underside
   (c) Underside
   (d) None of these

96. New design of solar still include

(a) Multiple solar distillation system

(b) Wick type solar distillation system

(c) Both (a) and (b)

(d) None of these

97. To improve the efficiency of solar distillation

(a) Keep the water depth in the basin low

(b) Use charcoal at the surface of water

(c) Make the side walls reflective

(d) Any or all of these

98. A shallow convective solar pond consists of pure water enclosed in a large narrow plastic bag so that

(a) It reduces convection

(b) It reduces evaporation

(c) It reduces both convection and evaporation

(d) None of these

99. Fresh water is used in (identify the most appropriate statement)

(a) Convective shallow solar pond

(b) Convective deep solar pond

(c) Both (a) and (b)

(d) Non-convective solar pond

100. Saline water is used in

(a) Non-convective solar pond

(b) Convective deep solar pond

(c) Convective shallow solar pond

(d) Both (b) and (c)

Chapter 12

101. Indirect type solar dryer with chimney comprises

(a) Solar air collector (b) Chimney

(c) Drying cabinet (d) All of these

102. A plenum chamber in an indirect type solar dryer with chimney is

(a) To allow the air to calm down

(b) Ensure an equal distribution of air pressure

(c) Ensure an equal distribution of air temperature before it enters the drying chamber

(d) All of these

103. In a mixed mode solar dryer

(a) The outlet vent is shifted to the side wall

(b) The roof is made transparent

(c) Energy is supplied through the solar radiation incident directly on the material to be dried and the circulation of pre-heated air from the solar air collector

(d) All of these

104. The volume of rock pile material required for thermal storage of required heat energy is dependent upon

(a) Specific heat of rock material

(b) Specific heat of air in case of air solar collector

(c) Both (a) and (b)

(d) None of these

105. Volume of air to be handled to dry 1 tonne of paddy depends upon

(a) Initial moisture content of the paddy

(b) Final moisture content of the paddy

(c) Latent heat of paddy

(d) All of these

Chapter 13

106. Solar cells made of Amorphous Silicon (a-Si) are used because these

(a) Have high efficiency

(b) Have low cost

(c) Have long life

(d) All of these

107. Out of 14 electrons in silicon, the outermost cell contains ______ electrons

(a) 4

(b) 5

(c) 14

(d) 8

108. The n-type silicon is made by doping it with atoms that have ________ electron in their outer shell than silicon.

(a) More

(b) Less

(c) Equal

(d) None of these

109. The p-type silicon is made by doping it with atoms that have ________ electron in their outer shell than silicon.

(a) More
(b) Less
(c) Equal
(d) None of these

110. The p-type silicon is a semiconductor rich in

(a) Holes
(b) Electrons
(c) Both (a) and (b)
(d) None of these

111. n-type silicon is a semiconductor rich in

(a) Holes
(b) Electrons
(c) Both (a) and (b)
(d) None of these

112. If more than two cells are connected in series, then

(a) Output current of the cell remains the same
(b) Voltage is doubled
(c) Both (a) and (b) are true
(d) None of these

113. If more than two cells are connected in parallel, then

(a) The voltage is doubled
(b) Magnitude of current is doubled
(c) Both (a) and (b)
(d) None of these

114. A PV module is a combination of

(a) PV cells
(b) PV panels
(c) Both (a) and (b)
(d) None of these

115. An inverter is used in a lantern, if.

(a) CFL bulbs are used
(b) DC LED bulbs are used
(c) AC LED bulbs are used
(d) Both (a) and (c)

116. Battery capacity is measured in

(a) kW
(b) kWh
(c) kW/h
(d) None of these

117. Main function of the energizer in a solar fence system is

(a) To produce short and sharp pulses of specified volts at regular intervals
(b) To charge the battery

(c) To save the battery from deep discharge

(d) None of these

118. In a all live wire fence system animals get the shock when

(a) They touch the live wires

(b) They touch the ground near the fencing system

(c) They touch the live wires and the ground at the same time

(d) None of these

119. In a earth or ground wire return system animals get the shock when

(a) They touch the live wires

(b) They touch the live wires and the ground at the same time

(c) They touch both the live and earth wires of the fence at the same time

(d) They touch the earth wire and ground at the same time

120. The intensity of shock ______ with increasing Joule rating of the energizer

(a) Increases

(b) Decreases

(c) Rating has no impact on the shock intensity

(d) None of these

Chapter 14

121. Oceanic hydropower comprises harnessing of

(a) Tidal energy

(b) Thermal gradients in the ocean

(c) Osmotic power

(d) All of these

122. The power output in a run-of-the-river scheme is proportional to the ______ of the velocity of the water.

(a) Square (b) Square root

(c) Cube (d) None of these

123. PHES stands for

(a) Pumped hydroelectric energy storage

(b) Potential hydroelectric energy storage

(c) Pumped head energy storage

(d) Pumped hydro electric storage

124. The tides are caused by the

(a) The spinning of the earth
(b) Gravitational pull of the moon
(c) Gravitational pull of the sun
(d) All of these

125. Tides have

(a) Only kinetic energy
(b) Only potential energy
(c) Both kinetic and potential energies
(d) Only electrical energy

126. Tidal energy is harnessed through

(a) Tidal barrages
(b) Tidal fences
(c) Tidal generators
(d) All of these

127. Support structure of the tidal stream generators can be

(a) Embedded in the bed of the sea
(b) It can be a piled structure
(c) Floating structure
(d) Any of these

128. Tides operate

(a) Strictly perpendicular to the shore
(b) Strictly parallel to the shore
(c) Both perpendicular and parallel directions
(d) None of these

129. Waves have

(a) Only kinetic energy
(b) Only potential energy
(c) Both kinetic and potential energies
(d) None of these

130. Abbreviation OTEC stands for

(a) Ocean temporary energy conversion
(b) Ocean Thermal Energy Conversion
(c) Optical thermal energy conversion
(d) Ocean thermal energy corporation

131. A working fluid in OTEC is used in

(a) Closed cycle
(b) Open cycle
(c) Hybrid cycle
(d) Both A and C

132. Desalinated water in OTEC is received as a by-product in

(a) Open cycle
(b) Closed cycle
(c) Both (a) and (b)
(d) None of these

133. A wave energy power plant is in operation in India in the state of

(a) Gujarat
(b) Kerala
(c) West Bengal
(d) Tamil Nadu

134. Oscillating water columns technology is used in harnessing

(a) Tidal energy
(b) Ocean thermal energy
(c) Wave energy
(d) Osmotic energy

135. How many high and low tides are guaranteed every day?

(a) 2, 2
(b) 3, 3
(c) 4, 4
(d) 1, 1

Chapter 15

136. Which is most appropriate statement when we say power in the wind is proportional to

(a) Area of windmill being swept by the wind
(b) Cube of the wind speed
(c) Air density
(d) All of these

137. If the wind velocity doubles then the power of the wind becomes

(a) Double of the original
(b) Four times of the original
(c) Eight times of the original
(d) Remains the same

138. Other conditions being similar, a wind generator produces _______ power in summer than in winter.

(a) Higher
(b) Lesser
(c) Equal, there is no change
(d) None of these

139. The theoretical maximum efficiency of a wind turbine is nearly

(a) 100 per cent
(b) 59 per cent
(c) 30 per cent
(d) 75 per cent

140. in a horizontal axis machine doubling the diameter of the turbine's rotor will increase the power output

(a) Double of the original
(b) Half of the original
(c) Four times of the original
(d) ¼ th of the original

141. The units of angular velocity are

(a) Radians/s
(b) 1/s
(c) Either (a) or (b)
(d) s/radians

142. Ratio of the coefficient of performance to coefficient of torque is equal to

(a) Betz coefficient (b) Tip speed ratio

(c) Furling wind speed ratio (d) None of these

143. Torque generated in a wind turbine is

(a) Proportional to radius of the rotor

(b) Proportional to square of the wind velocity

(c) Proportional to the rotor swept area

(d) All of these

144. if N is the rotation per second of a rotor of diameter, D meters, the tip speed (m/s) of the rotor is (R is the radius of the rotor)

(a) πND (b) πNR

(c) $\pi N/D$ (d) $\pi N/R$

145. Which of the following statement is incorrect

(a) Drag is in the direction of air flow

(b) Lift is perpendicular to the direction of air flow

(c) Generation of lift always develops a certain amount of drag

(d) Lift devices are generally less efficient than drag devices

146. Angle of attack can be decreased through

(a) Stalling a wind turbine (b) Furling a wind turbine

(c) Any of A and B (d) None of these

147. A tail vane is

(a) A passive yaw control mechanism

(b) An active yaw control mechanism

(c) Not a yaw control mechanism

(d) None of these

148. Which of the turbine is suitable for areas having low wind velocities

(a) Savonius wind turbine (b) Darrieus wind turbine

(c) Both are suitable (d) None of them is suitable

149. Mark the most appropriate statement for high-solidity machines

(a) Generate much higher starting torque

(b) Are less efficient than low-solidity machines

(c) Have low tip-speed ratio

(d) All of these

150. The wind speed at which rotor can be loaded is known as

(a) Cut-in wind speed
(b) Rated wind speed
(c) Furling wind speed
(d) Start-up wind speed

Chapter 16

151. Full form of LCCA is

(a) Life cycle cost analysis
(b) Long cycle cost analysis
(c) Life control cost analysis
(d) Life cycle conventional analysis

152. Discount factor, DF is given as (D_r is the discount rate in per cent and t is the year)

(a) $DF_t = (1+D_r/100)^t$
(b) $DF_t = 1/(1+D_r/100)^t$
(c) $DF_t = 1/(1+D_r)^t$
(d) $DF_t = (1+D_r)^t$

153. Negative value of net present value reveals

(a) Profit
(b) Neither loss nor profit
(c) Loss
(d) None of these

154. If discount rate is equal to internal rate of return then

(a) NPV = 0
(b) Benefit cost ratio =1
(c) Both (a) and (b)
(d) None of these

155. A company has invested Rs. 100000 in the base year. Annual cash flow is uniform @ Rs. 20000. The payback period is

(a) 5 years
(b) 4 years
(c) 0.2 years
(d) None of these

ANSWERS TO QUESTION 3

1. (d)
2. (a)
3. (b)
4. (c)
5. (b)
6. (a)
7. (c)
8. (d)
9. (d)
10. (a)
11. (c)
12. (d)
13. (c)
14. (a)
15. (b)
16. (d)
17. (c)
18. (a)
19. (d)
20. (d)
21. (c)
22. (a)
23. (b)
24. (c)
25. (d)
26. (a)
27. (d)
28. (d)
29. (b)
30. (a)
31. (d)
32. (d)
33. (a)
34. (b)
35. (c)
36. (d)
37. (b)
38. (c)
39. (a)
40. (b)
41. (c)
42. (d)
43. (a)
44. (b)
45. (c)
46. (d)
47. (c)
48. (d)
49. (a)
50. (b)
51. (c)
52. (d)
53. (a)
54. (c)
55. (b)
56. (a)
57. (d)
58. (d)
59. (b)
60. (c)
61. (d)
62. (c)
63. (a)
64. (b)
65. (c)
66. (d)
67. (d)
68. (a)
69. (d)
70. (a)
71. (d)
72. (c)
73. (a)
74. (c)
75. (b)
76. (c)
77. (b)
78. (c)
79. (c)
80. (a)
81. (a)
82. (b)
83. (c)
84. (a)
85. (d)
86. (c)
87. (b)
88. (d)
89. (c)
90. (a)
91. (b)
92. (a)
93. (c)
94. (b)
95. (c)
96. (c)
97. (d)
98. (b)
99. (c)
100. (a)
101. (d)
102. (d)
103. (d)
104. (c)
105. (d)
106. (d)
107. (a)
108. (a)
109. (b)
110. (a)
111. (b)
112. (c)
113. (b)
114. (a)
115. (d)
116. (b)
117. (a)
118. (c)
119. (c)
120. (a)

121. (d)
122. (c)
123. (a)
124. (d)
125. (c)
126. (d)
127. (d)
128. (c)
129. (c)
130. (b)
131. (d)
132. (b)
133. (b)
134. (c)
135. (a)
136. (d)
137. (c)
138. (b)
139. (b)
140. (c)
141. (c)
142. (b)
143. (d)
144. (a)
145. (d)
146. (b)
147. (a)
148. (a)
149. (d)
150. (a)
151. (a)
152. (b)
153. (c)
154. (c)
155. (a)

References

Abu-Aligah, M. 2011. Design of photovoltaic pumping system and compare it with diesel powered pumps. Jordan Journal of Mechanical and Industrial Engineering. 5: 273-280.

Acharya, S.K., Mohanty, M.K. and Swain, R.K. 2009. Kusum oil as a fuel for small horse power diesel engine. International Journal of Engineering and Technology. 1: 219-223.

Andrade, E., Hernández, J. and Arredondo, C. 2016. Osmotic energy, a little-known renewable energy source. Int. J. of Energy Prod. and Mgmt. 1: 233-244.

Asamoah, B., Nikiema, J., Gebrezgabher, S., Odonkor, E. and Njenga, M. 2016. A Review on Production, Marketing and Use of Fuel Briquettes. Resource Recovery and Reuse Series-7. International Water Management Institute, (IWMI), Colombo. 41 p

Basu, P. 2018. Biomass Gasification and Pyrolysis: Practical Design and Theory. Elsevier, Amsterdam, The Netherlands.

Bettis, B.L., Peterson, C.L., Driscoll, D.J. and Peterson, E.L. 1982. Fuel characteristics of vegetable oil from oilseed crops in the Pacific North-West. Agronomy Journal. 74: 335-339.

Bhavanam, A. and Sastry, R.C. 2011. Biomass gasification processes in downdraft fixed bed reactors: a review. International Journal of Chemical Engineering and Applications. 2. 425 133.

BIS. 2002. Biogas stove-specification (Second Revision). IS 8749:02. Bureau of Indian Standards (BIS), *Manak Bhavan*, New Delhi, India. 1-18.

BIS. 2016. Indian Standard Biodiesel (B100) Blend Stock for Diesel Fuel - Specification. IS 15607 (First revision). Bureau of Indian Standards, *Manak Bhawan*, New Delhi. 5 p.

BIS. 2022. Design, Construction, Installation and Operation of Biogas (Biomethane) Plant- Code of Practice. IS 9478 (Third Revision). Bureau of Indian Standards, *Manak Bhavan*, New Delhi. 33 p.

Buckman, N. 2015. Wind Turbine Aerodynamics: Theory of Drag and Power. https://nbuckman.scripts.mit.edu:444/homepage/wp-content/uploads/16/03/WindTurbine _Paper2-1.pdf

Demirbas, A. 2002. Biodiesel from vegetable oils via transesterification in supercritical methanol. Conversion Mgmt. 43: 2349–2356.

Goldemberg, J. (Ed.). 2000. World Energy Assessment: Energy and the Challenge of Sustainability. UNDP, New York.

Goutham, K. and Siva Krishna, C. 2019. Solar pond technology. International Journal of Engineering Research and General Science. 1: 1-10.

Gupta, S. K. 2019. Drainage Engineering: Principles and Practices. Scientific Publishers, Jodhpur. 531 p.

Hartung, H. and Pluschke, L. 2018. The benefits and risks of solar-powered irrigation - a global overview. FAO Land and Water Division. www.fao.org.

https://am.gallagher.com/-/media/Bynder/Animal-Management/Document/ALL-GGL-Brochures/Introduction-to-Electric-Fencing-Brochure-original.pdf

https://coabnau.in/uploads/16442346 _Ag.Engg4.3Lacture-Copy(2).pdf. Lecture Note: Renewable Energy and Green Technology.

https://mnre.gov.in/img/documents/uploads/file_f-1608040317211.pdf

https://snscourseware.org/snsctnew/files/1580286076.pdf. Construction details of KVIC biogas plant. Opened on 30.05.22.

https://www.canr.msu.edu › *Introduction to Biodiesel Production, Chapter 2: 22p.* opened on 30.06.2022.

https://www.ceew.in/sites/default/files/CEEW-Solar-for-Irrigation-Deployment-Report-7Jan18_0.pdf

https://courseware.cutm.ac.in/courses/renewable-energy-green-technology/ by Dash, A.R. Renewable Energy and Green Technology.

https://www.vssut.ac.in/lecture_notes/lecture1428910296.pdf. Lecture Notes on Renewable Energy Sources. Veer Surendra Sai University of Technology Burla.

ICAR (Indian Council of Agricultural Research). 2016. Renewable Energy. 124 p. www.agrimoon.com.

ICID. 2019. Solar Powered Irrigation Systems in India: Lessons for Africa through a FAO Study Tour: Draft Report. International Commission on Irrigation and Drainage, New Delhi. 136 p.

IEA (International Energy Agency). 2021. Hydropower Special Market Report: Analysis and Forecast to 30. International Energy Agency, France. www.iea.org, opened on 17.11.22.

IPCC (Intergovernmental Panel on Climate Change). 2011. Special Report on Renewable Energy Sources and Climate Change Mitigation. Cambridge University Press, Cambridge, United Kingdom and New York, NY, USA. 230 p.

Joshua O.S., Ejura G.J., Bako I.C., Gbaja I.S. and Yusuf Y. I. 2014. Fundamental principles of biogas product. International Journal of Scientific Engineering and Research (IJSER). 2: 47-50.

Kumar, S., Kumar, A. and Kumar, S. 2021. Renewable Energy and Green Technology. 112 p. www.agrifair.in Opened on 27.02.2023.

Lingayat, A.B., Chandramohan, V.P., Raju, V.R.K. and Meda, V. 2020. A review on indirect type solar dryers for agricultural crops – Dryer set-up, its performance, energy storage and important highlights. Applied Energy. 258: 1-22.

Loeb, S. and Norman, R. 1975. Osmotic power plants. Science. 189: 654–655.

Min, H.S. 2018. Renewable Energy and Wastewater Treatment. Ideal International E – Publication Pvt. Ltd. Indore (India)

Mishra, D. and Saikhedkar, N.K. 14. A study and theoretical analysis of evacuated tube collectors as solar energy conversion device for water heating. Advance Physics Letter. 1: 30-39.

MNRE. 2009. Ministry of New and Renewable Energy, Govt. of India, New Delhi. www.mnre.gov.in/biomassresources

Murthy, G.R.R. and Pandey, K.P. 2000. Solar Ponds: a Perspective from Indian agriculture. Department of Agricultural and Food Engineering, Indian institute of technology, Kharagpur, India.

NEED (National Energy Education Development Project). 2014. Energy of Moving Water: Student Guide. Manassas, VA, USA. 64 p.

Nijaguna, B.T. 2006. Biogas Technology. New Age International Private Limited, New Delhi, India. 231-246.

NREL. 2012. Preliminary Analysis of the Jobs and Economic Impacts of Renewable Energy Projects Supported by the 1603 Treasury Grant Program. Available: http://www.nrel.gov/docs/fy12osti/52739.pdf.

Patel, V., Pandit, S. and Chandrasekhar, K. 2017. Basics of methanogenesis in anaerobic digester. Microbial Applications. 2: 291-314.

Pathak, H., Bhatia, A. and Jain, N. 2010. Inventory of greenhouse gas emission from agriculture. Ministry of Environment and Forests. Govt. of India.

Punter, A. 2011. Solar Thermal Energy. Practical Action. The Schumacher Centre, Warwickshire, Reino Unido. 10 p.

Rajendran, K., Aslanzadeh, S. and Taherzadeh, M.J. 2012. Household, biogas digesters-a review. Energies. 5: 2911-2941.

Rajvanshi, A.K. 1986. Biomass gasification. In: Alternative Energy in Agriculture (Goswami, D.Y. Ed.). Vol. II. CRC Press. 83-102.

Ramachandra, T.V., Jain, R. and Krishnadas, G. 2011. Hotspots of solar potential in India. Renewable Sustainable Energy Rev. 15: 3178–3186.

Ranganath, N. and Sarkar, D. 2021. Life Cycle Costing Analysis of Solar Photo Voltaic Generation System in Indian Scenario. International Journal of Sustainable Engineering. 14: 1698-1713.

Reed, T.B. and Das, A. 1988. Handbook of Biomass Downdraft Gasifier Engine Systems. Solar Energy Research Institute, Colorado, USA. 140 p.

RFA (Renewable Fuel Association). 2017. Ethanol Industry Outlook: Pocket Guide. In: Building Partnerships and Growing Markets. Renewable Fuel Association, Washington, DC.

Senthil, M. and Jaikumar, M. 2014. A comprehensive study on performance, emission and combustion behavior of a compression ignition engine fuelled with WCO (waste cooking oil) emulsion as fuel. Journal of the Energy Institute. 87: 263-271.

Serrano, J.M. 2018. Other renewable energies. In: Energy and Telecommunications: Lecture Notes. University of Cantabria, Santander. 1-29.

Sharma, D. and Samar, K. 2016. Frequently asked questions on biogas technology. Biogas development and training centre, Department of Renewable Energy Engineering, College of Technology and Engineering, Maharana Pratap University of Agriculture and Technology, Udaipur (Raj.), 1–62.

Simeon, Iliev. 2018. Comparison of ethanol and methanol blending with gasoline using engine simulation. In: Bio-fuels - Challenges and Opportunities (Mansour Al Qubeissi, Ed.). IntechOpen, DOI: 10.5772/intechopen.81776.

Smedema, L.K. and Rycroft, D.W. 1983. Land Drainage: Planning and Design of Agricultural Drainage Systems. Cornell University Press, Ithaca, New York. 375p.

Spath, P.L. and Dayton, D.C. 2003. Preliminary Screening–Technical and Economic Assessment of Synthesis Gas to Fuels and Chemicals with Emphasis on the Potential for Biomass-Derived Syngas. National Renewable Energy Laboratory: Golden, CO, USA.

Subramanian, P. and Sampathrajan, A. 2020. Renewable Energy Technology. Agrimoon.com. 124 p.

TEDIC (Tata Energy Documentation and Information Centre). 1985. Biogas Technology: an Information Package. Tata Energy Research Institute, Bombay. 189 p.

TERI (The Energy Resources Institute). Undated, Salt-gradient solar ponds–Salt of the Earth. http://www.teriin.org/tech_solarponds.php/November 2022. Accessed December 2022.

Tripathi, R.S. 2007. Economic analysis of subsurface drainage projects. In: Agricultural Land Drainage: Reclamation of Waterlogged Saline Lands [Ambast, Gupta, and Singh (Eds.)]. Central Soil Salinity Research Institute, Karnal, India. 222-231.

Uppal, A. 2019. Geo-thermal and ocean energy technologies. Yojana. 24-28.

Venkatachalam, P. and Rao, S. 2020. Renewable Energy Sources. Agrimoon. com. 200 p.

Verma, D. K. and Pandey, A. Undated. Renewable Energy and Green Technology. Jawaharlal Nehru Krishi Vishwavidyalaya, Jabalpur. 106 p.

Verma. R.K. and Sahu, R.M. 2019. Deenbandhu and KVIC models of biogas (Gobar gas) plants under national biogas and manure management programme in Sehore district of Madhya Pradesh: An economic analysis. Journal of Pharmacognosy and Phytochemistry. 8: 1817-1821.

Vijayakumary, P. and Desai, S.R. 2021. Biomass Management for Fodder and Energy. Agrimoon.com. 92 p.

Index